MW00795216

An Introduction to Applied Multivariate Analysis

Tenko Raykov • George A. Marcoulides

Routledge
Taylor & Francis Group
711 Third Avenue
New York, NY 10017

Routledge
Taylor & Francis Group
2 Park Square
Milton Park, Abingdon
Oxon OX14 4RN

International Standard Book Number-13: 978-0-8058-6375-8 (Hardcover)

Library of Congress Cataloging-in-Publication Data

Introduction to applied multivariate analysis / by Tenko Raykov & George A. Marcoulides.
p. cm.
Includes bibliographical references and index.
ISBN-13: 978-0-8058-6375-8 (hardcover)
ISBN-10: 0-8058-6375-3 (hardcover)
1. Multivariate analysis. I. Raykov, Tenko. II. Marcoulides, George A.

QA278.I597 2008
519.5'35--dc22 2007039834

Visit the Taylor & Francis Web site at
http://www.taylorandfrancis.com

and the Psychology Press Web site at
http://www.psypress.com

Contents

Preface

Having taught applied multivariate statistics for a number of years, we have been impressed by the broad spectrum of topics that one may be expected to typically cover in a graduate course for students from departments outside of mathematics and statistics. Multivariate statistics has developed over the past few decades into a very extensive field that is hard to master in a single course, even for students aiming at methodological specialization in commonly considered applied fields, such as those within the behavioral, social, and educational disciplines. To meet this challenge, we tried to identify a core set of topics in multivariate statistics, which would be both of fundamental relevance for its understanding and at the same time would allow the student to move on to more advanced pursuits.

This book is a result of this effort. Our goal is to provide a coherent introduction to applied multivariate analysis, which would lay down the basics of the subject that we consider of particular importance in many empirical settings in the social and behavioral sciences. Our approach is based in part on emphasizing, where appropriate, analogies between univariate statistics and multivariate statistics. Although aiming, in principle, at a relatively nontechnical introduction to the subject, we were not able to avoid the use of mathematical formulas, but we employ these primarily in their definitional meaning rather than as elements of proofs or related derivations. The targeted audience who will find this book most beneficial consists primarily of graduate students, advanced undergraduate students, and researchers in the behavioral, social, as well as educational disciplines, who have limited or no familiarity with multivariate statistics. As prerequisites for this book, an introductory statistics course with exposure to regression analysis is recommended, as is some familiarity with two of the most widely circulated statistical analysis software: SPSS and SAS.

Without the use of computers, we find that an introduction to applied multivariate statistics is not possible in our technological era, and so we employ extensively these popular packages, SPSS and SAS. In addition, for the purposes of some chapters, we utilize the latent variable modeling program M*plus*, which is increasingly used across the social and behavioral sciences. On the book specific website, www.psypress.com/applied-multivariate-analysis, we supply essentially all data used in the text. (See Appendix for name of data file and of its variables, as well as their order as columns within it.) To aid with clarity, the software code (for SAS and M*plus*) or sequence of analytic/menu option selection (for SPSS) is also presented and discussed at appropriate places in the book.

We hope that readers will find this text offering them a useful introduction to and a basic treatment of applied multivariate statistics, as well as preparing them for more advanced studies of this exciting and comprehensive subject. A feature that seems to set apart the book from others in this field is our use of latent variable modeling in later chapters to address some multivariate analysis questions of special interest in the behavioral and social disciplines. These include the study of group mean differences on unobserved (latent) variables, testing of latent structure, and some introductory aspects of missing data analysis and longitudinal modeling.

Many colleagues have at least indirectly helped us in our work on this project. Tenko Raykov acknowledges the skillful introduction to multivariate statistics years ago by K. Fischer and R. Griffiths, as well as many valuable discussions on the subject with S. Penev and Y. Zuo. George A. Marcoulides is most grateful to H. Loether, B. O. Muthén, and D. Nasatir under whose stimulating tutelage many years ago he was first introduced to multivariate analysis. We are also grateful to C. Ames and R. Prawat from Michigan State University for their instrumental support in more than one way, which allowed us to embark on the project of writing this book. Thanks are also due to L. K. Muthén, B. O. Muthén, T. Asparouhov, and T. Nguyen for valuable instruction and discussions on applications of latent variable modeling. We are similarly grateful to P. B. Baltes, F. Dittmann-Kohli, and R. Kliegl for generously granting us access to data from their project "Aging and Plasticity in Fluid Intelligence," parts of which we adopt for our method illustration purposes in several chapters of the book. Many of our students provided us with very useful feedback on the lecture notes we first developed for our courses in applied multivariate statistics, from which this book emerged. We are also very grateful to Douglas Steinley, University of Missouri-Columbia; Spiridon Penev, University of New South Wales; and Tim Konold, University of Virginia for their critical comments on an earlier draft of the manuscript, as well as to D. Riegert and R. Larsen from Lawrence Erlbaum Associates, and R. Tressider of Taylor & Francis, for their essential assistance during advanced stages of our work on this project. Last but not least, we are more than indebted to our families for their continued support in lots of ways. Tenko Raykov thanks Albena and Anna, and George A. Marcoulides thanks Laura and Katerina.

Tenko Raykov
East Lansing, Michigan

George A. Marcoulides
Riverside, California

1

Introduction to Multivariate Statistics

One of the simplest conceptual definitions of multivariate statistics (MVS) is as a set of methods that deal with the simultaneous analysis of multiple outcome or response variables, frequently also referred to as dependent variables (DVs). This definition of MVS suggests an important relationship to univariate statistics (UVS) that may be considered a group of methods dealing with the analysis of a single DV. In fact, MVS not only exhibits similarities with UVS but can also be considered an extension of it, or conversely UVS can be viewed as a special case of MVS. At the same time, MVS and UVS have a number of distinctions, and this book deals with many of them whenever appropriate.

In this introductory chapter, our main objective is to discuss, from a principled standpoint, some of the similarities and differences between MVS and UVS. More specifically, we (a) define MVS; then (b) discuss some relationships between MVS and UVS; and finally (c) illustrate the use of the popular statistical software SPSS and SAS for a number of initial multiple variable analyses, including obtaining covariance, correlation, and sum-of-squares and cross-product matrices. As will be observed repeatedly throughout the book, these are three matrices of variable interrelationship indices, which play a fundamental role in many MVS methods.

1.1 Definition of Multivariate Statistics

Behavioral, social, and educational phenomena are often multifaceted, multifactorially determined, and exceedingly complex. Any systematic attempt to understand them, therefore, will typically require the examination of multiple dimensions that are usually intertwined in complicated ways. For these reasons, researchers need to evaluate a number of interrelated variables capturing specific aspects of phenomena under consideration. As a result, scholars in these sciences commonly obtain and have to deal with data sets that contain measurements on many interdependent dimensions, which are collected on subjects sampled from the studied populations. Consequently, in empirical behavioral and social studies, one is very often faced with data sets consisting of multiple interrelated variables that have been observed on a number of persons, and possibly on samples from different populations.

MVS is a scientific field, which for many purposes may be viewed a branch of mathematics and has been developed to meet these complex challenges. Specifically, MVS represents a multitude of statistical methods to help one analyze potentially numerous interrelated measures considered together rather than separately from one another (i.e., one at a time). Researchers typically resort to using MVS methods when they need to analyze more than one dependent (response, or outcome) variable, possibly along with one or more independent (predictor, or explanatory) variables, which are in general all correlated with each other.

Although the concepts of independent variables (IVs) and DVs are generally well covered in most introductory statistics and research methods treatments, for the aims of this chapter, we deem it useful to briefly discuss them here. IVs are typically different conditions to which subjects might be exposed, or reflect specific characteristics that studied persons bring into a research situation. For example, socioeconomic status (SES), educational level, age, gender, teaching method, training program or treatment are oftentimes considered IVs in various empirical settings. Conversely, DVs are those that are of main interest in an investigation, and whose examination is of focal interest to the researcher. For example, intelligence, aggression, college grade point average (GPA) or Graduate Record Exam (GRE) score, performance on a reading or writing test, math ability score or computer aptitude score can be DVs in a study aimed at explaining variability in any of these measures in terms of some selected IVs. More specifically, the IVs and DVs are defined according to the research question being asked. For this reason, it is possible that a variable that is an IV for one research query may become a DV for another one, or vice versa. Even within a single study, it is not unlikely that a DV for one question of interest changes status to an IV, or conversely, when pursuing another concern at a different point during the study.

To give an example involving IVs and DVs: suppose an educational scientist were interested in comparing the effectiveness of two teaching methods, a standard method and a new method of teaching number division. To this end, two groups of students are randomly assigned to the new and to the standard method. Assume that a test of number division ability was administered to all students who participated in the study, and that the researcher was interested in explaining the individual differences observed then. In this case, the score on the division test would be a DV. If the scientist had measured initial arithmetic ability as well as collected data on student SES or even hours watching television per week then all these three variables may be potential IVs. The particular posited question appears to be relatively simple and in fact may be addressed straightforwardly using UVS as it is phrased in terms of a single DV, namely score obtained on the number division test. However, if the study was carried out in such a way that measurements of division ability were collected for each student on each of three consecutive weeks after the two teaching methods were administered, and in addition data on hours watched television were gathered in each of these weeks, then this question becomes considerably more complicated. This is because there are now three measures of interest—division ability in each of the 3 weeks of measurement—and it may be appropriate to consider them all as DVs. Furthermore, because these measures are taken on the same subjects in the study, they are typically interrelated. In addition, when addressing the original question about comparative effectiveness of the new teaching method relative to the old one, it would make sense at least in some analyses to consider all three so-obtained division ability scores simultaneously. Under these circumstances, UVS cannot provide the sought answer to the research query. This is when MVS is typically used, especially where the goal is to address complicated research questions that cannot be answered directly with UVS.

A main reason for this MVS preference in such situations is that UVS represents a set of statistical methods to deal with just a single DV, while there is effectively no limit on the number of IVs that might be considered. As an example, consider the question of whether observed individual differences in average university freshman grades could be explained with such on their SAT score. In this case, the DV is freshman GPA, while the SAT score would play the role of an IV, possibly in addition to say gender, SES and type of high school attended (e.g., public vs. private), in case a pursued research question requires consideration of these as further IVs.

There are many UVS and closely related methods that can be used to address an array of queries varying in their similarity to this one. For example, for prediction goals, one could consider using regression analysis (simple or multiple regression, depending on the number of IVs selected), including the familiar *t* test. When examination of mean differences across groups is of interest, a traditionally utilized method would be analysis of variance (ANOVA) as well as analysis of covariance (ANCOVA)—either approach being a special case of regression analysis. Depending on the nature of available data, one may consider a chi-square analysis of say two-way frequency tables (e.g., for testing association of two categorical variables). When certain assumptions are markedly violated, in particular the assumption of normality, and depending on other study aspects, one may also consider nonparametric statistical methods. Most of the latter methods share the common feature that they are typically considered for application when for certain research questions one identifies single DVs.

By way of contrast, MVS may be viewed as an extension of UVS in the case where one is interested in studying multiple DVs that are interrelated, as they commonly would be in empirical research in the behavioral, social, and educational disciplines. For this reason, MVS typically deals in applications with fairly large data sets on potentially many subjects and in particular on possibly numerous interrelated variables of main interest. Due to this complexity, the underlying theme behind many MVS methods is also simplification, for instance, the reduction of the complexity of available data to several meaningful indices, quantities (parameters), or dimensions.

To give merely a sense of the range of questions addressed with MVS, let us consider a few simple examples—we of course return to these issues in greater detail later in this book. Suppose a researcher is interested in determining which characteristics or variables differentiate between achievers and nonachievers in an educational setting. As will be discussed in Chapter 10, these kinds of questions can be answered using a technique called discriminant function analysis (or discriminant analysis for short). Interestingly, discriminant analysis can also be used to predict group membership—in the currently considered example, achievers versus nonachievers—based on the knowledge obtained about differentiating characteristics or variables. Another research question may be whether there is a single underlying (i.e., unobservable, or so-called latent) dimension along which students differ and which is responsible for their observed interrelationships on some battery of intelligence tests. Such a question can be attended to using a method called factor analysis (FA), discussed in Chapters 8 and 9. As another example, one may be concerned with finding out whether a set of interrelated tests can be decomposed into groups of measures and accordingly new derived measures obtained so that they account for most of the observed variance of the tests. For these aims, a method called principal component analysis (PCA) is appropriate, to which we turn in Chapter 7. Further, if one were interested in whether there are mean differences between several groups of students exposed to different training programs, say with regard to their scores on a set of mathematical tasks—possibly after accounting for initial differences on algebra, geometry, and trigonometry tests—then multivariate analysis of variance (MANOVA) and multivariate analysis of covariance (MANCOVA) would be applicable. These methods are the subject of Chapters 4 and 6. When of concern are group differences on means of unobserved variables, such as ability, intelligence, neuroticism, or aptitude, a specific form of what is referred to as latent variable modeling could be used (Chapter 9). Last but not least, when studied variables have been repeatedly measured, application of special approaches of ANOVA or latent variable modeling can be considered, as covered in Chapters 5 and 13. All these examples are just a few of the kinds of questions that can be addressed using MVS, and the remaining chapters in this book are devoted to their discussion. The common theme unifying the research questions underlying these examples is the necessity to deal with potentially multiple correlated variables in such a way that their interrelationships are taken into account rather than ignored.

1.2 Relationship of Multivariate Statistics to Univariate Statistics

The preceding discussion provides leads to elaborate further on the relationship between MVS and UVS. First, as indicated previously, MVS may be considered an extension of UVS to the case of multiple, and commonly interrelated, DVs. Conversely, UVS can be viewed as a special case of MVS, which is obtained when the number of analyzed DVs is reduced to just one. This relationship is additionally highlighted by the observation that for some UVS methods, there is an MVS analog or multivariate generalization. For example, traditional ANOVA is extended to MANOVA in situations involving more than one outcome variable. Similarly, conventional ANCOVA is generalized to MANCOVA whenever more than a single DV is examined, regardless of number of covariates involved. Further, multiple regression generalizes to multivariate multiple regression (general linear model) in the case with more than one DVs. This type of regression analysis may also be viewed as path analysis or structural equation modeling with observed variables only, for which we refer to a number of alternative treatments in the literature (see Raykov & Marcoulides, 2006, for an introduction to the subject). Also, the idea underlying the widely used correlation coefficient, for example, in the context of a bivariate correlation analysis or simple linear regression, is extended to that of canonical correlation. In particular, using canonical correlation analysis (CCA) one may examine the relationships between sets of what may be viewed, for the sake of this example, as multiple IVs and multiple DVs. With this perspective, CCA could in fact be considered encompassing all MVS methods mentioned so far in this section, with the latter being obtained as specifically defined special cases of CCA.

With multiple DVs, a major distinctive characteristic of MVS relative to UVS is that the former lets one perform a single, simultaneous analysis pertaining to the core of a research question. This approach is in contrast to a series of univariate or even bivariate analyses, like regressions with a single DV, correlation estimation for all pairs of analyzed variables, or ANOVA/ANCOVA for each DV considered in turn (i.e., one at a time). Even though we often follow such a simultaneous multivariate test with further and more focused analyses, the benefit of using MVS is that no matter how many outcome variables are analyzed the overall Type I error rate is kept at a prespecified level, usually .05 or more generally the one at which the multivariate test is carried out. As a conceivable alternative, one might contemplate conducting multiple univariate analyses, one per DV. However, that approach will be associated with a higher (family-wise) Type I error rate due to the multiple testing involved. These are essentially the same reasons for which in a group mean comparison setup, carrying out a series of t tests for all pairs of groups would be associated with a higher than nominal error rate relative to an ANOVA, and hence make the latter a preferable analytic procedure.

At the same time, it is worth noting that with MVS we aim at the "big picture," namely analysis of more than one DV when considered together. This is why with MVS we rarely get "as close" to data as we can with UVS, because we typically do not pursue as focused an analysis in MVS as we do in UVS where a single DV is of concern. We emphasize however that the center of interest, and thus of analysis, depends on the specific research question asked. For example, at any given stage of an empirical study dealing with say a teaching method comparison, we may be interested in comparing two or more methods with regard only to a single DV. In such a case, the use of UVS will be quite appropriate. When alternatively the comparison is to be carried out with respect to several DVs simultaneously, an application of MVS is clearly indicated and preferable.

In conclusion of this section, and by way of summarizing much of the preceding discussion, multivariate analyses are conducted instead of univariate analyses for the following reasons:

1. With more than one DVs (say p in number), the use of p separate univariate tests inflates the Type I error rate, whereas a pertinent multivariate test preserves the significance level ($p > 1$).
2. Univariate tests, no matter how many in number, ignore the interrelationships possible among the DVs, unlike multivariate analyses, and hence potentially waste important information contained in the available sample of data.
3. In many cases, the multivariate test is more powerful than a corresponding univariate test, because the former utilizes the information mentioned in the previous point 2. In such cases, we tend to trust MVS more when its results are at variance with those of UVS (as we also do when of course our concern is primarily with a simultaneous analysis of more than one DV).
4. Many multivariate tests involving means have as a by-product the construction of a linear combination of variables, which provides further information (in case of a significant outcome) about how the variables unite to reject the hypothesis; we deal with these issues in detail later in the book (Chapter 10).

1.3 Choice of Variables and Multivariate Method, and the Concept of Optimal Linear Combination

Our discussion so far has assumed that we have already selected the variables to be used in a given multivariate analysis. The natural question that arises now is how one actually makes this variable choice. In general, main requirements for the selection of variables to be used in an MVS analysis—like in any univariate analysis—are those of high psychometric qualities (specifically, high validity and reliability) of the measures used as DVs and IVs, and that they pertain to the research questions being pursued, that is, are measures of aspects of a studied phenomenon that are relevant to these questions. Accordingly, throughout the rest of the book, we assume that the choice of considered variables has already been made in this way.

MVS encompasses an impressive array of analytic and modeling methods, each of which can be used to tackle certain research queries. Consequently, the next logical concern for any researcher is which one(s) of these methods to utilize. In addition to selecting from among those methods that allow answering the questions asked, the choice will also typically depend on the type of measurement of the involved DVs. Oftentimes, with continuous (or approximately so) DVs, such as reaction time, income, GPA, and intelligence test scores, a frequent choice may be made from among MANOVA or MANCOVA, FA, PCA, discriminant function analysis, CCA, or multivariate multiple regression. Alternatively, with discrete DVs—for example, answers to questions from a questionnaire that have limited number of response options, or items of another type of multiple-component measuring instrument—a choice may often be made from among logistic regression, contingency table analysis, log-linear models, latent variable modeling with categorical outcomes (e.g., latent class analysis), or item–response theory models. Obviously, even within a comprehensive textbook, only a limited number of these methods can be adequately addressed. Having to make such a choice, we elected the material in the book to center around what we found to be—at the level aimed in this text—most widely used multivariate methods for analysis of continuous DVs. For discussions of methods for analyzing discrete DVs, the reader is referred to a number of alternative texts (Agresti, 2002; Lord, 1980; Muthén, 2002; Skrondal & Rabe-Hesketh, 2004).

When choosing a statistical technique, in particular a multivariate method, whether the data to be analyzed are experimental (i.e., resulting after random assignment of subjects and manipulation of IVs, in addition to typically exercising control upon so-called extraneous variance) or observational (i.e., obtained from responses to questionnaires or surveys), is irrelevant. Statistical analysis will work in either case equally well. However, it is the resultant interpretation which will typically differ. In particular, potential causality attributions, if attempted, can be crucially affected by whether the data stem from an experimental or an observational (correlational) study. A researcher is in the strongest position to possibly make causal statements in the case of an experimental study. This fundamental matter is discussed at length in the literature, and we refer the reader to alternative sources dealing with experimental design, causality and related issues (Shadish, Cook, & Campbell, 2002). Throughout this book, we will not make or imply causal statements of any form or type.

Once the particular choice of variables and statistical method(s) of analysis is made, a number of MVS techniques will optimally combine the DVs and yield a special linear combination of them with certain features of interest. More specifically, these methods find that linear combination, Y^*, of the response measures $Y_1, Y_2, \ldots, Y_p$ ($p > 1$), which is defined as

$$Y^* = w_1 Y_1 + w_2 Y_2 + \cdots + w_p Y_p \tag{1.1}$$

and has special optimality properties (Y^* is occasionally referred to as supervariable). In particular, this constructed variable Y^* may be best at differentiating between groups of subjects that are built with respect to some IV(s). As discussed in Chapter 10, this will be the case when using the technique called discriminant function analysis. Alternatively, the variable Y^* defined in Equation 1.1 may possess the highest possible variance from all linear combinations of the measures $Y_1, Y_2, \ldots, Y_p$ that one could come up with. As discussed in Chapter 7, this will be the case when using PCA. As another option, Y^* may be constructed so as to possess the highest correlation with another appropriately obtained supervariable, that is, a linear combination of another set of variables. This will be the case in CCA. Such optimal linear combinations are typical for MVS, and in some sense parallel the search conducted in univariate regression analysis for that linear combination of a given set of predictors, which has the highest possible correlation with a prespecified DV. Different MVS methods use specific information about the relationship between the DVs and possibly IVs in evaluating the weights $w_1, w_2, \ldots, w_p$ in Equation 1.1, so that the resulting linear combination Y^* has the corresponding of the properties mentioned above.

1.4 Data for Multivariate Analyses

In empirical behavioral, social, and educational research, MVS methods are applied to data provided by examined subjects sampled from studied populations. For the analytic procedures considered in this book, these data are typically organized in what is commonly called data matrix. Because the notion of a data matrix is of special relevance both in UVS and MVS, we attend to it in some detail next.

TABLE 1.1

Data From Four Subjects in a General Mental Ability Study

Student	Test1	Test2	Test3	Gen	SES	MathAbTest
1	45	55	47	1	3	33
2	51	54	57	0	1	23
3	40	51	46	1	2	43
4	49	45	48	0	3	42

Note: Gen = gender; SES = socioeconomic status; MathAbTest = mathematics ability test score.

A data matrix is a rectangular array of collected (recorded) scores from studied subjects. The entries in this matrix are arranged in such a way that each row represents a given person's data, and each column represents a variable that may be either dependent or independent in a particular analysis, in accordance with a research question asked. For example, consider a study of general mental ability, and let us assume that data are collected from a sample of students on the following six variables: (a) three tests of intelligence, denoted below Test1, Test2, and Test3; (b) a mathematics ability test; (c) information about their gender; and (d) SES. For the particular illustrative purposes here, the data on only four subjects are provided in Table 1.1.

As seen from Table 1.1, for the continuous variables—in this case, the three tests of intelligence and the mathematics ability test—the actual performance scores are recorded in the data matrix. However, for discrete variables (here, gender and SES), codes for group membership are typically entered. For more details on coding schemes, especially in studies with multiple groups—a topic commonly discussed at length in most regression analysis treatments—we refer to Pedhazur (1997) or any other introductory to intermediate statistics text.

Throughout the rest of this book, we treat the elements of the data matrix as discussed so far in this section. We note, however, that in some repeated assessment studies (or mixed modeling contexts) data per measurement occasion may be recorded on a single line. Such data arrangements will not be considered in the book, yet we mention that they can be readily reordered in the form of data matrices outlined above that will be of relevance in the remainder. Additionally, this text will not be concerned with data that, instead of being in the form of subject performance or related scores, are determined as distances between stimuli or studied persons (e.g., data collected in psychophysics experiments or some areas of social psychology). For discussions on how to handle such settings, we refer to corresponding sources (Johnson & Wichern, 2002).

The above data-related remarks also lead us to a major assumption that will be made throughout the book, that of independence of studied subjects. This assumption may be violated in cases where there is a possibility for examined persons to interact with one another during the process of data collection, or to receive the same type of schooling, treatment instruction, or opportunities. The data resulting in such contexts typically exhibit a hierarchical nature. For example, data stemming from patients who are treated by the same doctor or in the same health care facility, or data provided by students taught by the same teachers, in certain classrooms or in schools (school districts), tend to possess this characteristic. This data property is at times also referred to as nestedness or clustering, and then scores from different subjects cannot be safely considered independent. While the methods discussed in this book may lead to meaningful results when the subject independence assumption is violated to a minor extent, beyond that they cannot be generally trusted. Instead, methods that have been specifically developed to deal with nested data, also referred to as hierarchical, multilevel, or clustered data, should be utilized. Such methods are available within the so-called mixed modeling methodology, and some of them are also known as hierarchical linear (or nonlinear) modeling. For a discussion of these methods, which are not covered in this text, the reader is referred to Heck and Thomas (2000), Hox (2002), or Raudenbush and Bryk (2002) and references therein. Later chapters of this book will, however, handle a special case of hierarchical data, which stem from repeated measure designs. In the latter, measurements from any given subject can be considered nested within that subject, that is, exhibit a two-level hierarchy. Beyond this relatively simple case of nested structure, however, alternative multilevel analysis approaches are recommended with hierarchical data.

While elaborating on the nature of the data utilized in applications of MVS methods considered in this book, it is also worthwhile to make the following notes. First, we stress that the rows of the data matrix (Table 1.1) represent a random sample from a studied population, or random samples from more than one population in the more general case of a multipopulation investigation. That is, the rows of the data matrix are independent of one another. On the other hand, the columns of the data matrix do not represent a random sample, and are in general not independent of one another but instead typically interrelated. MVS is in fact particularly concerned with this information about the interrelationship between variables (data matrix columns) of main interest in connection to a research question(s) being pursued.

Second, in many studies in the behavioral, social, and educational disciplines, some subjects do not provide data on all collected variables. Hence, a researcher has to deal with what is commonly referred to as missing data. Chapter 12 addresses some aspects of this in general difficult to deal with issue, but at this point we emphasize the importance of using appropriate and uniform representation of missing values throughout the entire data matrix. In particular, it is strongly recommended to use the

TABLE 1.2

Missing Data Declaration in an Empirical Study (cf. Table 1.1)

Student	Test1	Test2	Test3	Gen	SES	MathAbTest
1	45	55	47	1	3	33
2	51	54	57	0	1	23
3	40	51	46	1	2	43
4	49	45	48	0	3	42
5	−99	44	−99	1	−99	44
6	52	−99	44	−99	2	−99

Note: Gen = gender; SES = socioeconomic status; MathAbTest = mathematics ability test score.

same symbol(s) for denoting missing data, a symbol(s) that is not a legitimate value possible to take by any subject on a variable in the study. In addition, as a next step one should also insure that this value(s) is declared to the software used as being employed to denote missing value(s); failure to do so can cause severely misleading results. For example, if the next two subjects in the previously considered general mental ability study (cf. Table 1.1) had some missing data, the latter could be designated by the uniform symbol (−99) as illustrated in Table 1.2.

Dealing with missing data is in general a rather difficult and in part "technical" matter, and we refer the reader to Allison (2001) and Little and Rubin (2002) for highly informative and instructive treatments of this issue (see also Raykov, 2005, for a nontechnical introduction in a context of repeated measure analysis). In this book, apart from the discussion in Chapter 12, we assume that used data sets have no missing values (unless otherwise indicated).

1.5 Three Fundamental Matrices in Multivariate Statistics

A number of MVS methods may be conceptually thought of as being based on at least one of three matrices of variable interrelationship indices. For a set of variables to be analyzed, denoted $Y_1, Y_2, \ldots, Y_p$ $(p > 1)$, these matrices are

1. The covariance matrix, designated S for a given sample
2. The correlation matrix, symbolized R in the sample
3. The sum-of-squares and cross-products (SSCP) matrix, denoted Q in the sample

For the purpose of setting the stage for subsequent developments, we discuss each of these matrices in Sections 1.5.1 through 1.5.3.

1.5.1 Covariance Matrix

A covariance matrix is a symmetric matrix that contains the variable variances on its main diagonal and the variable covariances as remaining elements. The covariance coefficient represents a nonstandardized index of the linear relationship between two variables (in case it is standardized, this index is the correlation coefficient, see Section 1.5.2). For example, consider an intelligence study of $n = 500$ sixth-grade students, which used five intelligence tests. Assume that the following was the resulting empirical covariance matrix for these measures (for symmetry reasons, only main diagonal elements and those below are displayed, a practice followed throughout the remainder of the book):

$$S = \begin{bmatrix} 75.73 & & & & \\ 23.55 & 66.77 & & & \\ 33.11 & 37.22 & 99.54 & & \\ 29.56 & 33.41 & 37.41 & 74.44 & \\ 21.99 & 31.25 & 22.58 & 33.66 & 85.32 \end{bmatrix}. \quad (1.2)$$

We note that sample variances are always positive (unless of course on a given variable all subjects take the same value, a highly uninteresting case). In empirical research, covariances tend to be smaller, on average, in magnitude (absolute value) than variances. In fact, this is a main property of *legitimate* covariance matrices for observed variables of interest in behavioral and social research. In Chapter 2, we return to this type of matrices and discuss in a more formal way an important concept referred to as *positive definiteness*. For now, we simply note that unless there is a perfect linear relationship among a given set of observed variables (that take on real values), their covariance matrix will exhibit this feature informally mentioned here.

For a given data set on p variables ($p > 1$), there is a pertinent population covariance matrix (on the assumption of existence of their variances, which is practically not restrictive). This matrix, typically denoted Σ, consists of the population variances on its main diagonal and population covariances off it for all pairs of variables. In a random sample of n ($n > 1$) subjects drawn from a studied population, each element of this covariance matrix can be estimated as

$$s_{ij} = \sum_{k=1}^{n} \frac{(Y_{ki} - \bar{Y}_i)(Y_{kj} - \bar{Y}_j)}{n - 1} \quad (1.3)$$

($i, j = 1, \ldots, p$), where the bar stands for arithmetic mean for the variable underneath, while Y_{ki} and Y_{kj} denote the score of the kth subject on the ith and jth variable, respectively. (In this book, n generally denotes sample size.) Obviously, in the special case that $i = j$, from Equation 1.3 one estimates the variable variances in that sample as

$$s_i^2 = \sum_{k=1}^{n} \frac{(Y_{ki} - \bar{Y}_i)^2}{n - 1}, \tag{1.4}$$

where $(i = 1, \ldots, p)$. Equations 1.3 and 1.4 are utilized by statistical software to estimate an empirical covariance matrix, once the data matrix with observed raw data has been provided. If data were available on all members of a given finite population (the latter being the typical case in social and behavioral research), then using Equations 1.3 and 1.4 yet with the divisor (denominator) n would allow one to determine all elements of the population covariance matrix Σ.

We note that, as can be seen from Equations 1.3 and 1.4, variances and covariances depend on the specific units of measurement of the variables involved. In particular, the magnitude of either of these two indices is unrestricted. Hence, a change from one unit of measurement to another (such as from inches to centimeter measurements), which in empirical research is usually done via a linear transformation, can substantially increase or decrease the magnitude of variance and covariance coefficients.

1.5.2 Correlation Matrix

The last mentioned feature of scale dependence creates difficulties when trying to interpret variances and covariances. To deal with this problem for two random variables, one can use the correlation coefficient, which is obtained by dividing their covariance with the square-rooted product of their variances (i.e., the product of their standard deviations; see below for the case when this division would not be possible). For a given population, from the population covariance matrix Σ one can determine in this way the population correlation matrix, which is commonly denoted P (the capital Greek letter "rho"). Similar to the covariance coefficient, in a given sample the correlation coefficient r_{ij} between the variables Y_i and Y_j is evaluated as

$$r_{ij} = \frac{s_{ij}}{\sqrt{s_i^2 s_j^2}} = \frac{s_{ij}}{s_i s_j} = \frac{\sum_{k=1}^{n} (Y_{ki} - \bar{Y}_i)(Y_{kj} - \bar{Y}_j)}{\sqrt{\sum_{k=1}^{n} (Y_{ki} - \bar{Y}_i)^2 \sum_{k=1}^{n} (Y_{kj} - \bar{Y}_j)^2}} \tag{1.5}$$

$(i, j = 1, \ldots, p)$. A very useful property of the correlation coefficient is that the magnitude of the numerator in Equation 1.5 is never larger than that of the denominator, that is,

$$s_{ij} \leq s_i s_j, \tag{1.6}$$

and hence

$$-1 \leq r_{ij} \leq 1 \tag{1.7}$$

is always true $(i, j = 1, \ldots, p)$. As a matter of fact, in Inequalities 1.6 and 1.7 the equality sign is only then obtained when there is a perfect linear relationship between the two variables involved, that is, if and only there exist two numbers a_{ij} and b_{ij} such that $Y_i = a_{ij} + b_{ij}\ Y_j$ $(1 < i,\ j < p)$. As with the covariance matrix, if data were available on an entire (finite) population, using Equation 1.5 one could determine the population correlation matrix P. Also from Equation 1.5 it is noted that the correlation coefficient will not exist, that is, will not be defined, if at least one of the variables involved is constant across the sample or population considered (this is obviously a highly uninteresting and potentially unrealistic case in empirical research). Then obviously the commonly used notion of relationship is void as well.

As an example, consider a study examining the relationship between GPA, SAT scores, annual family income, and abstract reasoning test scores obtained from $n = 350$, 10th-grade students. Let us also assume that in this sample the following correlation matrix was obtained (positioning of variables in the data matrix is as just listed, from left to right and top to bottom):

$$R = \begin{bmatrix} 1 & & & \\ .69 & 1 & & \\ .48 & .52 & 1 & \\ .75 & .66 & .32 & 1 \end{bmatrix}. \tag{1.8}$$

As seen by examining the values in the right-hand side of Equation 1.8, most correlations are of medium size, with that between income and abstract reasoning test score being the weakest (.32). We note that while there are no strict rules to be followed when interpreting entries in a correlation matrix, it is generally easier to interpret their squared values (e.g., as in simple linear regression, where the squared correlation equals the R^2 index furnishing the proportion of explained or shared variance between the two variables in question).

We also observe that while in general only the sign of a covariance coefficient can be interpreted, both the sign and magnitude of the correlation coefficient are meaningful. Specifically, the closer the correlation coefficient is to 1 or -1, the stronger (more discernible) the linear relationship is between the two variables involved. In contrast, the closer this coefficient is to 0, being either a positive or negative number, the weaker the linear pattern of relationship between the variables. Like a positive covariance, a positive correlation suggests that persons with above mean (below mean) performance on one of the variables tend to be among those with above mean (below mean) performance on the other measure. Alternatively, a negative covariance (negative correlation) is indicative of the fact that subjects with above (below) mean performance on one of the variables tend to be among those with below (above) mean performance on the other measure. When the absolute value of the correlation coefficient is, for example, in the .90s, one may add that this tendency is strong, while otherwise it is moderate to weak (the closer to 0 this correlation is).

1.5.3 Sums-of-Squares and Cross-Products Matrix

An important feature of both the covariance and correlation matrices is that when determining or estimating their elements an averaging process takes place (see the summation sign and division by $(n-1)$ in Equations 1.3 through 1.5, and correspondingly for their population counterparts). This does not happen, however, when one is concerned with estimating the entries of another matrix of main relevance in MVS and especially in ANOVA contexts, the so-called sums-of-squares and cross-products (SSCP) matrix. For a sample from a studied population, this symmetric matrix contains along its diagonal the sums of squares, and off this diagonal the sums of cross products for all possible pairs of variables involved. As is well known from discussions of ANOVA in introductory statistics textbooks, the sum of squares for a given variable is

$$q_{ii} = \sum_{k=1}^{n} (Y_{ki} - \bar{Y}_i)^2 \tag{1.9}$$

$(i = 1, \ldots, p)$, which is sometimes also referred to as corrected or deviation sum of squares, due to the fact that the mean is subtracted from the observed score on the variable under consideration. Similarly, the sum of cross products is

$$q_{ij} = \sum_{k=1}^{n} (Y_{ki} - \bar{Y}_i)(Y_{kj} - \bar{Y}_j), \tag{1.10}$$

where $1 < i, j < p$. Obviously, Equation 1.9 is obtained from Equation 1.10 in the special case when $i = j$, that is, the sum of squares for a given variable equals the sum of cross products of this variable with itself $(i, j = 1, \ldots, p)$. As can be readily seen by a comparison of Equations 1.9

and 1.10 with the Equations 1.3 and 1.4, respectively, the elements of the SSCP matrix $Q = [q_{ij}]$ result by multiplying with $(n - 1)$ the corresponding elements of the empirical covariance matrix S. (In the rest of this chapter, we enclose in brackets the general element of a matrix; see Chapter 2 for further notation explication.) Hence, Q has as its elements measures of linear relationship that are not averaged over subjects, unlike the elements of the matrices S and R. As a result, there is no readily conceptualized population analogue of the SSCP matrix Q. This may in fact be one of the reasons why this matrix has been referred to explicitly less often in the literature (in particular applied) than the covariance or correlation matrix.

Another type of SSCP matrix that can also be considered and used to obtain the matrix Q is one that reflects the SSCP of the actual raw scores uncorrected for the mean. In this raw score SSCP matrix, $U = [u_{ij}]$, the sum of squares for a given variable, say Y_i is defined as

$$u_{ii} = \sum_{k=1}^{n} Y_{ki}^2, \tag{1.11}$$

and the sums of its cross products with the remaining variables are

$$u_{ij} = \sum_{k=1}^{n} Y_{ki} Y_{kj}, \tag{1.12}$$

where $1 \leq i, j \leq p$. We note that Equation 1.11 is obtained from Equation 1.12 in the special case when $i = j$ $(i, j = 1, \ldots, p)$.

As an example, consider a study concerning the development of aggression in middle-school students, with $p = 3$ consecutive assessments on an aggression measure. Suppose the SSCP matrix for these three measures, Q, is as follows:

$$Q = \begin{bmatrix} 1112.56 & & \\ 992.76 & 2055.33 & \\ 890.33 & 1001.36 & 2955.36 \end{bmatrix}. \tag{1.13}$$

We observe that similarly to the covariance matrix, the diagonal entries of Q are positive (unless a variable is a constant), and tend to be larger in magnitude than the off-diagonal ones. This feature can be readily explained with the fact mentioned earlier that the elements of Q equal $(n - 1)$ times those of the covariance matrix S, and the similar feature of the covariance matrix. We also note that the elements of Q may grow unrestrictedly, or alternatively decrease unrestrictedly (if negative), when sample size increases; this is due to their definition as nonaveraged sums across subjects. Similarly to the elements of any sample covariance matrix S, due to their nonstandardized feature, the elements of Q cannot in general be interpreted in magnitude, but only their sign can—in the same way as the sign of the elements of S. Further, the entries in Q also depend on the metric underlying the studied variables.

We stress that the matrices S, R, and Q will be very important in most MVS methods of interest in this book because they contain information on the linear interrelationships among studied variables. It is these interrelationships, which are essential for the multivariate methods considered later in the text. Specifically, MVS methods capitalize on this information and re-express it in their results, in addition to other features of the data. To illustrate, consider the following cases. A correlation matrix showing uniformly high variable correlations (e.g., for a battery of tests) may reveal a structure of relationships that is consistent with the assumption of a single dimension (e.g., abstract thinking ability) underlying a set of analyzed variables. Further, a correlation matrix showing two groups of similar (within-group) correlations with respect to size, may be consistent with two interrelated dimensions (e.g., reading ability and writing ability in an educational study). As it turns out, and preempting some of the developments to follow in subsequent chapters, we use the correlation and covariance matrix in FA and PCA; the SSCP matrix in MANOVA, MANCOVA, and discriminant function analysis; and the covariance matrix in confirmatory FA, in studies of group mean differences on unobserved variables, and in such with repeated measures. (In addition to the covariance matrix, also variable means will be of relevance in the latter two cases, as elaborated in Chapters 9 and 13.)

Hence, with some simplification, we may say that these three matrices of variable interrelationships—S, R, and Q—will often play the role of data in this text; that is, they will be the main starting points for applying MVS methods (with the raw data also remaining relevant in MVS in its own right). We also emphasize that, in this sense, for a given empirical data set, the covariance, correlation, and SSCP matrices are only the beginning and the means rather than the end of MVS applications.

1.6 Illustration Using Statistical Software

In this section, we illustrate the previous discussions using two of the most widely circulated statistical analysis software, SPSS and SAS. To achieve this goal, we use data from a sample of $n = 32$ freshmen in a study of the relationship between several variables that are defined below. Before we commence, we note that such relatively small sample size examples will occasionally be used in this book merely for didactic purposes, and stress that in empirical research it is strongly recommended to use large samples whenever possible. The desirability of large sample sizes is a topic that has received a considerable amount of attention in the literature because it is well recognized that the larger the sample, the more stable the parameter estimates will be, although there are no easy to apply general rules for sample size determination. (This is because the appropriate size of a sample depends in general on many factors, including psychometric properties of the variables selected, the strength of relationships among them, number of observed variables, amount of missing data, and the distributional characteristics of the analyzed variables; Marcoulides & Saunders, 2006; Muthén & Muthén, 2002). In this example data set, the following variables are considered: (a) GPA at the beginning of fall semester (called GPA in the data file ch1ex1.dat available from www.psypress.com/applied-multivariate-analysis), (b) an initial math ability test (called INIT_AB in that file), (c) an IQ test score (called IQ in the file), and (d) the number of hours the person watched television last week (called HOURS_TV in the file). For ease of discussion, we also assume that there are no anomalous values in the data on any of the observed variables. In Chapter 3, we revisit the issue of anomalous values and provide some guidance concerning how such values can be examined and assessed.

For the purposes of this illustration, we go through several initial data analysis steps. We begin by studying the frequencies with which scores occur for each of these variables across the studied sample. To accomplish this, in SPSS we choose the following menu options (in the order given next) and then click on the OK button:

Analyze → Descriptive Statistics → Frequencies

(Upon opening the data file, or reading in the data to be analyzed, the ''Analyze'' menu choice is available in the toolbar, with the ''Descriptive Statistics'' becoming available when ''Analyze'' is chosen, and similarly for ''Frequencies.'' Once the latter choice is clicked, the user must move over the variables of interest into the variable selection window.)

To obtain variable frequencies with SAS, the following set of commands can be used:

```
DATA CHAPTER1;
INFILE 'ch1ex1.dat';
INPUT GPA INIT_AB IQ HOURS_TV;
PROC FREQ;
RUN;
```

In general, SAS program files normally contain commands that describe the data to be analyzed (the so-called DATA statements), and the type of procedures to be performed (the so-called PROC statements). SAS PROC statements are just like computer programs that perform various manipulations, and then print the results of the analyses (for complete details, see the latest *SAS User's Guide* and related manuals). The INFILE command statement merely indicates the file name from which the data are to be read (see also specific software arrangements regarding accessing raw data), and the following INPUT statement invokes retrieval of the data from the named file, in the order of specified free-formatted variables.

Another way to inform SAS about the data to use is to include the entire data set (abbreviated below to the first five observations to conserve space) in the program file as follows:

```
DATA CHAPTER1;
INPUT GPA INIT_AB IQ HOURS_TV;
CARDS;
2.66  20  101  9
2.89  22  103  8
3.28  24   99  9
2.92  12  100  8
4     21  121  7
;
PROC  FREQ;
RUN;
```

Once either of these two command sequences is submitted to SAS, two types of files reporting results are created. One is called the *SAS log file* and the other is referred to as the *SAS output file*. The SAS log file contains all commands, messages, and information related to the execution of the program, whereas the SAS output file contains the actual statistical results.

The outputs created by running the software, SPSS and SAS, as indicated above are as follows. For ease of presentation, we separate them by program and insert clarifying comments after each output section accordingly.

SPSS output notes

Frequencies

Notes

Output Created		
Comments		
Input	Data	D:\Teaching\Multivariate.Statistics\Data\Lecture1.sav
	Filter	<none>
	Weight	<none>
	Split File	<none>
	N of Rows in Working Data File	32
Missing Value Handling	Definition of Missing	User-defined missing values are treated as missing.

	Cases Used	Statistics are based on all cases with valid data.
Syntax		FREQUENCIES VARIABLES = gpa init_ab iq hours_tv/ORDER = ANALYSIS.
Resources	Elapsed Time	0:00:00.09
	Total Values Allowed	149796

SAS output log file

```
NOTE: Copyright (c) 2002-2003 by SAS Institute Inc.,
      Cary, NC, USA.
NOTE: SAS (r) 9.1 (TS1M3)
      Licensed to CSU FULLERTON-SYSTEMWIDE-T/R, Site
      0039713013.
NOTE: This session is executing on the WIN_PRO platform.
NOTE: SAS 9.1.3 Service Pack 1
NOTE: SAS initialization used:
      real time       16.04 seconds
      cpu time        1.54 seconds
1     DATA CHAPTER1;
2     INPUT GPA INIT_AB IQ HOURS_TV;
3     CARDS;
NOTE: The data set WORK.CHAPTER1 has 32 observations
      and 5 variables.
NOTE: DATA statement used (Total process time):
      real time       0.31 seconds
      cpu time        0.06 seconds
36    ;
37    PROC FREQ;
38    RUN;
NOTE: There were 32 observations read from the data set
      WORK.CHAPTER1.
NOTE: PROCEDURE FREQ used (Total process time):
      real time       0.41 seconds
      cpu time        0.04 seconds
```

To save space, in the remainder of the book we dispense with these beginning output parts for each considered analytic session in both software, which echo back the input submitted to the program and/or contain information about internal arrangements that the latter invokes in order to meet the analytic requests. We also dispense with the output section titles, and supply instead appropriate headings and subheadings.

SPSS frequencies output

Statistics

		GPA	INIT_AB	PSI	IQ	HOURS_TV
N	Valid	32	32	32	32	32
	Missing	0	0	0	0	0

This section confirms that we are dealing with a complete data set, that is, one having no missing values, as indicated in Section 1.4.

Frequency Table

GPA

		Frequency	Percent	Valid Percent	Cumulative Percent
Valid	2.06	1	3.1	3.1	3.1
	2.39	1	3.1	3.1	6.3
	2.63	1	3.1	3.1	9.4
	2.66	1	3.1	3.1	12.5
	2.67	1	3.1	3.1	15.6
	2.74	1	3.1	3.1	18.8
	2.75	1	3.1	3.1	21.9
	2.76	1	3.1	3.1	25.0
	2.83	2	6.3	6.3	31.3
	2.86	1	3.1	3.1	34.4
	2.87	1	3.1	3.1	37.5
	2.89	2	6.3	6.3	43.8
	2.92	1	3.1	3.1	46.9
	3.03	1	3.1	3.1	50.0
	3.10	1	3.1	3.1	53.1
	3.12	1	3.1	3.1	56.3
	3.16	1	3.1	3.1	59.4
	3.26	1	3.1	3.1	62.5
	3.28	1	3.1	3.1	65.6
	3.32	1	3.1	3.1	68.8
	3.39	1	3.1	3.1	71.9
	3.51	1	3.1	3.1	75.0
	3.53	1	3.1	3.1	78.1

	Frequency	Percent	Valid Percent	Cumulative Percent
3.54	1	3.1	3.1	81.3
3.57	1	3.1	3.1	84.4
3.62	1	3.1	3.1	87.5
3.65	1	3.1	3.1	90.6
3.92	1	3.1	3.1	93.8
4.00	2	6.3	6.3	100.0
Total	32	100.0	100.0	

INIT_AB

		Frequency	Percent	Valid Percent	Cumulative Percent
Valid	12.0	1	3.1	3.1	3.1
	14.0	1	3.1	3.1	6.3
	17.0	3	9.4	9.4	15.6
	19.0	3	9.4	9.4	25.0
	20.0	2	6.3	6.3	31.3
	21.0	4	12.5	12.5	43.8
	22.0	2	6.3	6.3	50.0
	23.0	4	12.5	12.5	62.5
	24.0	3	9.4	9.4	71.9
	25.0	4	12.5	12.5	84.4
	26.0	2	6.3	6.3	90.6
	27.0	1	3.1	3.1	93.8
	28.0	1	3.1	3.1	96.9
	29.0	1	3.1	3.1	100.0
	Total	32	100.0	100.0	

IQ

		Frequency	Percent	Valid Percent	Cumulative Percent
Valid	97.00	2	6.3	6.3	6.3
	98.00	4	12.5	12.5	18.8
	99.00	4	12.5	12.5	31.3
	100.00	1	3.1	3.1	34.4
	101.00	7	21.9	21.9	56.3
	102.00	1	3.1	3.1	59.4
	103.00	2	6.3	6.3	65.6
	104.00	1	3.1	3.1	68.8
	107.00	1	3.1	3.1	71.9
	110.00	2	6.3	6.3	78.1
	111.00	1	3.1	3.1	81.3
	112.00	1	3.1	3.1	84.4
	113.00	2	6.3	6.3	90.6
	114.00	1	3.1	3.1	93.8
	119.00	1	3.1	3.1	96.9
	121.00	1	3.1	3.1	100.0
	Total	32	100.0	100.0	

HOURS_TV

		Frequency	Percent	Valid Percent	Cumulative Percent
Valid	6.000	4	12.5	12.5	12.5
	6.500	2	6.3	6.3	18.8
	7.000	6	18.8	18.8	37.5
	7.500	1	3.1	3.1	40.6
	8.000	10	31.3	31.3	71.9
	8.500	2	6.3	6.3	78.1
	9.000	6	18.8	18.8	96.9
	9.500	1	3.1	3.1	100.0
	Total	32	100.0	100.0	

SAS frequencies output

GPA	Frequency	Percent	Cumulative Frequency	Cumulative Percent
2.06	1	3.13	1	3.13
2.39	1	3.13	2	6.25
2.63	1	3.13	3	9.38
2.66	1	3.13	4	12.50
2.67	1	3.13	5	15.63
2.74	1	3.13	6	18.75
2.75	1	3.13	7	21.88
2.76	1	3.13	8	25.00
2.83	2	6.25	10	31.25
2.86	1	3.13	11	34.38
2.87	1	3.13	12	37.50
2.89	2	6.25	14	43.75
2.92	1	3.13	15	46.88
3.03	1	3.13	16	50.00
3.1	1	3.13	17	53.13
3.12	1	3.13	18	56.25
3.16	1	3.13	19	59.38
3.26	1	3.13	20	62.50
3.28	1	3.13	21	65.63
3.32	1	3.13	22	68.75
3.39	1	3.13	23	71.88
3.51	1	3.13	24	75.00
3.53	1	3.13	25	78.13
3.54	1	3.13	26	81.25
3.57	1	3.13	27	84.38
3.62	1	3.13	28	87.50
3.65	1	3.13	29	90.63
3.92	1	3.13	30	93.75
4	2	6.25	32	100.00

INIT_AB	Frequency	Percent	Cumulative Frequency	Cumulative Percent
12	1	3.13	1	3.13
14	1	3.13	2	6.25
17	3	9.38	5	15.63
19	3	9.38	8	25.00
20	2	6.25	10	31.25
21	4	12.50	14	43.75
22	2	6.25	16	50.00
23	4	12.50	20	62.50
24	3	9.38	23	71.88
25	4	12.50	27	84.38
26	2	6.25	29	90.63
27	1	3.13	30	93.75
28	1	3.13	31	96.88
29	1	3.13	32	100.00

IQ	Frequency	Percent	Cumulative Frequency	Cumulative Percent
97	2	6.25	2	6.25
98	4	12.50	6	18.75
99	4	12.50	10	31.25
100	1	3.13	11	34.38
101	7	21.88	18	56.25
102	1	3.13	19	59.38
103	2	6.25	21	65.63
104	1	3.13	22	68.75
107	1	3.13	23	71.88
110	2	6.25	25	78.13
111	1	3.13	26	81.25
112	1	3.13	27	84.38
113	2	6.25	29	90.63
114	1	3.13	30	93.75
119	1	3.13	31	96.88
121	1	3.13	32	100.00

HOURS_TV	Frequency	Percent	Cumulative Frequency	Cumulative Percent
6	4	12.50	4	12.50
6.5	2	6.25	6	18.75
7	6	18.75	12	37.50
7.5	1	3.13	13	40.63
8	10	31.25	23	71.88
8.5	2	6.25	25	78.13
9	6	18.75	31	96.88
9.5	1	3.13	100	100.00

These SPSS and SAS frequency tables provide important information regarding the distribution (specifically, the frequencies) of values that each of the variables is taking in the studied sample. In this sense, these output sections also tell us what the data actually are on each variable. In particular, we see that all variables may be considered as (approximately) continuous.

The next step is to examine what the descriptive statistics are for every measure used. From these statistics, we learn much more about the studied variables, namely their means, range of values taken, and standard deviations. To this end, in SPSS we use the following sequence of menu option selections:

Analyze → Descriptive statistics → Descriptives

In SAS, the procedure PROC MEANS would need to be used and stated instead of PROC FREQ (or in addition to the latter) in the second-to-last line of either SAS command file presented above.

Each of the corresponding software command sets provides identical output (up to roundoff error) shown below.

SPSS output

Descriptive Statistics

	N	Minimum	Maximum	Mean	Std. Deviation
GPA	32	2.06	4.00	3.1172	.46671
INIT_AB	32	12.0	29.0	21.938	3.9015
IQ	32	97.00	121.00	104.0938	6.71234
HOURS_TV	32	6.000	9.500	7.71875	1.031265
Valid N (listwise)	32				

SAS output

Variable	N	Mean	Std Dev	Minimum	Maximum
GPA	32	3.1171875	0.4667128	2.0600000	4.0000000
INIT_AB	32	21.9375000	3.9015092	12.0000000	29.0000000
IQ	32	104.0937500	6.7123352	97.0000000	121.0000000
HOURS_TV	32	7.7187500	1.0312653	6.0000000	9.5000000

To obtain standardized measures of variable interrelationships, we produce their correlation matrix. In SAS, this is accomplished by including the procedure PROC CORR at the end of the program file used previously (or instead of PROC FREQ or PROC MEANS), whereas in SPSS this is accomplished by requesting the following series of menu options:

Analyze → Correlations → Bivariate

These SPSS and SAS commands furnish the following output:

SPSS output

Correlations

		GPA	INIT_AB	IQ	HOURS_TV
GPA	Pearson Correlation	1	.387*	.659**	−.453**
	Sig. (2-tailed)	.	.029	.000	.009
	N	32	32	32	32
INIT_AB	Pearson Correlation	.387*	1	.206	−.125
	Sig. (2-tailed)	.029	.	.258	.496
	N	32	32	32	32
IQ	Pearson Correlation	.659**	.206	1	−.784**
	Sig. (2-tailed)	.000	.258	.	.000
	N	32	32	32	32
HOURS_TV	Pearson Correlation	−.453**	−.125	−.784**	1
	Sig. (2-tailed)	.009	.496	.000	.
	N	32	32	32	32

*. Correlation is significant at the 0.05 level (2-tailed).
**. Correlation is significant at the 0.01 level (2-tailed).

SAS output

Pearson Correlation Coefficients, N = 32
Prob > |r| under H0: Rho = 0

	GPA	INIT_AB	IQ	HOURS_TV
GPA	1.00000	0.38699	0.65910	−0.45276
		0.0287	<.0001	0.0093
INIT_AB	0.38699	1.00000	0.20594	−0.12477
	0.0287		0.2581	0.4963>
IQ	0.65910	0.20594	1.00000	−0.78362
	<.0001	0.2581		<.0001>
HOURS_TV	−0.45276	−0.12477	−0.78362	1.00000
	0.0093	0.4963	<.0001	

For the purposes of this introductory chapter, we will not place emphasis on the p-values that are associated with each of the numerical cells with estimated correlations in these tables (labeled in the SPSS output "Sig. (2-tailed)" and in the SAS output "Prob > |r| under H0: Rho = 0"). Instead, we primarily look at these correlations as descriptive measures of linear variable interrelationships, and thereby interpret mainly their sign and magnitude (i.e., distance from 1 or −1, or closeness to 0; see also earlier discussion in Section 1.4.).

In order to generate a covariance matrix and an SSCP matrix for the studied variables in SAS, instead of (or in addition to) PROC FREQ in the second-last line of either of the above SAS command files we use the procedure PROC CORR with its options "COV" and "SSCP," whereas in SPSS the following set of menu options can be used:

Analyze → Correlate → Bivariate (click Options and then check cross products and deviations)

These commands furnish the following output.

SAS output

SSCP Matrix

	GPA	INIT_AB	IQ	HOURS_TV
GPA	317.6919	2210.1100	10447.3600	763.1900
INIT_AB	2210.1100	15872.0000	73241.0000	5403.0000
IQ	10447.3600	73241.0000	348133.0000	25543.0000
HOURS_TV	763.1900	5403.0000	25543.0000	1939.5000

Covariance Matrix, DF = 31

	GPA	INIT_AB	IQ	HOURS_TV
GPA	0.21782087	0.70465726	2.06478831	−0.21791331
INIT_AB	0.70465726	15.22177419	5.39314516	−0.50201613
IQ	2.06478831	5.39314516	45.05544355	−5.42439516
HOURS_TV	−0.21791331	−0.50201613	−5.42439516	1.06350806

SPSS output

Correlations

		GPA	INIT_AB	PSI	IQ	HOURS_TV
GPA	Pearson Correlation	1	.387*	.040	.659**	−.453**
	Sig. (2-tailed)	.	.029	.829	.000	.009
	Sum of Squares and Cross-products	6.752	21.844	.289	64.008	−6.755
	Covariance	.218	.705	.009	2.065	−.218
	N	32	32	32	32	32
INIT_AB	Pearson Correlation	.387*	1	.113	.206	−.125
	Sig. (2-tailed)	.029	.	.539	.258	.496
	Sum of Squares and Cross-products	21.844	471.875	6.875	167.187	−15.562
	Covariance	.705	15.222	.222	5.393	−.502
	N	32	32	32	32	32
PSI	Pearson Correlation	.040	.113	1	.035	−.097
	Sig. (2-tailed)	.829	.539	.	.848	.598
	Sum of Squares and Cross-products	.289	6.875	7.875	3.688	−1.563
	Covariance	.009	.222	.254	.119	−.050
	N	32	32	32	32	32
IQ	Pearson Correlation	.659**	.206	.035	1	−.784**
	Sig. (2-tailed)	.000	.258	.848	.	.000
	Sum of Squares and Cross-products	64.008	167.187	3.688	1396.719	−168.156
	Covariance	2.065	5.393	.119	45.055	−5.424
	N	32	32	32	32	32
HOURS_TV	Pearson Correlation	−.453**	−.125	−.097	−.784**	1
	Sig. (2-tailed)	.009	.496	.598	.000	.
	Sum of Squares and Cross-products	−6.755	−15.562	−1.563	−168.156	32.969
	Covariance	−.218	−.502	−.050	−5.424	1.064
	N	32	32	32	32	32

*. Correlation is significant at the 0.05 level (2-tailed).

**. Correlation is significant at the 0.01 level (2-tailed).

On the basis of these tables, we can easily extract the needed empirical covariance matrix S. For example, using the output provided by SPSS, we look into each of the five panels pertaining to the analyzed variable, and find the fourth row (titled "Covariance"). Note that the respective diagonal elements are actually the squares of the entries in the column "Std. Deviations" of the output table obtained earlier in this section using the "Descriptives" procedure. We thus furnish the following sample covariance matrix S for the study under consideration:

$$S = \begin{bmatrix} .218 & & & & \\ .705 & 15.222 & & & \\ .009 & .222 & .254 & & \\ 2.065 & 5.393 & .119 & 45.055 & \\ -.218 & -.502 & -.050 & -5.424 & 1.064 \end{bmatrix}. \tag{1.14}$$

Finally, the above SPSS and SAS output tables can also be used to extract the SSCP matrix Q. For example, using the last considered SPSS output, we look into each of the five panels pertaining to the analyzed variables and extract its third row (titled "Sum of Squares and Cross Products"). It is important to note, however, that in the output provided by SAS it is the raw data SSCP matrix U that is rendered, whereas SPSS provides directly the (deviation) SSCP matrix Q. In order to obtain the deviation SSCP matrix, Q, using the output provided by SAS, one would have to compute the difference between each entry of the SSCP matrix for the raw data displayed in the SAS output, and the corresponding sum of squared mean or mean products (counted once for each subject). For example, with the value determined by SAS for the variable GPA (i.e., 317.6919), via subtraction of 32 times its squared mean (i.e., the squared mean 3.1171875 summed 32 times, resulting in 310.93945), one obtains the value 6.752, which is the one displayed as the corresponding SSCP matrix element in the SPSS output. In this way, either from the SPSS or SAS output, we can obtain the following deviation SSCP matrix for the five variables under consideration:

$$Q = \begin{bmatrix} 6.752 & & & & \\ 21.844 & 471.875 & & & \\ .289 & 6.875 & 7.875 & & \\ 64.008 & 167.187 & 3.688 & 1396.719 & \\ -6.755 & -15.562 & -1.563 & -168.156 & 32.969 \end{bmatrix}. \tag{1.15}$$

Note that as mentioned earlier, the elements of Q are much larger than the respective ones of S—the reason is that the former are 31 times (i.e., $n - 1$ times in general) those of the latter.

We conclude this chapter with the following general cautionary remark. In applied research in the behavioral, social, and educational sciences, often all elements of a correlation matrix are evaluated for significance. However, if their number is large, even if in the population all correlation coefficients (between all pairs of variables) are 0, it is likely that some will turn out to be significant purely by chance. Therefore, when faced with a correlation matrix a researcher should better not evaluate more than a limited number of correlations for significance, unless he or she has an a priori idea exactly which very few of them to examine.

2

Elements of Matrix Theory

Multivariate statistical methods have been developed to deal with the inherent complexity of studies based on multiple interrelated variables. To handle this complexity, and still keep clarity about the analytic ideas underlying the methods, it is essential that one makes use of systematic means that aim at reducing matters to more manageable levels. Many such means are provided by matrix theory, and a discussion of some of its elements is the subject of this chapter. In particular, we introduce the reader to the compact language of matrices, and familiarize them with a number of basic matrix operations, including the important notions of determinant and trace that will be of special relevance for later developments in the book. In addition, we define the concept of distance and particularly the so-called Mahalanobis distance (MD), which will play an instrumental role in the next chapters.

2.1 Matrix Definition

Matrix theory (MT) is a compact language that we will use in this text in part as a means to underscore important aspects of multivariate statistics (MVS). MT readily enables one to understand a number of multivariate techniques using their univariate analogs and counterparts. In addition, MT allows us to handle efficiently the complexity of MVS in settings with multiple observed variables.

A matrix is in general a rectangular set, a table, or an array of numbers. These often, but not always, are in this book the scores of subjects on some set of observed variables—e.g., a set of intelligence tests—whereby subjects are represented by their rows and variables by their columns. For most of the text, matrices will (a) consist of the data collected in a study (i.e., will have the form of a data matrix—see the pertinent discussion in the preceding chapter), or (b) represent indices of interrelationships between studied variables, such as the covariance, correlation, and sum-of-squares and cross-products (SSCP) matrices S, R, and Q from Chapter 1. For example, the data matrix provided below (also described previously as Table 1.1 in the last chapter, apart from its first column) would represent a 4×6 matrix (read "four by six matrix"):

$$X = \begin{bmatrix} 45 & 55 & 47 & 1 & 3 & 33 \\ 51 & 54 & 57 & 0 & 1 & 23 \\ 40 & 51 & 46 & 1 & 2 & 43 \\ 49 & 45 & 48 & 0 & 3 & 42 \end{bmatrix}.$$

Other examples of matrices considered in Chapter 1 included those representing the matrix S in Equation 1.2, which was a 5×5 matrix; the matrix R in Equation 1.8 that was a 4×4 matrix; and the matrix Q in Equation 1.3, which was of size 3×3. We will use capital letters to denote matrices in this text.

Being defined as a two-dimensional entity means that any matrix has size. This size is determined by the number of its rows (say r, with $r \geq 1$) as well as that of its columns (say c, with $c \geq 1$), and we denote it as $r \times c$. There are no restrictions on r and c, as long as they are whole positive numbers. In particular, as indicated above when considering the correlation, covariance and SSCP matrices, r and c can (but need not) be equal to one another, in which case one is dealing with a square matrix. Furthermore, the case in which $r = 1$ is also possible, when the matrix is called a row vector. Similarly, the case of $c = 1$ is possible, when the matrix is called a column vector. In fact, both $r = 1$ and $c = 1$ are possible, and thus any number can be viewed as a 1×1 matrix, also called scalar.

For example, taking only the second row of the matrix X above, we could represent it as the following 1×6 row vector

$$\underline{x} = [51, 54, 57, 0, 1, 23], \tag{2.1}$$

where we place the elements of the vector within brackets, a convention followed throughout the book; similarly, we will underline the lowercase symbol used to denote the vector, as in the left-hand side of Equation 2.1.

Alternatively, the data of the four subjects displayed in the second column of the above matrix X (i.e., their values obtained on the Reading test; see Chapter 1 and Table 1.1) would represent a 4×1 column vector:

$$\underline{y} = \begin{bmatrix} 55 \\ 54 \\ 51 \\ 45 \end{bmatrix}. \tag{2.2}$$

In the methodological literature, a commonly accepted rule is to assume a column vector when referring to a vector; otherwise the reference "row vector" is explicitly used.

2.2 Matrix Operations, Determinant, and Trace

In order to use matrix theory effectively, certain operations need to be carried out on matrices. Below we provide a discussion of some essential operations needed to enhance one's understanding of MVS. We begin with the notion of matrix equality.

Matrix equality. Two (or more) matrices are said to be equal if (a) they are of the same size, and (b) each element (i.e., each entry) of one of them is identical to the element in the other, that is, in exactly the same position. That is, equal matrices have identical corresponding elements. For example, if we define the matrix A as

$$\mathrm{A} = \begin{bmatrix} 2 & 3 \\ 3 & 2 \\ 4 & 2 \end{bmatrix} \tag{2.3}$$

and the matrix B as

$$\mathrm{B} = \begin{bmatrix} 2 & 3 \\ 3 & 2 \\ 4 & 2 \end{bmatrix}, \tag{2.4}$$

then these are two equal matrices, which is commonly written as A = B. On the other hand, if we consider the matrix C

$$\mathrm{C} = \begin{bmatrix} 2 & 3 \\ 2 & 3 \\ 4 & 2 \end{bmatrix}, \tag{2.5}$$

then A does not equal C. That is, even though the matrices A and C consist of the same elements (numbers in this case), they are positioned at different places, and for this reason A is not equal to C, which is written as A $\neq$ C.

Throughout this book, we will use a special notation for a matrix with real numbers as its elements—specifically, we will place the general matrix element within brackets. For example, the matrix

$$\mathrm{A} = \begin{bmatrix} a_{11} & a_{12} & \dots & a_{1c} \\ a_{21} & a_{22} & \dots & a_{2c} \\ \cdot & \cdot & \cdot & \cdot \\ a_{r1} & a_{r2} & \dots & a_{rc} \end{bmatrix}, \tag{2.6}$$

where a_{ij} are some real numbers, will be symbolized in this notation as $\mathrm{A} = [a_{ij}]$ $(i = 1, 2, \dots, r; j = 1, 2, \dots, c)$, with the last statement in parentheses indicating the range which i and j cover. (In the literature, this statement is sometimes omitted, if the range is clear from the context or not of relevance.) Thus, the matrices $A=[a_{ij}]$ and $B=[b_{ij}]$ that are of the same size will be equal if and only if $a_{ij}=b_{ij}$ for all i and j; otherwise, if for some i and j it is the case that $a_{ij}\neq b_{ij}$, then $A\neq B$ ($i=1,2,\ldots,r; j=1,2,\ldots,c$).

Now that we have defined when two matrices are equal, we can move on to a discussion of matrix operations.

Matrix addition and subtraction. These operations are defined only for matrices that are compatible (also sometimes referred to as conformable or simply conform), i.e., matrices that are of the same size. Under such conditions, the sum and the difference of two matrices is obtained by simply adding or subtracting one by one their corresponding elements. That is, if $C=[c_{ij}]$ is the sum of the conform matrices $A=[a_{ij}]$ and $B=[b_{ij}]$, in other words if $C=A+B$, then $c_{ij}=a_{ij}+b_{ij}$. Similarly, if $D=[d_{ij}]$ denotes the difference between the matrices A and B, that is $D=A-B$, then $d_{ij}=a_{ij}-b_{ij}$ ($i=1, 2,\ldots, r$; $j=1, 2,\ldots, c$). For example, if A and B are the first and second matrix in the middle part of the following Equation 2.7, respectively, their sum and difference are readily obtained as follows:

$$C=A+B=\begin{bmatrix}3 & 7 & 4\\ 2 & 4 & 5\\ 1 & 4 & 3\end{bmatrix}+\begin{bmatrix}11 & 13 & 14\\ 2 & 4 & 5\\ 2 & 3 & 4\end{bmatrix}=\begin{bmatrix}14 & 20 & 18\\ 4 & 8 & 10\\ 3 & 7 & 7\end{bmatrix} \qquad (2.7)$$

and

$$D=A-B=\begin{bmatrix}3 & 7 & 4\\ 2 & 4 & 5\\ 1 & 4 & 3\end{bmatrix}-\begin{bmatrix}11 & 13 & 14\\ 2 & 4 & 5\\ 2 & 3 & 4\end{bmatrix}=\begin{bmatrix}-8 & -6 & -10\\ 0 & 0 & 0\\ -1 & 1 & -1\end{bmatrix}. \qquad (2.8)$$

Addition and subtraction operations of conform matrices are straightforwardly extended to any number of matrices. We also note that matrices do not need to be square, as they happen to be in the last two examples, in order for their sum and difference to be defined (which is readily seen from their general definition given above), as long as they are of the same size.

Matrix multiplication with a number (scalar). A matrix is multiplied with a number (scalar) by multiplying each element of the matrix with that number. That is, if the matrix $A=[a_{ij}]$ is considered, then $gA=[ga_{ij}]$, where g is a scalar ($i=1, 2,\ldots, r$; $j=1, 2,\ldots, c$). For example, if $g=3$, and

$$A=\begin{bmatrix}2 & 4\\ 3 & 2\\ 1 & 5\end{bmatrix},$$

then

$$gA = 3\begin{bmatrix} 2 & 4 \\ 3 & 2 \\ 1 & 5 \end{bmatrix} = \begin{bmatrix} 6 & 12 \\ 9 & 6 \\ 3 & 15 \end{bmatrix}. \qquad (2.9)$$

We note that the order of matrix multiplication with a scalar does not matter. Accordingly, the following Equality holds for a given scalar g:

$$gA = [ga_{ij}] = Ag = [a_{ij}g] \qquad (2.10)$$

($i = 1, 2, \ldots, r; j = 1, 2, \ldots, c$). In addition, there is no requirement on matrix size, in order for either addition, subtraction, or multiplication with a scalar to be carried out; however, as indicated previously, for addition and subtraction to be possible the matrices involved must be of the same size.

Matrix transposition. Oftentimes it is useful to consider the rows of a given matrix as columns of another one, and the columns of the former as rows of the latter. In other words, a new matrix of interest can be formed by interchanging the rows and columns of a given matrix. Such an exchange of the rows with the columns in a matrix is called matrix transposition, and is commonly denoted by a prime ($'$); sometimes, instead of a prime, a superscripted letter T is used in the literature to denote the transpose of a matrix. That is, if $A = [a_{ij}]$, then its transpose matrix is denoted by A' (or A^T) and is defined as $A' = [a_{ji}]$ ($i = 1, 2, \ldots, r; j = 1, 2, \ldots, c$). For example, if

$$A = \begin{bmatrix} 4 & 3 & 5 \\ 2 & 6 & 7 \\ 3 & 5 & 6 \end{bmatrix},$$

then its transpose is

$$A' = \begin{bmatrix} 4 & 2 & 3 \\ 3 & 6 & 5 \\ 5 & 7 & 6 \end{bmatrix}.$$

A matrix that remains the same after transposition is called *symmetric*. As is well known from introductory statistics courses (see also Chapter 1), all covariance, correlation, and SSCP matrices are symmetric. (For this reason, whenever a symmetric matrix is displayed we will usually present only its lower "half," i.e., only its elements on the main diagonal and below it; alternative representations of only its elements on the main diagonal and above it are also found in the literature.)

There is no requirement on the size of a matrix in order for the transposition operation to be executed. In particular, vectors (whether they are row or column) are readily transposed, as are square or rectangular matrices. Note that the transpose of a row vector is a column vector, and vice versa. Also note that $(A')' = A$, for any matrix A; in particular, for any scalar a, the equality $a' = a$ holds.

A couple of examples with transposition of vectors follow below. These will turn out to be quite useful throughout the rest of the book. First, let us consider the case of a row vector. If

$$\underline{x} = [3, 5, 7],$$

then its transpose is

$$\underline{x}' = \begin{bmatrix} 3 \\ 5 \\ 7 \end{bmatrix}. \tag{2.11}$$

Alternatively, if the following column vector is given,

$$\underline{y} = \begin{bmatrix} 6 \\ 1 \\ 4 \\ 5 \end{bmatrix},$$

then

$$\underline{y}' = [6, 1, 4, 5]. \tag{2.12}$$

We note that only square matrices can possibly be symmetric (but need not necessarily be symmetric just because they are square); a rectangular matrix by definition cannot be symmetric. Indeed, if

$$A = \begin{bmatrix} 3 & 5 & 7 \\ 5 & 4 & 8 \end{bmatrix},$$

then its transpose would be

$$A' = \begin{bmatrix} 3 & 5 \\ 5 & 4 \\ 7 & 8 \end{bmatrix},$$

and hence is unequal to A since the two matrices are of different size (one is a 2×3, while the other is a 3×2; see previous section on matrix equality). That is, for a given matrix A, even though the matrices A and A′ consist of the same elements, they are not equal in general.

Many matrices will be of relevance when discussing various topics in this book, but it is the correlation, covariance, and SSCP matrices that will be of special importance. This is because they reflect the degree of interrelationship between all studied variables, in particular among the dependent variables that are of main relevance in MVS. We will be interested in these interrelationship indices, and the multivariate methods discussed in the remainder of the text will take them appropriately into account. This is a major difference between multivariate and univariate statistical methods, as we emphasized throughout Chapter 1. While multivariate methods take into account the interrelation between the dependent variables, univariate methods do not, i.e., in some circumstances (depending on research question asked) they waste that important information.

Matrix multiplication. One of the most frequently used matrix operations in this book is matrix multiplication. In order for this operation to be executable on a given pair of matrices, they must be multiplication conform. The notion of two matrices being multiplication conform requires that the number of columns of the first matrix is equal to the number of rows in the other matrix. That is to say, if a matrix A is of size $u \times v$ and a matrix B is of size $v \times w$ (where u, v, and w are integer positive numbers), then the two matrices are multiplication conform. For example, a matrix of size 2×3 and another of size 3×2 are multiplication conform. There is no limitation on u and w, but note that the same number v appears "between" them when the size of the two matrices is considered, that is, the number of columns in the first matrix equals the number of rows in the second matrix. Hence, two matrices of size $u \times v$ and $w \times q$ are in general not multiplication conform, unless of course $v = w$ (with q being also a whole positive number).

The reason for the requirement of identical number of columns in the first matrix with number of rows in the second has to do with the particular way in which matrix multiplication is carried out. In order to perform matrix multiplication on two multiplication conform matrices, the elements of each consecutive row in the first matrix are multiplied by the corresponding elements of each consecutive column in the second matrix and the resulting products are added, for each pair of a row and a column. For example, if $A = [a_{ij}]$ and $B = [b_{ij}]$ are the above-mentioned multiplication conform matrices, then to find the ijth element of their product $C = AB = [c_{ij}]$ we take the sum of products of the corresponding elements in the ith row of A with those of the jth column of B: $c_{ij} = \sum_{k=1}^{v} a_{ik}b_{kj}$ $(i = 1, \ldots, u, j = 1, \ldots, w)$. To illustrate, if the following matrices A and B are given:

$$A = \begin{bmatrix} 3 & 5 \\ 4 & 7 \end{bmatrix} \quad \text{and} \quad B = \begin{bmatrix} 7 & 4 \\ 5 & 3 \end{bmatrix}, \tag{2.13}$$

then their product is equal to the following matrix C with 2 rows and as many columns:

$$C = AB = \begin{bmatrix} 46 = 3 \times 7 + 5 \times 5 & 27 = 3 \times 4 + 5 \times 3 \\ 63 = 4 \times 7 + 7 \times 5 & 37 = 4 \times 4 + 7 \times 3 \end{bmatrix}, \tag{2.14}$$

where "×" denotes multiplication. Note that in order to determine the elements of the product matrix C, the elements in each consecutive row of the matrix A are multiplied by each element of every consecutive column in B (e.g., the first element of the matrix C is obtained as $c_{11} = a_{11}b_{11} + a_{12}b_{21} = 3 \times 7 + 5 \times 5 = 46$, etc.). We stress that unlike the multiplication of numbers (scalars) that is commutative—i.e., for any two numbers a and b, $a \times b = b \times a$ is true—matrix multiplication is in general not commutative. Indeed, for a given pair of matrices A and B, the product AB need not equal the product BA. In fact, if AB is defined (that is, exists) then the product BA need not exist at all because B and A, in this order, may not be multiplication conform.

As an example, using the two matrices A and B in Equation 2.13, the product BA is readily shown not to equal the product AB (compare Equation 2.14 with Equation 2.15 next):

$$BA = \begin{bmatrix} 37 & 63 \\ 27 & 46 \end{bmatrix} \neq AB = C, \tag{2.15}$$

even though the matrices $D = BA$ and C consist of the same elements (but at different positions; see earlier discussion of matrix equality). As another example, consider the situation where the product AB may exist while the product BA may not even be defined. Indeed, if the following two matrices were considered

$$A = \begin{bmatrix} 4 & 33 \\ 21 & 23 \\ 32 & 3 \end{bmatrix} \quad \text{and} \quad B = \begin{bmatrix} 12 & 3 & 4 & 5 \\ 3 & 66 & 5 & 79 \end{bmatrix},$$

then it is easily seen that while the product AB exists (as the number of columns in the first matrix equals the number of rows in the second), the product BA is not defined since B and A are not multiplication conform (in this order—because the number of columns in B, the first matrix now, is not the same as the number of rows in the second, A). Hence, when speaking of matrix multiplication, we should actually specifically say that we premultiply B with A if we intend to obtain the matrix $C = AB$, or that we postmultiply B with A if we mean to obtain the matrix $D = BA$ (assuming of course the pertinent matrix product exists).

An interesting property that we only mention in passing is that the transpose of a matrix product, when the latter exists, is actually the reverse product of the transposes of the matrices involved:

$$(AB)' = B'A'. \tag{2.16}$$

Note that B′A′ exists as long as AB does. Equation 2.16 is readily generalized to the case of more than two matrices involved in the product. For instance, with three matrices A, B, and C

$$(ABC)' = C'B'A', \tag{2.17}$$

as long as the matrix product in either side of Equation 2.17 exists. (For three matrices, say A, B and C, their triple product is defined as ABC = (AB)C or A(BC); the last two products of what is eventually pairs of matrices are equal to one another due to the so-called associative law of matrix multiplication. Similarly one can define the product of any number of appropriately sized matrices.)

There are two specific kinds of matrix multiplication that are very important in MVS: (a) the multiplication of a matrix with a vector, and (b) the multiplication of a row vector with a matrix. Let us consider a few examples. First, we illustrate the (post) multiplication of a matrix A with a vector $\underline{x}$ (i.e., a column vector); to this end, we use the following example. If

$$A = \begin{bmatrix} 3 & 7 \\ 5 & 11 \\ 44 & 6 \end{bmatrix}$$

and $\underline{x} = [9, 4]'$, then

$$A\underline{x} = \begin{bmatrix} 3 & 7 \\ 5 & 11 \\ 44 & 6 \end{bmatrix} \begin{bmatrix} 9 \\ 4 \end{bmatrix} = \begin{bmatrix} 55 \\ 89 \\ 420 \end{bmatrix}. \tag{2.18}$$

Note the size of both matrix A and the vector $\underline{x}$ that it is postmultiplied with, as well as the size of their product that has as many rows as the matrix and as many columns as the vector, i.e., is of size 3×1.

Second, we illustrate the premultiplication of a matrix B by a row vector $\underline{y}$ with the following example:

$$\underline{y}B = [7, 4, 6] \begin{bmatrix} 2 & 8 \\ 5 & 2 \\ 4 & -2 \end{bmatrix} = [58, 52]. \tag{2.19}$$

Once again we note the size of the matrix B and the vector $\underline{y}$ that the former is being premultiplied with, and the size of their product that has as many rows as the vector and as many columns as the matrix, i.e., is of size 1×2.

We could have also written the last two examples as follows, using row and column vector transposition (notice the priming next):

$$\begin{bmatrix} 3 & 7 \\ 5 & 11 \\ 44 & 6 \end{bmatrix} [9, 4]' = \begin{bmatrix} 55 \\ 89 \\ 420 \end{bmatrix} \tag{2.20}$$

and

$$\begin{bmatrix} 7 \\ 4 \\ 6 \end{bmatrix}' \begin{bmatrix} 2 & 8 \\ 5 & 2 \\ 4 & -2 \end{bmatrix} = [58, 52], \tag{2.21}$$

respectively.

The last two forms of notation for multiplication of a row vector with a matrix, and of a matrix with a vector, will be frequently used in this text, viz. as $A\underline{x}'$ and $\underline{y}'B$, where $\underline{x}$ and $\underline{y}$ are vectors and A and B are matrices such that these products exist.

We complete our discussion of matrix multiplication with the following "combined" example. This type of "double" multiplication of a matrix with a vector, namely pre- and postmultiplication of a matrix with the transpose of a vector and the latter, will be oftentimes used later in the book. Let us consider the following matrix A and vector $\underline{x}$:

$$A = \begin{bmatrix} 2 & 6 & 3 \\ 4 & 5 & 8 \\ 8 & 4 & 3 \end{bmatrix}$$

and

$$\underline{x} = \begin{bmatrix} 3 \\ 5 \\ 6 \end{bmatrix}.$$

Then the "combined" product we had in mind is $\underline{x}'\, A\, \underline{x}$, that is,

$$\underline{x}'\, A\, \underline{x} = \begin{bmatrix} 3 \\ 5 \\ 6 \end{bmatrix}' \begin{bmatrix} 2 & 6 & 3 \\ 4 & 5 & 8 \\ 8 & 4 & 3 \end{bmatrix} \begin{bmatrix} 3 \\ 5 \\ 6 \end{bmatrix} = [3 \quad 5 \quad 6] \begin{bmatrix} 2 & 6 & 3 \\ 4 & 5 & 8 \\ 8 & 4 & 3 \end{bmatrix} \begin{bmatrix} 3 \\ 5 \\ 6 \end{bmatrix} = 959. \tag{2.22}$$

We stress that the matrix A involved in this type of product must be square, since it is premultiplied by the transpose of the same vector $\underline{x}$ with which it is postmultiplied. In addition, we emphasize that the resulting combined product is a scalar, as it is of size 1×1.

This product, $\underline{x}'$ A $\underline{x}$, when A is a square matrix, is so frequently used in MVS and other cognate areas that it has received a special name, viz. quadratic form (in particular, when A is symmetric). Quadratic forms will be very useful in later discussions, especially when we consider the notion of multivariate distance (the extension of the "normal" distance we are used to, in the multivariate case), which distance is exactly of this form, $\underline{x}'$ A $\underline{x}$. As can be readily verified, when $A = [a_{ij}]$ is symmetric and of size $p \times p$ say ($p > 1$), and $\underline{x} = [x_1, x_2, \ldots, x_p]'$ is a $p \times 1$ vector, then the quadratic form $\underline{x}'$ A $\underline{x}$ is the following scalar:

$$\underline{x}' \, A \, \underline{x} = [x_1, x_2, \ldots, x_p] \begin{bmatrix} a_{11} & a_{12} & \ldots & a_{1p} \\ a_{21} & a_{22} & \ldots & a_{2p} \\ . & . & . & . \\ a_{p1} & a_{p2} & \ldots & a_{pp} \end{bmatrix} [x_1, x_2, \ldots, x_p]'$$
$$= a_{11} x_1^2 + a_{22} x_2^2 + \ldots + a_{pp} x_p^2$$
$$+ 2a_{12}x_1x_2 + 2a_{13}x_1x_3 + \ldots + 2a_{p-1,p} x_{p-1}x_p. \quad (2.23)$$

In words, a quadratic form (with a symmetric matrix, as throughout the rest of this text) is the scalar that results as the sum of all squares of successive vector elements with the corresponding diagonal elements of the matrix involved, plus the product of different vector elements multiplied by pertinent elements of that matrix (with subindexes being those of the vector elements involved).

Matrix inversion. When using multivariate statistical methods, we will often need to be concerned with a (remote) analog of number division. This is the procedure of matrix inversion. Only square matrices may have inverses (for the concerns of this book), although not all square matrices will have inverses. In particular, a matrix with the property that there is a linear relationship between its columns (or rows)—e.g., one of the columns being a linear combination of some or all of the remaining columns (rows)—does not have an inverse. Such a matrix is called singular, as opposed to matrices that have inverses and which are called nonsingular or invertible. Inversion is denoted by the symbol $(.)^{-1}$, whereby A^{-1} denotes the inverse of the matrix A (when A^{-1} exists), and is typically best carried out by computers. Interestingly, similarly to transposition, matrix inversion works like a "toggle": the inverse of an inverse of a given matrix is the original matrix itself, that is, $(A^{-1})^{-1} = A$, for any invertible matrix A. In addition, as shown in more advanced treatments, the inverse of a matrix is unique when it exists.

How does one work out a matrix inverse? To begin with, there is a recursive rule that allows one to start with an obvious inverse of a 1×1 matrix (i.e., a scalar), and define those of higher-order matrices. For a given scalar a, obviously $a^{-1} = 1/a$ is its inverse (assuming $a \neq 0$); that is, inversion of a 1×1 matrix is the ordinary division of 1 by this scalar. To determine the inverse of a 2×2 matrix, we must first introduce the notion of a "determinant"—which we do in the next subsection—and then move on to the topic of matrix inversion for higher-order matrices.

An important type of matrix that is needed for this discussion is that of an identity matrix. The identity matrix, for a given size, is a square matrix which has 1's along its main diagonal and 0's off it. For example, the identity matrix with 4 rows and 4 columns, i.e., of size 4×4, is

$$I_4 = \begin{bmatrix} 1 & 0 & 0 & 0 \\ 0 & 1 & 0 & 0 \\ 0 & 0 & 1 & 0 \\ 0 & 0 & 0 & 1 \end{bmatrix}. \tag{2.24}$$

We note that for each integer number, say m ($m > 0$), there is only one identity matrix, namely I_m. The identity matrix plays a similar role to that of the unity (the number 1) among the real numbers. That is, any matrix multiplied with the identity matrix (with which the former is matrix conform) will remain unchanged, regardless whether it has been pre- or postmultiplied with that identity matrix. As an example, if

$$A = \begin{bmatrix} 4 & 5 & 6 \\ 66 & 5 & 55 \\ 45 & 32 & 35 \end{bmatrix},$$

then

$$AI_3 = I_3A = A. \tag{2.25}$$

(Note that this matrix A cannot be pre- or postmultiplied with an identity matrix other than that of size 3×3.)

Returning now to the issue of matrix inverse, we note its following characteristic feature: if A is an arbitrary (square) matrix that is nonsingular, then $AA^{-1} = I = A^{-1}A$, where I stands for the identity matrix of the size of A. An interesting property of matrix inverses, like of transposition, is that the inverse of a matrix product is the reverse product of the inverses of the matrices involved (assuming all inverses exist). For example, the following would be the case with the product of two multiplication conform matrices: $(AB)^{-1} = B^{-1}A^{-1}$; in case of more than two matrices involved in the product, say A, B, ..., Z, where they are multiplication conform in this order, this feature looks as follows:

$$(\mathrm{A\ B\ C}\ldots\mathrm{Y\ Z})^{-1} = \mathrm{Z}^{-1}\,\mathrm{Y}^{-1}\ldots\mathrm{C}^{-1}\,\mathrm{B}^{-1}\,\mathrm{A}^{-1} \tag{2.26}$$

(assuming all involved matrix inverses exist).

Determinant of a matrix. Matrices in empirical research typically contain many numerical elements, and as such are hard to remember or take note of. Further, a table of numbers—which a matrix is—cannot be readily manipulated. It would therefore be desirable to have available a single number that characterizes a given matrix.

One such characterization is the so-called determinant. Only square matrices have determinants. The determinant of a matrix is sometimes regarded as a measure of generalized variance for a given set of random variables, i.e., a random vector. Considering for example the determinant of a sample covariance matrix, it indicates the variability (spread) of the individual raw data it stems from—the larger the determinant of that matrix, the more salient the individual differences on the studied variables as reflected in the raw data, and vice versa. Unfortunately, determinants do not uniquely characterize a matrix; in fact, different matrices can have the same number as their determinant.

The notion of a determinant can also be defined in a recursive manner. To start, let us consider the simple case of a 1×1 matrix, i.e., a scalar, say a. In this case, the determinant is that number itself, a. That is, using vertical bars to symbolize a determinant, $|a| = a$, for any scalar a. In case of a 2×2 matrix, the following rule applies when finding the determinant: subtract from the product of its main diagonal elements, the product of the remaining two off-diagonal elements. That is, for the 2×2 matrix

$$\mathrm{A} = \begin{bmatrix} a & b \\ c & d \end{bmatrix},$$

its determinant is

$$|\mathrm{A}| = ad - bc. \tag{2.27}$$

For higher-order matrices, their determinants are worked out following a rule that reduces their computation to that of determinants of lower order. Without getting more specific, fortunately this recursive algorithm for determinant computation is programmed into most widely available statistical software, including SPSS and SAS, and we will leave their calculation to the computer throughout the rest of this text. (See discussion later in this chapter concerning pertinent software instructions and resulting output.) For illustrative purposes, the following are two simple numerical examples:

$$|3| = 3$$

and

$$\left|\begin{bmatrix} 2 & 5 \\ 6 & 7 \end{bmatrix}\right| = 2 \times 7 - 5 \times 6 = -16.$$

Now that we have introduced the notion of a determinant, let us return to the issue of matrix inversion. We already know how to render the inverse of a 1×1 matrix. In order to find the inverse of a 2×2 matrix, one more concept concerning matrices needs to be introduced. This is the concept of the so-called adjoint of a quadratic matrix, $A = [a_{ij}]$, denoted *adj(A)*. To furnish the latter, in a first step one obtains the matrix consisting of the determinants, for each element of A, pertaining to the matrix resulting after deleting the row and column of that element in A. In a second step, one multiplies each of these determinants by $(-1)^q$, where q is the sum of the numbers corresponding to the row and column of the corresponding element of A (i.e., $q = i + j$ for its general element, a_{ij}). To exemplify, consider the following 2×2 matrix

$$B = \begin{bmatrix} a & b \\ c & d \end{bmatrix},$$

for which the adjoint is

$$adj(B) = \begin{bmatrix} d & -c \\ -b & a \end{bmatrix}.$$

That is, the adjoint of B is the matrix that results by switching position of elements on the main diagonal, as well as on the diagonal crossing it, and adding the negative sign to the off-diagonal elements of the newly formed matrix.

In order to find the inverse of a nonsingular matrix A, i.e., A^{-1}, the following rule needs to be applied (as we will indicate below, the determinant of an invertible matrix is distinct from 0, so the following division is possible):

$$A^{-1} = [adj(A)]'/|A|. \tag{2.28}$$

That is, for the last example,

$$B^{-1} = \begin{bmatrix} d & -b \\ -c & a \end{bmatrix} \Big/ |B| = \begin{bmatrix} d & -b \\ -c & a \end{bmatrix} \Big/ (ad - bc).$$

We stress that the rule stated in Equation 2.28 is valid for any size of a square invertible matrix A. When its size is higher than 2×2, the computation of the elements of the adjoint matrix *adj(A)* is obviously more tedious, though following the above described steps, and is best left to the computer. That is, finding the inverse of higher-order matrices proceeds via use of determinants of matrices of lower-order and of the same order. (See discussion later in this chapter for detailed software instructions and resulting output.)

Before finishing this section, let us note the following interesting properties concerning matrix determinant. For any two multiplication conform square matrices A and B,

$$|AB| = |A|\ |B|, \tag{2.29}$$

that is, the determinant of the product of the two matrices is the product of their determinants. Further, if c is a constant, then the determinant of the product of c with a matrix A is given by $|c\,A| = c^p\,|A|$, where the size of A is $p \times p$ $(p \geq 1)$. Last but not least, if the matrix A is singular, then $|A| = 0$, while its determinant is nonzero if it is invertible. Whether a particular matrix is singular or nonsingular is very important in MVS because as mentioned previously only nonsingular matrices have inverses—when a matrix of interest is singular, the inverse fails to exist. For example, in a correlation matrix for a set of observed variables, singularity will occur whenever there is a linear relationship between the variables (either all of them or a subset of them). In fact, it can be shown in general that for a square matrix A, $|A| = 0$ (i.e., the matrix is singular) if and only if there is a linear relationship between its columns (or rows; e.g., Johnson & Wichern, 2002).

Trace of a matrix. Another candidate for a single number that could be used to characterize a given matrix is its trace. Again, like determinant, we define trace only for square matrices—the trace of a square matrix is the sum of its diagonal elements. That is, if

$$A = [a_{ij}],$$

then

$$\text{tr}(A) = a_{11} + a_{22} + \ldots + a_{pp}, \tag{2.30}$$

where tr(.) denotes trace and A is of order $p \times p$ $(p \geq 1)$. For example, if

$$A = \begin{bmatrix} 3 & 5 & 7 & 6 \\ 8 & 7 & 9 & 4 \\ 0 & 23 & 34 & 35 \\ 34 & 23 & 22 & 1 \end{bmatrix},$$

then its trace is $\text{tr}(A) = 3 + 7 + 34 + 1 = 45$.

We spent considerable amount of time on the notions of determinant and trace in this section because they are used in MVS as generalizations of the concept of variance for a single variable. In particular, for a given covariance matrix, the trace reflects the overall variability in a studied data set since it equals the sum of the variances of all involved variables. Further, the determinant of a covariance matrix may be seen as representing the generalized variance of a random vector with this covariance matrix. Specifically, large values of the generalized variance tend to go together with a broad data scatter around their mean (mean vector) and conversely, as well as with large amounts of individual differences on studied variables. Similarly, for a correlation matrix, R, small values of $|R|$ signal high degree of intercorrelation among the variables, whereas large values of $|R|$ are indicative of a limited extent of intercorrelation.

2.3 Using SPSS and SAS for Matrix Operations

As mentioned on several occasions in this chapter, matrix operations are best left to the computer. Indeed, with just a few simple instructions, one can readily utilize for instance either of the widely circulated software packages SPSS and SAS for these purposes.

The illustration given next shows how one can employ SPSS for computing matrix sum, difference, product, inversion, and determinant. (This is followed by examples of how one can use SAS for the same aims.) We insert comments preceded by a star to enhance comprehensibility of the following input files. Note the definition of the vectors and matrices involved, which happens with the COMPUTE statement (abbreviated to COMP); elements within a row are delineated by a comma, whereas successive rows in a matrix are so by a semicolon.

```
TITLE 'USING SPSS TO CARRY OUT SOME MATRIX OPERATIONS'.
* BELOW WE UTILIZE '*' TO INITIATE A COMMENT RE. PRECEEDING COMMAND IN LINE.
* DO NOT CONFUSE IT WITH THE SIGN '*' USED FURTHER BELOW FOR MULTIPLICATION!
* FIRST WE NEED TO TELL SPSS WE WANT IT TO CARRY OUT MATRIX OPERATIONS FOR US.
MATRIX. * THIS IS HOW WE START THE SPSS MODULE FOR MATRIX OPERATIONS.
COMP X = {1,3,6,8}.      * NEED TO ENCLOSE MATRIX IN CURLY BRACKETS.
* USE COMMAS TO SEPARATE ELEMENTS.
COMP Y = {6,8,7,5}.
COMP Z1 = X*T(Y).        * USE 'T' FOR TRANSPOSE AND '*' FOR MULTIPLICATION.
PRINT Z1.                * PRINT EACH RESULT SEPARATELY.
COMP Z2 = T(X)*Y.
PRINT Z2.
```

```
COMP A=X+Y.
COMP B=Y-X.
COMP C=3*X.
PRINT A.
PRINT B.
PRINT C.
COMP DET.Z1=DET(Z1).      * USE 'DET(.)' FOR EVALUATING A DETERMINANT.
COMP DET.Z2=DET(Z2).
PRINT DET.Z1.
PRINT DET.Z2.
COMP TR.Z1=TRACE(Z1).     * USE TRACE(.) TO COMPUTE TRACE.
COMP TR.Z2=TRACE(Z2).
PRINT TR.Z1.
PRINT TR.Z2.
END MATRIX.    * THIS IS HOW TO QUIT THE SPSS MATRIX OPERATIONS MODULE.
```

We hint here to the fact that the resulting matrices Z1 and Z2 will turn out to be of different size, even though they are the product of the same constituent matrices (vectors). The reason is, as mentioned before, that Z1 and Z2 result when matrix multiplication is performed in different orders.

To accomplish the same matrix operations with SAS, the following program file utilizing the Interactive Matrix Language procedure (called PROC IML) must be submitted to that software.

```
proc iml; /* THIS IS HOW WE START MATRIX OPERATIONS WITH SAS*/
X={1 3 6 8}; /* NEED TO ENCLOSE MATRIX IN CURLY BRACKETS*/
Y={6 8 7 5}; /* ELEMENTS ARE SEPARATED BY SPACES*/
Z1=X*T(Y); /* USE 'T' FOR TRANSPOSE AND '*' FOR MULTIPLICATION*/
print Z1; /* PRINT EACH RESULT SEPARATELY*/
Z2=T(X)*Y;
print Z2;
A=X+Y;
B=Y-X;
C=3*X;
print A;
print B;
print C;
DETZ1=det(Z1); /* USE 'det(.)' FOR EVALUATING DETERMINANT*/
DETZ2=det(Z2);
print DETZ1;
print DETZ2;
TRZ1=TRACE(Z1); /* USE 'TRACE(.)' TO COMPUTE TRACE*/
```

```
TRZ2=TRACE(Z2);
print TRZ1;
print TRZ2;
FINISH; /* END OF MATRIX MANIPULATIONS*/
```

The outputs created by submitting the above SPSS and SAS command files are given next. For clarity, we present them in a different font from that of the main text, and provide comments at their end.

SPSS output

```
USING SPSS TO CARRY OUT SOME MATRIX OPERATIONS

Run MATRIX procedure:

Z1
  112


Z2
    6   8   7   5
   18  24  21  15
   36  48  42  30
   48  64  56  40


A
    7   11   13   13


B
    5   5   1   -3


C
    3   9   18   24

DET.Z1
  112.0000000

DET.Z2
  0

TR.Z1
  112

TR.Z2
  112

—— END MATRIX ——
```

SAS output

```
                 The SAS System

                      Z1
                         112

                      Z2

   6             8                 7          5
  18            24                21         15
  36            48                42         30
  48            64                56         40

                      A
   7            11                13         13

                      B
   5             5                 1         -3

                      C
   3             9                18         24

                 DETZ1
                         112

                 DETZ2
                           0

                  TRZ1
                         112

                  TRZ2
                         112
```

As can be deduced from the preceding discussion in this chapter, multiplication of the same vectors in each of the two possible orders renders different results—in the first instance a scalar (number) as in the case of Z1, and in the other a matrix as in case of Z2. Indeed, to obtain Z1 we (post)multiply a row vector with a column vector, whereas to obtain Z2 we (post)multiply a column vector with a row vector. Hence, following the earlier discussion on matrix multiplication, the resulting two matrices are correspondingly of size 1×1 and 4×4: the product $Z1 = \underline{x}' \ \underline{y}$ is a single number, often called inner product, whereas $Z2 = \underline{xy}'$ is a matrix, sometimes called outer product.

We also note that the above matrix Z2 also possesses a determinant equal to 0. The reason is that it is a singular matrix, since any of its rows is a constant multiple of any other row. (In general, a singular matrix usually exhibits a more subtle, but still linear, relationship between some or all of its rows or columns.)

2.4 General Form of Matrix Multiplications With Vector, and Representation of the Covariance, Correlation, and Sum-of-Squares and Cross-Product Matrices

Much of the preceding discussion about matrices was mainly confined to a number of specific operations. Nevertheless, it would be of great benefit to the reader to get a sense of how these matrix manipulations can actually be considered in a more general form. In what follows, we move on to illustrate the use of symbols to present matrices (and vectors) and operations with them in their most general form of use for our purposes in this book.

2.4.1 Linear Modeling and Matrix Multiplication

Suppose we are given the vector

$$\underline{y} = [y_1, y_2, \ldots, y_n]'$$

consisting of n elements (that can be any real numbers), where $n > 1$, and the vector

$$\underline{b} = \begin{bmatrix} b_0 \\ b_1 \\ b_2 \\ \cdots \\ b_p \end{bmatrix} = [b_0, b_1, \ldots, b_p]',$$

which is a vector of $p + 1$ elements (that can be arbitrary numbers as well), with $0 < p < n$. Let us also assume that

$$X = \begin{bmatrix} 1 & x_{11} & \ldots & x_{1p} \\ 1 & x_{21} & \ldots & x_{2p} \\ . & . & . & . \\ 1 & x_{n1} & \ldots & x_{np} \end{bmatrix}$$

is a matrix with data from n subjects on p variables in its last p columns. We can readily observe that the equation

$$\underline{y} = X\underline{b} \tag{2.31}$$

in actual fact states that each consecutive element of $\underline{y}$ equals a linear combination of the elements of that row of X, which corresponds to the location of the element in question within the vector $\underline{y}$. For instance, the sth element of $\underline{y}$ $(1 \leq s \leq n)$ equals from Equation 2.31

$$y_s = b_0 + b_1 x_{s1} + b_2 x_{s2} + \ldots + b_p\, x_{sp}, \tag{2.32}$$

that is, represents a linear combination of the elements of the sth row of the matrix X. We stress that since Equation 2.32 holds for each s $(s = 1, \ldots, n)$, this linear combination utilizes the same weights for each element of $\underline{y}$. (These weights are the successive elements of the vector $\underline{b}$.)

Now let us think of $\underline{y}$ as a set of individual (sample) scores on a dependent variable of interest, and of X as a matrix consisting of the subjects' scores on a set of p independent variables, with the added first column consisting of 1's only. Equation 2.31, and in particular Equation 2.32, is then in fact the equation of how one obtains predicted scores for the dependent variable in a multiple regression analysis session, if the b's were the estimated partial regression weights. Further, if we now add in Equation 2.31 an $n \times 1$ vector of error scores, denoted e, for the considered case of a single dependent variable we get the general equation of the multiple linear regression model:

$$\underline{y} = X\underline{b} + \underline{e}. \tag{2.33}$$

Hence, already when dealing with univariate regression analysis, one has in fact been implicitly carrying out matrix multiplication (of a matrix by vector) any time when obtaining predicted values for a response variable.

2.4.2 Three Fundamental Matrices of Multivariate Statistics in Compact Form

Recall from introductory statistics how we estimate the variance of a single (unidimensional) random variable X with observed sample values $x_1, x_2, \ldots, x_n$ $(n > 1)$: if we denote that estimator $s^2{}_X$, then

$$s^2{}_X = \frac{1}{n \quad 1}[(x_1 - \bar{x})^2 + \ldots + (x_n - \bar{x})^2] = \frac{1}{n-1}\sum_{i=1}^{n}(x_i - \bar{x})^2. \tag{2.34}$$

From Equation 2.34, the sum of squares for this random variable, X, is also based on its realizations $x_1, x_2, \ldots, x_n$ and is given by

$$SS_X = (n-1)s^2{}_X = \sum_{i=1}^{n}(x_i - \bar{x})^2, \tag{2.35}$$

where $\bar{x}$ is their sample average.

As one may also recall from univariate statistics (UVS), sums of squares play a very important part in a number of its methods (in particular, analysis of variance). It is instructive to mention here that the role played by sums of squares in UVS is played by the SSCP matrix Q in MVS. Further, as we mentioned in Chapter 1, the elements of the SSCP matrix Q are $n-1$ times the corresponding elements of the covariance matrix S. Hence, using the earlier discussed rule of matrix multiplication with a scalar, it follows that the SSCP matrix can be written as

$$Q = (n-1)S. \tag{2.36}$$

Comparing now Equations 2.35 and 2.36, we see that the latter is a multivariate analog of the former, with Equation 2.35 resulting from Equation 2.36 in the special case of a single variable.

This relationship leads us to a more generally valid analogy that facilitates greatly understanding the conceptual ideas behind a number of multivariate statistical methods. To describe it, notice that we could carry out the following two operations in the right-hand side of Equation 2.35 in order to obtain Equation 2.36:

(i) exchange single variables with vectors, and
(ii) exchange the square with the product of the underlying expression (the one being squared) with its transpose.

In this way, following steps (i) and (ii), from Equation 2.35 one directly obtains the formula for the SSCP matrix from that of sum of squares for a given random variable:

$$Q = \sum_{i=1}^{n} (\underline{x}_i - \bar{\underline{x}})(\underline{x}_i - \bar{\underline{x}})'. \tag{2.37}$$

where $\underline{x}_i$ is the vector of scores of the ith subject on a set of p studied variables, and $\bar{\underline{x}}$ is the vector with elements being the means of these variables ($i = 1, \ldots, n$). Note that in the right-hand side of Equation 2.37 we have a sum of n products of a $p \times 1$ column vector with a $1 \times p$ row vector, i.e., a sum of n matrices each of size $p \times p$. With this in mind, it follows that

$$S = (n-1)^{-1}Q = \frac{1}{n-1}\sum_{i=1}^{n} (\underline{x}_i - \bar{\underline{x}})(\underline{x}_i - \bar{\underline{x}})', \tag{2.38}$$

which is the "reverse" relationship to that in Equation 2.36, and one that we emphasized earlier in this chapter as well as in Chapter 1. We also note in passing that Equation 2.36 follows as a special case from Equation 2.37, when the dimensionality of the observed variable vector $\underline{x}$ in the latter is 1; in that case, from Equation 2.38 follows also the formula Equation 2.34 for estimation of variance for a given random variable (based on its random realizations in a sample).

Next recall the definitional relationship between the correlation coefficient of two random variables X_1 and X_2 (denoted $Corr(X_1,X_2)$), their covariance coefficient $Cov(X_1,X_2)$, and their standard deviations s_{X_1} and s_{X_2} (assuming the latter are not zero; see Chapter 1):

$$Corr(X_1,X_2) = \frac{Cov(X_1,X_2)}{s_{X_1}s_{X_2}}. \tag{2.39}$$

In the case of more than two variables, Equation 2.39 would relate just one element of the correlation matrix R with the corresponding element of the covariance matrix S and the reciprocal of the product of the involved variables' standard deviations. Now, for a given random vector $\underline{x}$, that is a set of random variables $X_1, X_2, \ldots, X_p$, one can define the following diagonal matrix:

$$D = \begin{bmatrix} s_{X_1} & 0 & 0 & 0 \\ 0 & s_{X_2} & 0 & 0 \\ . & . & . & . \\ 0 & 0 & 0 & s_{X_p} \end{bmatrix},$$

which has as its only nonzero elements the standard deviations of the corresponding elements of the vector $\underline{x}$ along its main diagonal ($p > 1$). The inverse of the matrix D (i.e., D^{-1}) would simply be a diagonal matrix with the reciprocals of the standard deviations along its main diagonal (as can be found out by direct multiplication of D and D^{-1}, which renders the unit matrix; recall earlier discussion in this chapter on uniqueness of matrix inverse). Thus, the inverse of D is

$$D^{-1} = \begin{bmatrix} 1/s_{X_1} & 0 & 0 & 0 \\ 0 & 1/s_{X_2} & 0 & 0 \\ . & . & . & . \\ 0 & 0 & 0 & 1/s_{X_p} \end{bmatrix}.$$

Based on these considerations, and using the earlier discussed rules of matrix multiplication, one can readily find out that the correlation matrix R can now be written as

$$R = D^{-1}SD^{-1} = D^{-1}\left[\frac{1}{n-1}\sum_{i=1}^{n}(\underline{x}_i - \bar{\underline{x}})(\underline{x}_i - \bar{\underline{x}})'\right]D^{-1}. \tag{2.40}$$

Hence, in Equations 2.37, 2.38, and 2.40 we have expressed in compact form three fundamental matrices in MVS. Thereby, we have instrumentally used the "uni-to-multivariate analogy" indicated previously on several occasions (Rencher, 1995).

2.5 Raw Data Points in Higher Dimensions, and Distance Between Them

Data points. If we wanted to take a look at the data of a particular person from a given data matrix, e.g., that in Table 1.1 (see Chapter 1), we can just take his/her row and represent it separately, i.e., as a row vector. For example, if we were interested in examining the data for the third person in Table 1.1., they would be represented as the following 1×7 matrix (row vector; i.e., a horizontal "slice" of the data matrix):

$$\underline{x}' = [3, 40, 51, 46, 1, 2, 43],$$

where we use the prime symbol in compliance with the widely adopted convention to imply a column vector from a simple reference to a vector.

Similarly, if we wanted to look at all subjects' data on only 1 variable, say Test 3 in the same data table, we would obtain the following 4×1 matrix (column vector; a vertical "slice" of the matrix):

$$\underline{y} = \begin{bmatrix} 47 \\ 57 \\ 46 \\ 48 \end{bmatrix}.$$

Note that we can also represent this vector, perhaps more conveniently, by stating its transpose:

$$\underline{y} = [47, 57, 46, 48]'.$$

We could also think of both vectors $\underline{x}$ and $\underline{y}$ as representing two data points in a multidimensional space that we next turn to.

Multivariable Distance (Mahalanobis Distance)

Many notions underlying MVS cannot be easily or directly visualized because they are typically related to a q-dimensional space, where $q > 3$. Regrettably, we are only three-dimensional creatures, and thus so is also our immediate imagination. It will therefore be quite helpful to utilize whenever possible extensions of our usual notions of three-dimensional space, especially if these can assist us in understanding much of MVS, at least at a conceptual level.

Two widely used spaces in MVS. To accomplish such extensions, it will be beneficial to think of each subject's data (on all studied variables) as a point in a p-dimensional space, where each row of the data matrix represents a corresponding point. At times it will be similarly useful to think of all subjects' data on each separate variable as a point in an n-dimensional space, where each column of the data matrix is represented by a corresponding point. Notice the difference between these two spaces. Specifically, the meaning and coordinates of the actual data point indicated are different in both cases. This difference stems from what the coordinate axes in these two spaces are supposed to mean, as well as their number.

In the first case (p-dimensional space), one can think of the studied variables being the axes and individuals being points in that space, with these points corresponding to their data. This p-dimensional space will be quite useful for most multivariate techniques discussed in this text and is perhaps "more natural," but the second mentioned space above is also quite useful. In the latter, n-dimensional space, one may want to think of individuals being the axes while variables are positioned within it according to the columns in an observed data matrix. Preemptying some of the discussion in a later chapter, this space will be particularly helpful when considering the technique of factor analysis, and especially when interested in factor rotation (Chapter 8).

In addition to reference to a multivariate space, many MVS procedures can be understood by using the notion of a distance, so we move now to this concept.

Distance in a multivariable space. For a point that is say in a q-dimensional space, $\underline{x} = [x_1, x_2, \ldots, x_q]'$ $(q > 1)$, we can define its distance to the origin as its length (denoted by $\|\underline{x}\|$) as follows:

$$\|\underline{x}\| = \sqrt{x_1^2 + x_2^2 + \ldots + x_q^2} \tag{2.41}$$

(We will refer below to $x_1, x_2, \ldots, x_q$ at times also as components of $\underline{x}$.)

Now, if there are two points in that q-dimensional space, say $\underline{x}$ and $\underline{y}$, we can define their distance $D(\underline{x}, \underline{y})$ as the length of their difference, $\underline{x} - \underline{y}$ (i.e., as the distance of $\underline{x} - \underline{y}$ to the origin), where $\underline{x} - \underline{y}$ is obviously the vector having as components the corresponding differences in the components of $\underline{x}$ and $\underline{y}$:

$$D(\underline{x}, \underline{y}) = \|\underline{x} - \underline{y}\| = \sqrt{(x_1 - y_1)^2 + (x_2 - y_2)^2 + \ldots + (x_q - y_q)^2}. \tag{2.42}$$

For example, if $\underline{x} = (2, 8)'$ and $\underline{y} = (4, 2)'$, then their distance would be as follows (note that $q = 2$ here, i.e., these points lie in a two-dimensional space):

$$D(\underline{x}, \underline{y}) = \|\underline{x} - \underline{y}\| = \sqrt{(2-4)^2 + (8-2)^2} = \sqrt{4+36} = 6.235.$$

Using the earlier discussed rules of matrix multiplication, it is readily observed that the expression (2.42) can also be written as

$$\|x - y\|^2 = (x - y)' I_q (x - y). \tag{2.43}$$

Equation 2.43 defines what is commonly known as ''Euclidean distance'' between the vectors $\underline{x}$ and $\underline{y}$. This is the conventional concept used to represent distance between points in a multivariable space. In this definition, each variable (component, or coordinate) participates with the same weight, viz. 1. Note that Equation 2.41 is a special case of Equation 2.42, which results from the latter when $\underline{y} = \underline{0}$ (the last being the zero vector consisting of only 0's as its elements).

Most of the time in empirical behavioral and social research, however, some observed variables may have larger variances than others, and in addition some variables are more likely to be stronger related with one another than with other variables. These facts are not taken into account in the Euclidean distance, but are so in what is called multivariable (statistical) distance. In other words, the Euclidean distance depends on the units in which variables (components) are measured, and thus can be influenced by whichever variable takes numerically larger values. This effect of differences in units of measurement is counteracted in the multivariate distance by particular weighting that can be given to different components (studied variables or measures). This is accomplished by employing a prespecified weight matrix W that is an appropriate square matrix, with which the multivariable distance is more compactly defined as

$$D_W(\underline{x}, \underline{y}) = (\underline{x} - \underline{y})' \, W(\underline{x} - \underline{y}). \tag{2.44}$$

The product in the right-hand side of Equation 2.44 is also denoted by $\|\underline{x} - \underline{y}\|_W^2$.

The weight matrix W can be in general an appropriately sized symmetric matrix with the property that $\underline{z}' \, W \, \underline{z} > 0$ for any nonzero vector $\underline{z}$ (of dimension making it multiplication conform with W). Such a matrix is called positive definite, typically denoted by $W > 0$. Any covariance, correlation, or SSCP matrix of a set of random variables (with real-valued realizations) has this property, if the variables are not linearly related. An interesting feature of a positive definite matrix is that only positive numbers can lie along its main diagonal; that is, if $W = [w_{ii}]$ is positive definite, then $w_{ii} > 0$ ($i = 1, 2, \ldots, q$, where q is the size of the matrix; $q \geq 1$).

The distance $D_W(\underline{x},\underline{y})$ is also called generalized distance between the points $\underline{x}$ and $\underline{y}$ with regard to the weight matrix W. In MVS, the W matrices typically take into account individual differences on the observed variables as well as their interrelationships. In particular, the most important distance measure used in this book will be the so-called Mahalanobis distance (MD; sometimes also called "statistical distance"). This is the distance between an observed data point $\underline{x}$ (i.e., the vector of an individual's scores on p observed variables, $p > 1$), to the mean for a sample, $\underline{\bar{x}}$, weighted by the inverse of the sample covariance matrix S of the p variables, and is denoted by Mah($\underline{x}$):

$$\text{Mah}(x) = (\underline{x} - \underline{\bar{x}})' S^{-1} (\underline{x} - \underline{\bar{x}}). \tag{2.45}$$

Note that this is a distance in the p-space, i.e., the variable space, where the role of axes is played by the measured variables.

As can be easily seen from Equation 2.45, the MD is a direct generalization or extension of the intuitive notion of univariate distance to the multivariate case. A highly useful form of univariate distance is the well-known z-score:

$$z = (x - \bar{x})/s, \tag{2.46}$$

where s denotes the standard deviation of the variable x. Specifically, if we were to write out Equation 2.46 as follows:

$$z^2 = (x - \bar{x})'(s^2)^{-1}(x - \bar{x}) \tag{2.47}$$

and then compare it with Equation 2.45, the analogy is readily apparent. As we have indicated before, this type of "uni-to-multivariate" analogy will turn out to be highly useful as we discuss more complex multivariate methods.

The MD evaluates the distance of each individual data point, in the p-dimensional space, to the centroid of the sample data. The centroid is the point with coordinates being the means of the observed variables, i.e., the vector with elements being the means of the observed variables. Data points $\underline{x}$ with larger Mah($\underline{x}$) are further out from the sample mean than points with smaller MD. In fact, points with very large Mah are potentially abnormal observations. (We discuss this issue in greater detail in the next chapter.)

So how can one evaluate in practice the MD for a given point $\underline{x}$ with respect to the mean of a data set? To accomplish this aim, one needs to find out the means on all variables and their covariance matrix for the particular data set (sample). Then, using either SPSS or SAS, the distance defined in Equation 2.45 of that point, $\underline{x}$, to the centroid of the data set can be evaluated.

To illustrate, consider a study involving three tests (i.e., $p = 3$) of writing, reading, and general mental ability, which were administered to a sample of

$n = 200$ elementary school children. Suppose one were interested in finding out the distance between the point $\underline{x} = [77, 56, 88]'$ of scores obtained by one of the students, to the centroid of the sample data, $\bar{\underline{x}} = [46, 55, 65]'$, with the sample covariance matrix of these three variables being

$$S = \begin{bmatrix} 994.33 & & \\ 653.3 & 873.24 & \\ 554.12 & 629.88 & 769.67 \end{bmatrix}. \tag{2.48}$$

To work out the MD for this point, Mah($\underline{x}$), either the following SPSS command file or subsequent SAS command file can be used. (To save space, only comments are inserted pertaining to operations and entities not used previously in the chapter.)

SPSS command file

```
TITLE 'HOW TO USE SPSS TO WORK OUT MAHALANOBIS DISTANCE'.
MATRIX.
COMP X={77, 56, 88}.
* NOTE THAT WE DEFINE X AS A ROW VECTOR HERE.
COMP X.BAR={46, 55, 65}.
COMP S =  {994.33, 653.30, 554.12;
           653.30, 873.24, 629.88;
           554.12, 629.88, 769.67}.
PRINT S.
COMP S.INV=INV(S).
PRINT S.INV.
COMP MAH.DIST= (X-X.BAR)*S.INV*T(X-X.BAR).
* SEE NOTE AT THE BEGINNING OF THIS INPUT FILE.
PRINT MAH.DIST.
END MATRIX.
```

SAS command file

```
proc iml;
X={77 56 88};
XBAR={46 55 65};
S={994.33 653.30 554.2,
   653.30 873.24 629.88,
   554.12 629.88 769.67}; /* NOTE THE USE OF COMMA TO SEPARATE ROWS*/
print S;
INVS=INV(S);
Print INVS;
DIFF=X-XBAR; /* THIS IS THE DEVIATION SCORE*/
```

```
TRANDIFF=T(DIFF); /* THIS IS THE TRANSPOSE OF THE DEVIATION SCORE*/
MAHDIST=DIFF*INVS*TRANDIFF;
print MAHDIST;
QUIT;
```

The following output results are furnished by these SPSS and SAS program files:

SPSS output

```
HOW TO USE SPSS TO WORK OUT MAHALANOBIS DISTANCE

Run MATRIX procedure:

S
  994.3300000  653.3000000  554.1200000
  653.3000000  873.2400000  629.8800000
  554.1200000  629.8800000  769.6700000

S.INV
  10 ** -3  X
   2.067023074  -1.154499684   -.543327095
  -1.154499684   3.439988733  -1.984030478
   -.543327095  -1.984030478   3.314108030

MAH.DIST
  2.805383491

— END MATRIX —
```

SAS output

```
                 The SAS System

                       S
     994.33          653.3          554.2
     653.3           873.24         629.88
     554.12          629.88         769.67

                      INVS

   0.0020671      -0.001154      -0.000544
  -0.001155        0.0034398     -0.001984
  -0.000543       -0.001984       0.0033143

                    MAHDIST

                   2.8051519
```

It is important to note that the inverse of the empirical covariance matrix provided in Equation 2.48—which is denoted S.INV in the SPSS command file and INVS in the SAS file—is the matrix that is instrumentally needed for calculating the MD value of 2.805. Further, in SPSS the matrix S.INV is provided in scientific notation, whereas it is not so in SAS.

We finalize this chapter by emphasizing once again that the concept of MD renders important information about the distance between points in a p-dimensional variable space, for any given sample of data from a multivariable study. In addition, as we will see in more detail in the next chapter, it allows us to evaluate whether some observations may be very different from the majority in a given sample. Last but not least, it helps us conceptually understand a number of multivariate statistical procedures, especially those related to group differences, which are dealt with in later chapters.

3

Data Screening and Preliminary Analyses

Results obtained through application of univariate or multivariate statistical methods will in general depend critically on the quality of the data and on the numerical magnitude of the elements of the data matrix as well as variable relationships. For this reason, after data are collected in an empirical study and before they are analyzed using a particular method(s) to respond to a research question(s) of concern, one needs to conduct what is typically referred to as data screening. These preliminary activities aim (a) to ensure that the data to be analyzed represent correctly the data originally obtained, (b) to search for any potentially very influential observations, and (c) to assess whether assumptions underlying the method(s) to be applied subsequently are plausible. This chapter addresses these issues.

3.1 Initial Data Exploration

To obtain veridical results from an empirical investigation, the data collected in it must have been accurately entered into the data file submitted to the computer for analysis. Mistakes committed during the process of data entry can be very costly and can result in incorrect parameter estimates, standard errors, and test statistics, potentially yielding misleading substantive conclusions. Hence, one needs to spend as much time as necessary to screen the data for entry errors, before proceeding with the application of any uni- or multivariate method aimed at responding to the posited research question(s). Although this process of data screening may be quite time consuming, it is an indispensable prerequisite of a trustworthy data analytic session, and the time invested in data screening will always prove to be worthwhile.

Once a data set is obtained in a study, it is essential to begin with proofreading the available data file. With a small data set, it may be best to check each original record (i.e., each subject's data) for correct entry. With larger data sets, however, this may not be a viable option, and so one may instead arrange to have at least two independent data entry sessions followed by a comparison of the resulting files. Where discrepancies are found, examination of the raw (original) data records must then be carried out in order to correctly represent the data into a computer file to be analyzed subsequently using particular statistical methods. Obviously, the use of independent data entry sessions can prove to be expensive and time consuming. In addition, although such checks may resolve noted discrepancies when entering the data into a file, they will not detect possible common errors across all entry sessions or incorrect records in the original data. Therefore, for any data set once entered into a computer file and proofread, it is recommended that a researcher carefully examine frequencies and descriptive statistics for each variable across all studied persons. (In situations involving multiple-population studies, this should also be carried out within each group or sample.) Thereby, one should check, in particular, the range of each variable, and specifically whether the recorded maximum and minimum values on it make sense. Further, when examining each variable's frequencies, one should also check if all values listed in the frequency table are legitimate. In this way, errors at the data-recording stage can be spotted and immediately corrected.

To illustrate these very important preliminary activities, let us consider a study in which data were collected from a sample of 40 university freshmen on a measure of their success in an educational program (referred to below as "exam score" and recorded in a percentage correct metric), and its relationship to an aptitude measure, age in years, an intelligence test score, as well as a measure of attention span. (The data for this study can be found in the file named ch3ex1.dat available from www.psypress.com/applied-multivariate-analysis.) To initially screen the data set, we begin by examining the frequencies and descriptive statistics of all variables.

To accomplish this initial data screening in SPSS, we use the following menu options (in the order given next) to obtain the variable frequencies:

Analyze → Descriptive statistics → Frequencies,

and, correspondingly, to furnish their descriptive statistics:

Analyze → Descriptive statistics → Descriptives.

In order to generate the variable frequencies and descriptive statistics in SAS, the following command file can be used. In SAS, there are often a number of different ways to accomplish the same aim. The commands provided below were selected to maintain similarity with the structure of the output rendered by the above SPSS analysis session. In particular, the order of the options in the SAS PROC MEANS statement is structured to create similar output (with the exception of fw=6, which requests the field width of the displayed statistics be set at 6—alternatively, the command "maxdec=6" could be used to specify the maximum number of decimal places to output).

```
DATA CHAPTER3;
INFILE 'ch3ex1.dat';
INPUT id Exam_Score Aptitude_Measure Age_in_Years
  Intelligence_Score Attention_Span;
PROC MEANS n range min max mean std fw=6;
  var Exam_Score Aptitude_Measure Age_in_Years
  Intelligence_Score Attention_Span;
RUN;
PROC FREQ;
TABLES Exam_Score Aptitude_Measure Age_in_Years
  Intelligence_Score Attention_Span;
RUN;
```

The resulting outputs produced by SPSS and SAS are as follows:

SPSS descriptive statistics output

Descriptive Statistics

	N	Range	Minimum	Maximum	Mean	Std. Deviation
Exam Score	40	102	50	152	57.60	16.123
Aptitude Measure	40	24	20	44	23.12	3.589
Age in Years	40	9	15	24	18.22	1.441
Intelligence Score	40	8	96	104	99.00	2.418
Attention Span	40	7	16	23	20.02	1.349
Valid N (listwise)	40					

SAS descriptive statistics output

The SAS System

The MEANS Procedure

Variable	N	Range	Min	Max	Mean	Std Dev
Exam_Score	40	102.0	50.00	152.0	57.60	16.12
Aptitude _Measure	40	24.00	20.00	44.00	23.13	3.589
Age_in_Years	40	9.000	15.00	24.00	18.23	1.441
Intelligence _Score	40	8.000	96.00	104.0	99.00	2.418
Attention _Span	40	7.000	16.00	23.00	20.03	1.349

By examining the descriptive statistics in either of the above tables, we readily observe the high range on the dependent variable Exam Score. This apparent anomaly is also detected by looking at the frequency distribution of each measure, in particular of the same variable. The pertinent output sections are as follows:

SPSS frequencies output

Frequencies

Exam Score

		Frequency	Percent	Valid Percent	Cumulative Percent
Valid	50	5	12.5	12.5	12.5
	51	3	7.5	7.5	20.0
	52	8	20.0	20.0	40.0
	53	5	12.5	12.5	52.5
	54	3	7.5	7.5	60.0
	55	3	7.5	7.5	67.5
	56	1	2.5	2.5	70.0
	57	3	7.5	7.5	77.5
	62	1	2.5	2.5	80.0
	63	3	7.5	7.5	87.5
	64	1	2.5	2.5	90.0
	65	2	5.0	5.0	95.0
	69	1	2.5	2.5	97.5
	152	1	2.5	2.5	100.0
	Total	40	100.0	100.0	

Note how the score 152 "sticks out" from the rest of the values observed on the Exam Score variable—there is no one else having a score even close to 152; the latter finding is also not unexpected because as mentioned this variable was recorded in the metric of percentage correct responses. We continue our examination of the remaining measures in the study and return later to the issue of discussing and dealing with found anomalous, or at least apparently so, values.

Aptitude Measure

		Frequency	Percent	Valid Percent	Cumulative Percent
Valid	20	2	5.0	5.0	5.0
	21	6	15.0	15.0	20.0
	22	8	20.0	20.0	40.0
	23	14	35.0	35.0	75.0
	24	8	20.0	20.0	95.0
	25	1	2.5	2.5	97.5
	44	1	2.5	2.5	100.0
	Total	40	100.0	100.0	

Here we also note a subject whose aptitude score tends to stand out from the rest: the one with a score of 44.

Age in Years

		Frequency	Percent	Valid Percent	Cumulative Percent
Valid	15	1	2.5	2.5	2.5
	16	1	2.5	2.5	5.0
	17	9	22.5	22.5	27.5
	18	15	37.5	37.5	65.0
	19	9	22.5	22.5	87.5
	20	4	10.0	10.0	97.5
	24	1	2.5	2.5	100.0
	Total	40	100.0	100.0	

On the age variable, we observe that a subject seems to be very different from the remaining persons with regard to age, having a low value of 15. Given that this is a study of university freshmen, although not a common phenomenon to encounter someone that young, such an age per se does not seem really unusual for attending college.

Intelligence Score

		Frequency	Percent	Valid Percent	Cumulative Percent
Valid	96	9	22.5	22.5	22.5
	97	4	10.0	10.0	32.5
	98	5	12.5	12.5	45.0
	99	5	12.5	12.5	57.5
	100	6	15.0	15.0	72.5
	101	5	12.5	12.5	85.0
	102	2	5.0	5.0	90.0
	103	2	5.0	5.0	95.0
	104	2	5.0	5.0	100.0
	Total	40	100.0	100.0	

The range of scores on this measure also seems to be well within what could be considered consistent with expectations in a study involving university freshmen.

Attention Span

		Frequency	Percent	Valid Percent	Cumulative Percent
Valid	16	1	2.5	2.5	2.5
	18	6	15.0	15.0	17.5
	19	2	5.0	5.0	22.5
	20	16	40.0	40.0	62.5
	21	12	30.0	30.0	92.5
	22	2	5.0	5.0	97.5
	23	1	2.5	2.5	100.0
	Total	40	100.0	100.0	

Finally, with regard to the variable attention span, there is no subject that appears to have an excessively high or low score compared to the rest of the available sample.

SAS frequencies output

Because the similarly structured output created by SAS would obviously lead to interpretations akin to those offered above, we dispense with inserting comments in the next presented sections.

The SAS System

The FREQ Procedure

Exam_Score	Frequency	Percent	Cumulative Frequency	Cumulative Percent
50	5	12.50	5	12.50
51	3	7.50	8	20.00
52	8	20.00	16	40.00
53	5	12.50	21	52.50
54	3	7.50	24	60.00
55	3	7.50	27	67.50
56	1	2.50	28	70.00
57	3	7.50	31	77.50
62	1	2.50	32	80.00
63	3	7.50	35	87.50
64	1	2.50	36	90.00
65	2	5.00	38	95.00
69	1	2.50	39	97.50
152	1	2.50	40	100.00

Aptitude_Measure	Frequency	Percent	Cumulative Frequency	Cumulative Percent
20	2	5.00	2	5.00
21	6	15.00	8	20.00
22	8	20.00	16	40.00
23	14	35.00	30	75.00
24	8	20.00	38	95.00
25	1	2.50	39	97.50
44	1	2.50	40	100.00

Age_in_Years	Frequency	Percent	Cumulative Frequency	Cumulative Percent
15	1	2.50	1	2.50
16	1	2.50	2	5.00
17	9	22.50	11	27.50
18	15	37.50	26	65.00
19	9	22.50	35	87.50
20	4	10.00	39	97.50
24	1	2.50	40	100.00

Intelligence_Score	Frequency	Percent	Cumulative Frequency	Cumulative Percent
96	9	22.50	9	22.50
97	4	10.00	13	32.50
98	5	12.50	18	45.00
99	5	12.50	23	57.50
100	6	15.00	29	72.50
101	5	12.50	34	85.00
102	2	5.00	36	90.00
103	2	5.00	38	95.00
104	2	5.00	40	100.00

Attention_Span	Frequency	Percent	Cumulative Frequency	Cumulative Percent
16	1	2.50	1	2.50
18	6	15.00	7	17.50
19	2	5.00	9	22.50
20	16	40.00	25	62.50
21	12	30.00	37	92.50
22	2	5.00	39	97.50
23	1	2.50	40	100.00

Although examining the descriptive statistics and frequency distributions across all variables is highly informative, in the sense that one learns what the data actually are (especially when looking at their frequency tables), it is worthwhile noting that these statistics and distributions are only available for each variable when considered separately from the others. That is, like the descriptive statistics, frequency distributions provide only univariate information with regard to the relationships among the values that subjects give rise to on a given measure. Hence, when an (apparently) anomalous value is found for a particular variable, neither descriptive statistics nor frequency tables can provide further information about the person(s) with that anomalous score, in particular regarding their scores on some or all of the remaining measures. As a first step toward obtaining such information, it is helpful to extract the data on all variables for any subject exhibiting a seemingly extreme value on one or more of them. For example, to find out who the person was with the exam score of 152, its extraction from the file is accomplished in SPSS by using the following menu options/sequence (the variable Exam Score is named "exam_score" in the data file):

Data → Select cases → If condition "exam_score=152" (check "delete unselected cases").

To accomplish the printing of apparently aberrant data records, the following command line would be added to the above SAS program:

```
IF Exam_Score=152 THEN LIST;
```

Consequently, each time a score of 152 is detected (in the present example, just once) SAS prints the current input data line in the SAS log file.

When this activity is carried out and one takes a look at that person's scores on all variables, it is readily seen that apart from the screening results mentioned, his/her values on the remaining measures are unremarkable (i.e., lie within the variable-specific range for meaningful scores; in actual fact, reference to the original data record would reveal that this subject had an exam score of 52 and his value of 152 in the data file simply resulted from a typographical error).

After the data on all variables are examined for each subject with anomalous value on at least one of them, the next question that needs to be addressed refers to the reason(s) for this data abnormality. As we have just seen, the latter may result from an incorrect data entry, in which case the value is simply corrected according to the original data record. Alternatively, the extreme score may have been due to a failure to declare to the software a missing value code, so that a data point is read by the computer program as a legitimate value while it is not. (Oftentimes, this may be the result of a too hasty move on to the data analysis phase, even a preliminary one, by a researcher skipping this declaration step.) Another possibility could be that the person(s) with an out-of-range value may actually not be a member of the population intended to be studied, but happened to be included in the investigation for some unrelated reasons. In this case, his/her entire data record would have to be deleted from the data set and following analyses. Furthermore, and no less importantly, an apparently anomalous value may in fact be a legitimate value for a sample from a population where the distribution of the variable in question is highly skewed. Because of the potential impact such situations can have on data analysis results, these circumstances are addressed in greater detail in a later section of the chapter. We move next to a more formal discussion of extreme scores, which helps additionally in the process of handling abnormal data values.

3.2 Outliers and the Search for Them

As indicated in Section 3.1, the relevance of an examination for extreme observations, or so-called outliers, follows from the fact that these may exert very strong influence upon the results of ensuing analyses. An outlier is a case with (a) such an extreme value on a given variable, or (b) such an abnormal combination of values on several variables, which may render it having a substantial impact on the outcomes of a data analysis and modeling session. In case (a), the observation is called univariate outlier, while in case (b) it is referred to as multivariate outlier. Whenever even a single outlier (whether univariate or multivariate) is present in a data set, results generated with and without that observation(s) may be very different, leading to possibly incompatible substantive conclusions. For this reason, it is critically important to also consider some formal means that can be used to routinely search for outliers in a given data set.

3.2.1 Univariate Outliers

Univariate outliers are usually easier to spot than multivariate outliers. Typically, univariate outliers are to be sought among those observations with the following properties: (a) the magnitude of their z-scores is greater than 3 or smaller than -3; and (b) their z-scores are to some extent "disconnected" from the z-scores of the remaining observations. One of the easiest ways to search for univariate outliers is to use descriptive methods and/or graphical methods. The essence of using the descriptive methods is to check for individual observations with the properties (a) and (b) just mentioned. In contrast, graphical methods involve the use of various plots, including boxplots, steam-and-leaf plots, and normal probability (detrended) plots for studied variables. Before we discuss this topic further, let us mention in passing that often with large samples (at least in the hundreds), there may occasionally be a few apparent extreme observations that need not necessarily be outliers. The reason is that large samples have a relatively high chance of including extreme cases in a studied population that are legitimate members of it and thus need not be removed from the ensuing analyses.

To illustrate, consider the earlier study of university freshmen on the relationship between success in an educational program, aptitude, age, intelligence, and attention span (see data file ch3ex1.dat available from www.psypress.com/applied-multivariate-analysis). To search for univariate outliers, we first obtain the z-scores for all variables. This is readily achieved with SPSS using the following menu options/sequence:

Analyze → Descriptive statistics → Descriptives (check "save standardized values").

With SAS, the following PROC STANDARD command lines could be used:

```
DATA CHAPTER3;
INFILE 'ch3ex1.dat';
INPUT id Exam_Score Aptitude_Measure Age_in_Years
   Intelligence_Score Attention_Span;
zscore=Exam_Score;
PROC STANDARD mean=0 std=1 out=newscore;
  var zscore;
RUN;
PROC print data=newscore;
  var Exam_Score zscore;
  title 'Standardized Exam Scores';
RUN;
```

In these SAS statements, PROC STANDARD standardizes the specified variable from the data set (for our illustrative purposes, in this example only the variable exam_score was selected), using a mean of 0 and standard deviation of 1, and then creates a new SAS data set (defined here as the outfile "newscore") that contains the resulting standardized values. The PROC PRINT statement subsequently prints the original values alongside the standardized values for each individual on the named variables.

As a result of these software activities, SPSS and SAS generate an extended data file containing both the original variables plus a "copy" of each one of them, which consists of all subjects' z-scores; to save space, we only provide next the output generated by the above SAS statements (in which the variable "Exam Score" was selected for standardization).

Standardized Test Scores

Obs	Exam_Score	zscore
1	51	−0.40936
2	53	−0.28531
3	50	−0.47139
4	63	0.33493
5	65	0.45898
6	53	−0.28531
7	52	−0.34734
8	50	−0.47139
9	57	−0.03721
10	54	−0.22329
11	65	0.45898
12	50	−0.47139
13	52	−0.34734
14	63	0.33493
15	52	−0.34734
16	52	−0.34734
17	51	−0.40936
18	52	−0.34734
19	55	−0.16126

20	55	−0.16126
21	53	−0.28531
22	54	−0.22329
23	152	5.85513
24	50	−0.47139
25	63	0.33493
26	57	−0.03721
27	52	−0.34734
28	62	0.27291
29	52	−0.34734
30	55	−0.16126
31	54	−0.22329
32	56	−0.09924
33	52	−0.34734
34	53	−0.28531
35	64	0.39696
36	57	−0.03721
37	50	−0.47139
38	51	−0.40936
39	53	−0.28531
40	69	0.70708

Looking through the column labeled "zscore" in the last output table (and in general each of the columns generated for the remaining variables under consideration), we try to spot the *z*-scores that are larger than 3 or smaller than −3 and at the same time "stick out" of the remaining values in that column. (With a larger data set, it is also helpful to request the descriptive statistics for each variable along with their corresponding *z*-scores, and then look for any extreme values.) In this illustrative example, subject #23 clearly has a very large *z*-score relative to the rest of the observations on exam score (viz. larger than 5, although as discussed above this was clearly a data entry error). If we similarly examined the *z*-scores on the other variables (not tabled above), we would observe no apparent univariate outliers with respect to the variables Intelligence and Attention Span; however, we would find out that subject #40 had a large *z*-score on the Aptitude measure (*z*-score = 5.82), like subject #8 on age (*z*-score – 4.01).

Once possible univariate outliers are located in a data set, the next step is to search for the presence of multivariate outliers. We stress that it may be premature to make a decision for deleting a univariate outlier before examination for multivariate outliers is conducted.

3.2.2 Multivariate Outliers

Searching for multivariate outliers is considerably more difficult to carry out than examination for univariate outliers. As mentioned in the preceding section, a multivariate outlier is an observation with values on several variables that are not necessarily abnormal when each variable is considered separately, but are unusual in their combination. For example, in a study concerning income of college students, someone who reports an income of \$100,000 per year is not an unusual observation per se. Similarly, someone who reports that they are 16 years of age would not be considered an unusual observation. However, a case with these two measures in combination is likely to be highly unusual, that is, a possible multivariate outlier (Tabachnick & Fidell, 2007).

This example shows the necessity of utilizing such formal means when searching for multivariate outliers, which capitalize in an appropriate way on the individual variable values for each subject and at the same time also take into consideration their interrelationships. A very useful statistic in this regard is the Mahalanobis distance (MD) that we have discussed in Chapter 2. As indicated there, in an empirical setting, the MD represents the distance of a subject's data to the centroid (mean) of all cases in an available sample, that is, to the point in the multivariate space, which has as coordinates the means of all observed variables. That the MD is so instrumental in searching for multivariate outliers should actually not be unexpected, considering the earlier mentioned fact that it is the multivariate analog of univariate distance, as reflected in the *z*-score (see pertinent discussion in Chapter 2). As mentioned earlier, the MD is also frequently referred to as statistical distance since it takes into account the variances and covariances for all pairs of studied variables. In particular, from two variables with different variances, the one with larger variability will contribute less to the MD; further, two highly correlated variables will contribute less to the MD than two nearly uncorrelated ones. The reason is that the inverse of the empirical covariance matrix participates in the MD, and in effect assigns in this way weights of "importance" to the contribution of each variable to the MD.

In addition to being closely related to the concept of univariate distance, it can be shown that with multinormal data on a given set of variables and a large sample, the Mahalanobis distance follows approximately a chi-square distribution with degrees of freedom being the number of these variables (with this approximation becoming much better with larger samples) (Johnson & Wichern, 2002). This characteristic of the MD helps considerably in the search for multivariate outliers. Indeed, given this distributional property, one may consider an observation as a possible multivariate outlier if its MD is larger than the critical point (generally specified at a conservative recommended significance level of $\alpha = .001$) of the chi-square distribution with degrees of freedom being the number of variables participating in the MD. We note that the MDs for different observations are not unrelated to one another, as can be seen from their formal definition in Chapter 2. This suggests the need for some caution when using the MD in searching for multivariate outliers, especially with samples that cannot be considered large.

We already discussed in Chapter 2 a straightforward way of computing the MD for any particular observation from a data set. Using it for examination of multivariate outliers, however, can be a very tedious and time-consuming activity especially with large data sets. Instead, one can use alternative approaches that are readily applied with statistical software. Specifically, in the case of SPSS, one can simply regress a variable of no interest (e.g., subject ID, or case number) upon all variables participating in the MD; requesting thereby the MD for each subject yields as a byproduct this distance for all observations (Tabachnick & Fidell, 2007). We stress that the results of this multiple regression analysis are of no interest and value per se, apart from providing, of course, each individual's MD.

As an example, consider the earlier study of university freshmen on their success in an educational program in relation to their aptitude, age, intelligence, and attention span. (See data file ch3ex1.dat available from www.psypress.com/applied-multivariate-analysis.) To obtain the MD for each subject, we use in SPSS the following menu options/sequence:

Analyze → Regression → Linear → (ID as DV; all others as IVs)
→ Save "Mahalanobis Distance"

At the end of this analysis, a new variable is added by the software to the original data file, named MAH_1, which contains the MD values for each subject. (We note in passing that a number of SPSS macros have also been proposed in the literature for the same purposes, which are readily available.) (De Carlo, 1997).

In order to accomplish the same goal with SAS, several options exist. One of them is provided by the following PROC IML program:

```
title 'Mahalanobis Distance Values';
DATA CHAPTER3;
INFILE 'ch3ex1.dat';
INPUT id $ y1 y2 y3 y4 y5;
%let id=id; /* THE %let IS A MACRO STATEMENT*/
%let var=y1 y2 y3 y4 y5; /* DEFINES A VARIABLE */
PROC iml;
  start dsquare;
    use _last_;
    read all var {&var} into y [colname=vars rowname=&id];
    n=nrow(y);
    p=ncol(y);
    r1=&id;
    mean=y[:,];
```

```
    d=y-j(n,1)*mean;
    s=d'* d/(n-1);
    dsq=vecdiag(d* inv(s) * d');
    r=rank(dsq); /* ranks the values of dsq */
    val=dsq; dsq[r, ]=val;
    val=r1; &id [r]=val;
    result=dsq;
    cl={'dsq'};
  create dsquare from result [colname=cl rowname=&id];
  append from result [rowname=&id];
finish;
print dsquare;
run dsquare;
quit;
PROC print data=dsquare;
    var id dsq;
run;
```

The following output results would be obtained by submitting this command file to SAS (since the resulting output from SPSS would lead to the same individual MDs, we only provide next those generated by SAS); the column headings "ID" and "dsq" below correspond to subject ID number and MD, respectively. (Note that the observations are rank ordered according to their MD rather than their identification number.)

Mahalanobis Distance Values

Obs	ID	dsq
1	6	0.0992
2	34	0.1810
3	33	0.4039
4	3	0.4764
5	36	0.6769
6	25	0.7401
7	16	0.7651
8	38	0.8257
9	22	0.8821
10	32	1.0610
11	27	1.0714
12	21	1.1987
13	7	1.5199
14	14	1.5487
15	1	1.6823
16	2	2.0967
17	30	2.2345
18	28	2.5811
19	18	2.7049

```
20          13          2.8883
21          10          2.9170
22          31          2.9884
23           5          3.0018
24          29          3.0367
25           9          3.1060
26          19          3.1308
27          35          3.1815
28          12          3.6398
29          26          3.6548
30          15          3.8936
31           4          4.1176
32          17          4.4722
33          39          4.5406
34          24          4.7062
35          20          5.1592
36          37         13.0175
37          11         13.8536
38           8         17.1867
39          40         34.0070
40          23         35.7510
```

Mahalanobis distance measures can also be obtained in SAS by using the procedure PROC PRINCOMP along with the STD option. (These are based on computing the uncorrected sum of squared principal component scores within each output observation; see pertinent discussion in Chapters 1 and 7.) Accordingly, the following SAS program would generate the same MD values as displayed above (but ordered by subject ID instead):

```
PROC PRINCOMP std out=scores noprint;
  var Exam_Score Aptitude_Measure Age_in_Years
    Intelligence_Score Attention_Span;
RUN;
DATA mahdist;
    set scores;
    md=(uss(of prin1-prin5));
RUN;
PROC PRINT;
  var md;
RUN;
```

Yet another option available in SAS is to use the multiple regression procedure PROC REG and, similarly to the approach utilized with SPSS above, regress a variable of no interest (e.g., subject ID) upon all variables participating in the MD. The information of relevance to this discussion is obtained using the INFLUENCE statistics option as illustrated in the next program code.

```
PROC REG;
  model id=Exam_Score Aptitude_Measure Age_in_Years
           Intelligence_Score Attention_Span/INFLUENCE;
RUN;
```

This INFLUENCE option approach within PROC REG does not directly provide the values of the MD but a closely related individual statistic called leverage—commonly denoted by h_i and labeled in the SAS output as HAT DIAG H (for further details, see Belsley, Kuh, & Welsch, 1980). However, the leverage statistic can easily be used to determine MD values for each observation in a considered data set. In particular, it has been shown that MD and leverage are related (in the case under consideration) as follows:

$$\text{MD} = (n - 1)(h_i - 1/n), \tag{3.1}$$

where n denotes sample size and h_i is the leverage associated with the ith subject (i=1, . . . , n) (Belsley et al., 1980).

Note from Equation 3.1 that MD and leverage are directly proportional to one another—as MD grows (decreases) so does leverage.

The output resulting from submitting these PROC REG command lines to SAS is given below:

The SAS System

The REG Procedure
Model: MODEL1
Dependent Variable: id
Output Statistics

Obs	Residual	RStudent	Hat Diag H	Cov Ratio	DFFITS
1	−14.9860	−1.5872	0.0681	0.8256	−0.4292
2	−14.9323	−1.5908	0.0788	0.8334	−0.4652
3	−18.3368	−1.9442	0.0372	0.6481	−0.3822
4	−9.9411	−1.0687	0.1306	1.1218	−0.4142
5	−13.7132	−1.4721	0.1020	0.9094	−0.4961
6	−14.5586	−1.5039	0.0275	0.8264	−0.2531
7	−6.2042	−0.6358	0.0640	1.1879	−0.1662

8	−2.2869	−0.3088	0.4657	2.2003	−0.2882
9	−7.0221	−0.7373	0.1046	1.2112	−0.2521
10	−7.2634	−0.7610	0.0998	1.1971	−0.2534
11	3.0439	0.3819	0.3802	1.8796	0.2991
12	−4.5687	−0.4812	0.1183	1.3010	−0.1763
13	4.0729	0.4240	0.0991	1.2851	0.1406
14	−9.3569	−0.9669	0.0647	1.0816	−0.2543
15	−1.8641	−0.1965	0.1248	1.3572	−0.0742
16	−4.7932	−0.4850	0.0446	1.1998	−0.1048
17	−0.5673	−0.0603	0.1397	1.3894	−0.0243
18	0.9985	0.1034	0.0944	1.3182	0.0334
19	−11.8243	−1.2612	0.1053	1.0079	−0.4326
20	2.4913	0.2677	0.1573	1.4011	0.1157
21	6.9400	0.7092	0.0557	1.1569	0.1723
22	4.7030	0.4765	0.0476	1.2053	0.1066
23	0.2974	0.1214	0.9417	20.4599	0.4880
24	−8.1462	−0.8786	0.1457	1.2187	−0.3628
25	1.8029	0.1818	0.0440	1.2437	0.0390
26	10.6511	1.1399	0.1187	1.0766	0.4184
27	12.1511	1.2594	0.0525	0.9525	0.2964
28	7.7030	0.8040	0.0912	1.1715	0.2547
29	0.6869	0.0715	0.1029	1.3321	0.0242
30	6.6844	0.6926	0.0823	1.1953	0.2074
31	2.0881	0.2173	0.1016	1.3201	0.0731
32	9.5648	0.9822	0.0522	1.0617	0.2305
33	12.4692	1.2819	0.0354	0.9264	0.2454
34	14.2581	1.4725	0.0296	0.8415	0.2574
35	4.2887	0.4485	0.1066	1.2909	0.1549
36	15.9407	1.6719	0.0424	0.7669	0.3516
37	4.0544	0.5009	0.3588	1.7826	0.3746
38	19.1304	2.0495	0.0462	0.6111	0.4509
39	6.8041	0.7294	0.1414	1.2657	0.2961
40	−0.4596	−0.1411	0.8970	11.5683	−0.4165

As can be readily seen, using Equation 3.1 with, say, the obtained leverage value of 0.0681 for subject #1 in the original data file, his/her MD is computed as

$$MD = (40 - 1)(0.0681 - 1/40) = 1.681, \tag{3.2}$$

which corresponds to his or her MD value in the previously presented output.

By inspection of the last displayed output section, it is readily found that subjects #23 and #40 have notably large MD values—above 30—that may fulfill the above-indicated criterion of being possible multivariate outliers. Indeed, since we have analyzed simultaneously $p = 5$ variables, we are dealing with 5 degrees of freedom for this evaluation, and at a significance level of $\alpha = .001$, the corresponding chi-square cutoff is 20.515 that is exceeded by the MD of these two cases. Alternatively, requesting extraction from the data file of all subjects' records for whom their MD value is larger than 20.515 (see preceding section) would yield only these two subjects with values beyond this cutoff that can be, thus, potentially considered as multivariate outliers.

With respect to examining leverage values, we note in passing that they range from 0 to 1 with $(p+1)/n$ being their average (in this empirical example, 0.15). Rules of thumb concerning high values of leverage have also been suggested in the literature, whereby in general observations with leverage greater than a certain cutoff may be considered multivariate outliers (Fung, 1993; Huber, 1981). These cutoffs are based on the above-indicated MD cutoff at a specified significance level α (denoted MD_α). Specifically, the leverage cutoffs are

$$h_{\text{cutoff}} = (\text{MD}_\alpha)/(n-1) + 1/n, \tag{3.3}$$

which yields $20.515/39 + 1/40 = .551$ for the currently considered example. With the use of Equation 3.3, if one were to utilize the output generated by PROC REG, there is no need to convert to MD the then reported leverage values to determine the observations that may be considered multivariate outliers. In this way, it can be readily seen that only subjects #23 and #40 could be suggested as multivariate outliers.

Using diagnostic measures to identify an observation as a possible multivariate outlier depends on a potentially rather complicated correlational structure among a set of studied variables. It is therefore quite possible that some observations may have a masking effect upon others. That is, one or more subjects may appear to be possible multivariate outliers, yet if one were to delete them, other observations might emerge then as such. In other words, the former group of observations, while being in the data file, could mask the latter ones that, thus, could not be sensed at an initial inspection as possible outliers. For this reason, if one eventually decides to delete outliers masked by previously removed ones, ensuing analysis findings must be treated with great caution since there is a potential that the latter may have resulted from capitalization on chance fluctuations in the available sample.

3.2.3 Handling Outliers: A Revisit

Multivariate outliers may be often found among those that are univariate outliers, but there may also be cases that do not have extreme values on separately considered variables (one at a time). Either way, once an observation is deemed to be a possible outlier, a decision needs to be made with respect to handling it. To this end, first one should try to use all available information, or information that it is possible to obtain, to determine what reason(s) may have led to the observation appearing as an outlier. Coding or typographical errors, instrument malfunction or incorrect instructions during its administration, or being a member of another population that is not of interest are often sufficient grounds to correspondingly correct or consider removing the particular observation(s) from further analyses. Second, when there is no such relatively easily found reason, it is important to assess to what degree the observation(s) in question may be reflecting legitimate variability in the studied population. If the latter is the case, instead of subject removal variable transformations may be worth considering, a topic that is discussed later in this chapter.

There is a growing literature on robust statistics that deals with methods aimed at down-weighting the contribution of potential outliers to the results of statistical analyses (Wilcox, 2003). Unfortunately, at present there are still no widely available and easily applicable multivariate robust statistical methods. For this reason, we only mention here this direction of current methodological developments that is likely to contribute in the future readily used procedures for differential weighting of observations in multivariate analyses. These procedures will also be worth considering in empirical settings with potential outliers.

When one or more possible outliers are identified, it should be borne in mind that any one of these may unduly influence the ensuing statistical analysis results, but need not do so. In particular, an outlier may or may not be an influential observation in this sense. The degree to which it is influential is reflected in what are referred to as influence statistics and related quantities (such as the leverage value discussed earlier) (Pedhazur, 1997). These statistics have been developed within a regression analysis framework and made easily available in most statistical software. In fact, it is possible that keeping one or more outliers in the subsequent analyses will not change their results appreciably, and especially their substantive interpretations. In such a case, the decision regarding whether to keep them in the analysis or not does not have a real impact upon the final conclusions. Alternatively, if the results and their interpretation depend on whether the outliers are retained in the analyses, while a clear-cut decision for removal versus no removal cannot be reached, it is important to provide the results and interpretations in both cases. For the case where the outlier is removed, it is also necessary that one explicitly mentions, that is, specifically reports, the characteristics of the deleted outlier(s), and then restricts the final substantive conclusions to a population that does not contain members with the outliers' values on the studied variables. For example, if one has good reasons to exclude the subject with ID = 8 from the above study of university freshmen, who was 15 years old, one should also explicitly state in the substantive result interpretations of the following statistical analyses that they do not necessarily generalize to subjects in their mid-teens.

3.3 Checking of Variable Distribution Assumptions

The multivariate statistical methods we consider in this text are based on the assumption of multivariate normality for the dependent variables. Although this assumption is not used for parameter estimation purposes, it is needed when statistical tests and inference are performed. Multivariate normality (MVN) holds when and only when any linear combination of the individual variables involved is univariate normal (Roussas, 1997). Hence, testing for multivariate normality per se is not practically possible, since it involves infinitely many tests. However, there are several implications of MVN that can be empirically tested. These represent necessary conditions, rather than sufficient conditions, for multivariate normality. That is, these are implied by MVN, but none of these conditions by itself or in combination with any other(s) condition(s) entails multivariate normality.

In particular, if a set of p variables is multivariate normally distributed, then each of them is univariate normal ($p > 1$). In addition, any pair or subset of k variables from that set is bivariate or k-dimensional normal, respectively ($2 < k < p$). Further, at any given value for a single variable (or values for a subset of k variables), the remaining variables are jointly multivariate normal, and their variability does not depend on that value (or values, $2 < k < p$); moreover, the relationship of any of these variables, and a subset of the remaining ones that are not fixed, is linear.

To examine univariate normality, two distributional indices can be judged: skewness and kurtosis. These are closely related to the third and fourth moments of the underlying variable distribution, respectively. The skewness characterizes the symmetry of the distribution. A univariate normally distributed variable has a skewness index that is equal to zero. Deviations from this value on the positive or negative side indicate asymmetry. The kurtosis characterizes the shape of the distribution in terms of whether it is peaked or flat relative to a corresponding normal distribution (with the same mean and variance). A univariate normally distributed variable has a kurtosis that is (effectively) equal to zero, whereby positive values are indicative of a leptokurtic distribution and negative values of a platykurtic distribution. Two statistical tests for evaluating univariate normality are also usually considered, the Kolmogorov–Smirnov Test and the Shapiro–Wilk Test. If the sample size cannot be considered large, the Shapiro–Wilk Test may be preferred, whereas if the sample size is large the Kolmogorov–Smirnov Test is highly trustworthy. In general terms, both tests consider the following null hypothesis H_0: "The sampled data have been drawn from a normally distributed population." Rejection of this hypothesis at some prespecified significance level is suggestive of the data not coming from a population where the variable in question is normally distributed.

To examine multivariate normality, two analogous measures of skewness and kurtosis—called Mardia's skewness and kurtosis—have been developed (Mardia, 1970). In cases where the data are multivariate normal, the skewness coefficient is zero and the kurtosis is equal to $p(p+2)$; for example, in case of bivariate normality, Mardia's skewness is 0 and kurtosis is 8. Consequently, similar to evaluating their univariate counterparts, if the distribution is, say, leptokurtic, Mardia's measure of kurtosis will be comparatively large, whereas if it is platykurtic, the coefficient will be small. Mardia (1970) also showed that these two measures of multivariate normality can be statistically evaluated. Although most statistical analysis programs readily provide output of univariate skewness and kurtosis (see examples and discussion in Section 3.4), multivariate measures are not as yet commonly evaluated by software. For example, in order to obtain Mardia's coefficients with SAS, one could use the macro called %MULTNORM. Similarly, with SPSS, the macro developed by De Carlo (1997) could be utilized. Alternatively, structural equation modeling software may be employed for this purpose (Bentler, 2004; Jöreskog & Sörbom, 1996).

In addition to examining normality by means of the above-mentioned statistical tests, it can also be assessed by using some informal methods. In case of univariate normality, the so-called normal probability plot (often also referred to as Q–Q plot) or the detrended normal probability plot can be considered. The normal probability plot is a graphical representation in which each observation is plotted against a corresponding theoretical normal distribution value such that the points fall along a diagonal straight line in case of normality. Departures from the straight line indicate violations of the normality assumption. The detrended probability plot is similar, with deviations from that diagonal line effectively plotted horizontally. If the data are normally distributed, the observations will be basically evenly distributed above and below the horizontal line in the latter plot (see illustrations considered in Section 3.4).

Another method that can be used to examine multivariate normality is to create a graph that plots the MD for each observation against its ordered chi-square percentile value (see earlier in the chapter). If the data are multivariate normal, the plotted values should be close to a straight line, whereas points that fall far from the line may be multivariate outliers (Marcoulides & Hershberger, 1997). For example, the following PROC IML program could be used to generate such a plot:

```
TITLE 'Chi-Square Plot';
DATA CHAPTER3;
INFILE 'ch3ex1.dat';
INPUT id $ y1 y2 y3 y4 y5;
%let id=id;
%let var=y1 y2 y3 y4 y5;
PROC iml;
  start dsquare;
    use_last_;
    read all var {&var} into y [colname=vars rowname=&id];
    n=nrow(y);
    p=ncol(y);
    r1=&id;
    mean=y[ :,];
    d=y - j(n,1)*mean;
    s=d'* d / (n - 1);
    dsq=vecdiag(d* inv(s) * d');
    r=rank(dsq);
    val=dsq; dsq[r, ]=val;
    val=r1; &id [r]=val;
    z=((1:n)' - .5)/n;
    chisq=2 * gaminv(z, p/2);
    result=dsq||chisq;
    cl={'dsq' 'chisq'};
  create dsquare from result [colname=cl rowname=&id];
  append from result [rowname=&id];
finish;
print dsquare; /* THIS COMMAND IS ONLY NEEDED IF YOU WISH TO PRINT THE MD */
RUN dsquare;
quit;
PROC print data=dsquare;
   var id dsq chisq;
RUN;
PROC gplot data=dsquare;
plot chisq*dsq;
RUN;
```

This command file is quite similar to that presented earlier in Section 3.2.2, with the only difference being that now, in addition to the MD values, ordered chi-square percentile values are computed. Submitting this PROC IML program to SAS for the last considered data set generates the

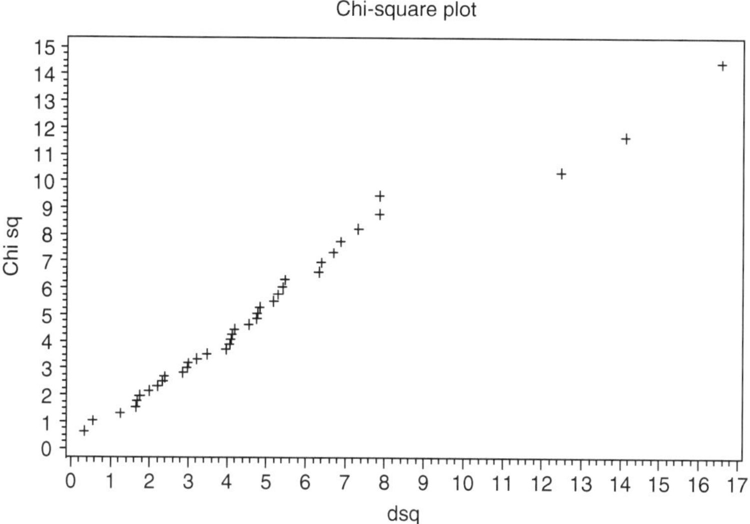

FIGURE 3.1
Chi-square plot for assessing multivariate normality.

above multivariate probability plot (if first removing the data lines for subjects #23 and #40 suggested previously as multivariate outliers).

An examination of Figure 3.1 reveals that the plotted values are reasonably close to a diagonal straight line, indicating that the data do not deviate considerably from normality (keeping in mind, of course, the relatively small sample size used for this illustration).

The discussion in this section suggests that examination of MVN is a difficult yet important topic that has been widely discussed in the literature, and there are a number of excellent and accessible treatments of it (Mardia, 1970; Johnson & Wichern, 2002). In conclusion, we mention that most MVS methods that we deal with in this text can tolerate minor nonnormality (i.e., their results can be viewed also then as trustworthy). However, in empirical applications it is important to consider all the issues discussed in this section, so that a researcher becomes aware of the degree to which the normality assumption may be violated in an analyzed data set.

3.4 Variable Transformations

When data are found to be decidedly nonnormal, in particular on a given variable, it may be possible to transform that variable to be closer to normally distributed whereupon the set of variables under consideration would likely better comply with the multivariate normality assumption. (There is no guarantee for multinormality as a result of the transformation, however, as indicated in Section 3.3.) In this section, we discuss a class of transformations that can be used to deal with the lack of symmetry of individual variables, an important aspect of deviation from the normal distribution that as well known is symmetric. As it often happens, dealing with this aspect of normality deviation may also improve variable kurtosis and make it closer to that of the normal distribution. Before we begin, however, let us emphasize that asymmetry or skewness as well as excessive kurtosis—and consequently nonnormality in general—may be primarily the result of outliers being present in a given data set. Hence, before considering any particular transformation, it is recommended that one first examines the data for potential outliers. In the remainder of this section, we assume that the latter issue has been already handled.

We start with relatively weak transformations that are usually applicable with mild asymmetry (skewness) and gradually move on to stronger transformations that may be used on distributions with considerably longer and heavier tails. If the observed skewness is not very pronounced and positive, chances are that the square root transformation, $Y' = \sqrt{Y}$, where Y is the original variable, will lead to a transformed measure Y' with a distribution that is considerably closer to the normal (assuming that all Y scores are positive). With SPSS, to obtain the square-rooted variable Y', we use

Transform → Compute,

and then enter in the small left- and right-opened windows correspondingly

SQRT_Y=SQRT(Y),

where Y is the original variable. In the syntax mode of SPSS, this is equivalent to the command

COMPUTE SQRT_Y=SQRT(Y).

(which as mentioned earlier may be abbreviated to COMP SQRT_Y=SQRT(Y).)

With SAS, this can be accomplished by inserting the following general format data-modifying statement immediately after the INPUT statement (but before any PROC statement is invoked):

New-Variable-Name=Formula-Specifying-Manipulation-of-an-Existing-Variable

For example, the following SAS statement could be used in this way for the square root transformation:

SQRT_Y=SQRT(Y),

which is obviously quite similar to the above syntax with SPSS.

If for some subjects $Y < 0$, since a square root cannot be taken then, we first add the absolute value of the smallest of them to all scores, and then proceed with the following SPSS syntax mode command that is to be executed in the same manner as above:

```
COMP SQRT_Y=SQRT(Y + |MIN(Y)|).
```

where |MIN(Y)| denotes the absolute value of the smallest negative Y score, which may have been obtained beforehand, for example, with the descriptives procedure (see discussion earlier in the chapter). With SAS, the same operation could be accomplished using the command:

```
SQRT_Y=SQRT(Y + ABS(min(Y)),
```

where ABS(min(Y)) is the absolute value of the smallest negative Y score (which can either be obtained directly or furnished beforehand, as mentioned above).

For variables with more pronounced positive skewness, the stronger logarithmic transformation may be more appropriate. The notion of "stronger" transformation is used in this section to refer to a transformation with a more pronounced effect upon a variable under consideration. In the presently considered setting, such a transformation would reduce more notably variable skewness; see below. The logarithmic transformation can be carried out with SPSS using the command:

```
COMP LN_Y=LN(Y).
```

or with SAS employing the command:

```
LN_Y=log(Y);
```

assuming all Y scores are positive since otherwise the logarithm is not defined. If for some cases $Y = 0$ (and for none $Y < 0$ holds), we add 1 first to Y and then take the logarithm, which can be accomplished in SPSS and SAS using respectively the following commands:

```
COMP LN_Y=LN(Y + 1).

LN_Y=log(Y + 1);
```

If for some subjects $Y < 0$, we first add to all scores $1 + |MIN(Y)|$, and then take the logarithm (as indicated above).

A stronger yet transformation is the inverse, which is more effective on distributions with larger skewness, for which the logarithm does not render them close to normality. This transformation is obtained as follows using either of the following SPSS or SAS commands, respectively:

```
COMP INV_Y=1/Y.

INV_Y=1/Y;
```

in cases where there are no zero scores. Alternatively, if for some cases $Y = 0$, we add first 1 to Y before taking inverse:

COMPUTE INV_Y=1/(Y + 1).

or

INV_Y=1/(Y + 1);

(If there are zero and negative scores in the data, we add first to all scores 1 plus the absolute value of their minimum, and then proceed as in the last two equations.) An even stronger transformation is the inverse squared, which under the assumption of no zero scores in the data can be obtained using the commands:

COMPUTE INVSQ_Y=1/Y^2.

or

INV_Y=1/(Y**2);

If there are some cases with negative scores, or zero scores, first add the constant 1 plus the absolute value of their minimum to all subjects' data, and then proceed with this transformation.

When a variable is negatively skewed (i.e., its left tail is longer than its right one), then one needs to first "reflect" the distribution before conducting any further transformations. Such a reflection of the distribution can be accomplished by subtracting each original score from 1 plus their maximum, as illustrated in the following SPSS statement:

COMPUTE Y_NEW=MAX(Y) + 1 – Y.

where MAX(Y) is the highest score in the sample, which may have been obtained beforehand (e.g., with the descriptives procedure). With SAS, this operation is accomplished using the command:

SQRT_Y=max(Y) + 1 – Y;

where max(Y) returns the largest value of Y (obtained directly, or using instead that value furnished beforehand via examination of variable descriptive statistics). Once reflected in this way, the variable in question is positively skewed and all above discussion concerning transformations is then applicable.

In an empirical study, it is possible that a weaker transformation does not render a distribution close to normality, for example, when the transformed distribution still has a significant and substantial skewness (see below for a pertinent testing procedure). Therefore, one needs to examine the transformed variable for normality before proceeding with it in any analyses that assume normality. In this sense, if one transformation is not strong enough, it is recommendable that a stronger transformation be chosen. However, if one applies a stronger than necessary transformation, the sign of the skewness may end up being changed (e.g., from positive to negative). Hence, one might better start with the weakest transformation that appears to be worthwhile trying (e.g., square root). Further, and no less important, as indicated above, it is always worthwhile examining whether excessive asymmetry (and kurtosis) may be due to outliers. If the transformed variable exhibits substantial skewness, it is recommendable that one examines it, in addition to the pretransformed variable, also for outliers (see Section 3.3).

Before moving on to an example, let us stress that caution is advised when interpreting the results of statistical analyses that use transformed variables. This is because the units and possibly origin of measurement have been changed by the transformation, and thus those of the transformed variable(s) are no longer identical to the variables underlying the original measure(s). However, all above transformations (and the ones mentioned at the conclusion of this section) are monotone, that is, they preserve the rank ordering of the studied subjects. Hence, when units of measurement are arbitrary or irrelevant, a transformation may not lead to a considerable loss of substantive interpretability of the final analytic results. It is also worth mentioning at this point that the discussed transformed variables result from other than linear transformations, and hence their correlational structure is in general different from that of the original variables. This consequence may be particularly relevant in settings where one considers subsequent analysis of the structure underlying the studied variables (such as factor analysis; see Chapter 8). In those cases, the alteration of the relationships among these variables may contribute to a decision perhaps not to transform the variables but instead to use subsequently specific correction methods that are available within the general framework of latent variable modeling, for which we refer to alternative sources (Muthén, 2002; Muthén & Muthén, 2006; for a nontechnical introduction, see Raykov & Marcoulides, 2006).

To exemplify the preceding discussion in this section, consider data obtained from a study in which $n = 150$ students were administered a test of inductive reasoning ability (denoted IR1 in the data file named ch3ex2.dat available from www.psypress.com/applied-multivariate-analysis). To examine the distribution of their scores on this intelligence measure, with SPSS we use the following menu options/sequence:

Analyze → Descriptive statistics → Explore,

whereas with SAS the following command file could be used:

```
DATA Chapter3EX2;
INFILE 'ch3ex2.dat';
INPUT ir1 group gender sqrt_ir1 ln_ir1;
PROC UNIVARIATE plot normal;
```

```
/* Note that instead of the "plot" statement, additional
 commands like "QQPLOT", "PROBPLOT" or "HISTOGRAM" can be
 provided in a line below to create separate plots */
var ir1;
RUN;
```

The resulting outputs produced by SPSS and SAS are as follows (provided in segments to simplify the discussion).

SPSS descriptive statistics output

Descriptives

			Statistic	Std. Error
IR1	Mean		30.5145	1.20818
	95% Confidence Interval for Mean	Lower Bound	28.1272	
		Upper Bound	32.9019	
	5% Trimmed Mean		29.9512	
	Median		28.5800	
	Variance		218.954	
	Std. Deviation		14.79710	
	Minimum		1.43	
	Maximum		78.60	
	Range		77.17	
	Interquartile Range		18.5700	
	Skewness		.643	.198
	Kurtosis		.158	.394

Extreme Values

			Case Number	Value
IR1	Highest	1	100	78.60
		2	60	71.45
		3	16	64.31
		4	107	61.45
		5	20	60.02[a]
	Lowest	1	22	1.43
		2	129	7.15
		3	126	7.15
		4	76	7.15
		5	66	7.15[b]

a. Only a partial list of cases with the value 60.02 are shown in the table of upper extremes.

b. Only a partial list of cases with the value 7.15 are shown in the table of lower extremes.

SAS descriptive statistics output

The SAS System
The UNIVARIATE Procedure
Variable: ir1

Moments

N	150	Sum Weights	150
Mean	30.5145333	Sum Observations	4577.18
Std Deviation	14.7971049	Variance	218.954312
Skewness	0.64299511	Kurtosis	0.15756849
Uncorrected SS	172294.704	Corrected SS	32624.1925
Coeff Variation	48.4919913	Std Error Mean	1.20817855

Basic Statistical Measures

Location		Variability	
Mean	30.51453	Std Deviation	14.79710
Median	28.58000	Variance	218.95431
Mode	25.72000	Range	77.17000
		Interquartile Range	18.57000

Extreme Observations

----Lowest----		----Highest----	
Value	Obs	Value	Obs
1.43	22	60.02	78
7.15	129	61.45	107
7.15	126	64.31	16
7.15	76	71.45	60
7.15	66	78.60	100

As can be readily seen by examining the skewness and kurtosis in either of the above sections with descriptive statistics, skewness of the variable under consideration is positive and quite large (as well as significant, since the ratio of its estimate to standard error is larger than 2; recall that at α=.05, the cutoff is $\pm$1.96 for this ratio that follows a normal distribution). Such a finding is not the case for its kurtosis, however. With respect to the listed extreme values, at this point, we withhold judgment about any of these 10 cases since their being apparently extreme may actually be due to lack of normality. We turn next to this issue.

SPSS tests of normality

Tests of Normality

	Kolmogoroy-Smirnov[a]			Shapiro-Wilk		
	Statistic	**df**	**Sig.**	**Statistic**	**df**	**Sig.**
IR1	.094	150	.003	.968	150	.002

a. Lilliefors Significance Correction

SAS tests of normality

```
                         Tests for Normality
Test                    ----Statistic----     ------p Value------

Shapiro-Wilk            W         0.96824     Pr < W         0.0015
Kolmogorov-Smirnov      D        0.093705     Pr > D        <0.0100
Cramer-von Mises        W-Sq     0.224096     Pr > W-Sq     <0.0050
Anderson-Darling        A-Sq     1.348968     Pr > A-Sq     <0.0050
```

As mentioned in Section 3.3, two statistical means can be employed to examine normality, the Kolmogorov–Smirnov (K–S) and Shapiro–Wilk (S–W) tests. (SAS also provides the Cramer–von Mises and the Anderson–Darling tests, which may be viewed as modifications of the K–S Test.) Note that both the K–S and S–W tests indicate that the normality assumption is violated.

The graphical output created by SPSS and SAS would lead to essentially identical plots. To save space, below we only provide the output generated by invoking the SPSS commands given earlier in this section.

Consistent with our earlier findings regarding skewness, the positive tail of the distribution is considerably longer, as seen by examining the following histogram, stem-and-leaf plot, and box plot. This can also be noticed when inspecting the normal probability plots provided next. The degree of skewness is especially evident when examining the detrended plot next, in which the observations are not close to evenly distributed following and below the horizontal line.

So far, we have seen substantial evidence for pronounced skewness of the variable in question to the right. In an attempt to deal with this skewness, which does not appear to be excessive, we try first the square root transformation on this measure, which is the weakest from the ones

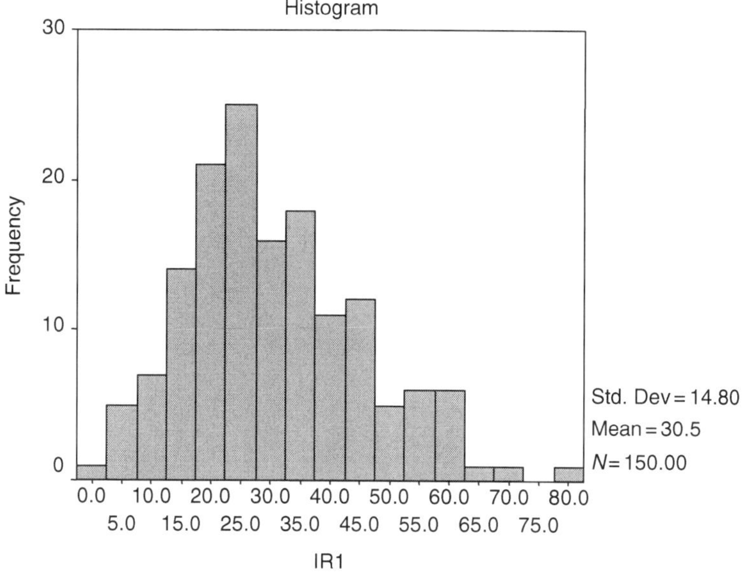

discussed above. To this end, we use with SPSS the following menu options/command (which as illustrated earlier, could also be readily implemented with SAS):

```
IR1 Stem-and-Leaf Plot
Frequency   Stem & Leaf

  1.00        0 . 1
  7.00        0 . 7777788
 11.00        1 . 00011222244
 17.00        1 . 55555777888888888
 23.00        2 . 00000011111122224444444
 21.00        2 . 555555555557778888888
 22.00        3 . 0000111112222222444444
 13.00        3 . 5577788888888
 10.00        4 . 0112222444
  7.00        4 . 5577788
  5.00        5 . 01122
  7.00        5 . 5577888
  4.00        6 . 0014
  2.00 Extremes   (>=71)
Each leaf:        1 case(s)
Stem width:          10.00
```

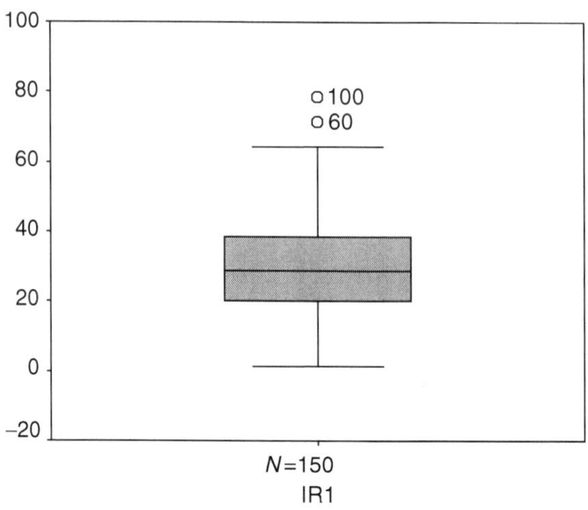

100
80
60
40
20
0
–20
100
60
N=150
IR1

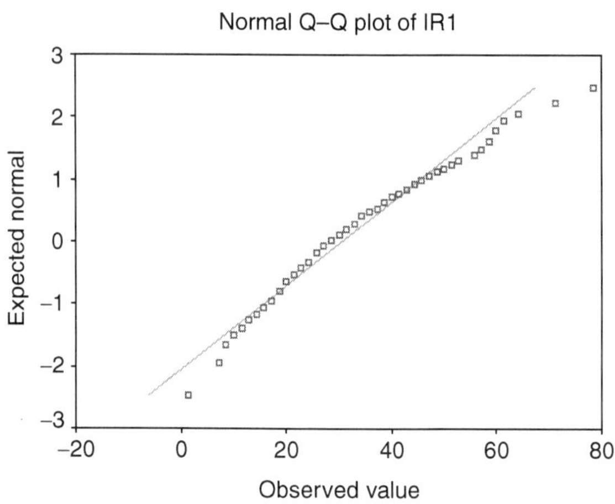

Normal Q–Q plot of IR1
Expected normal
Observed value

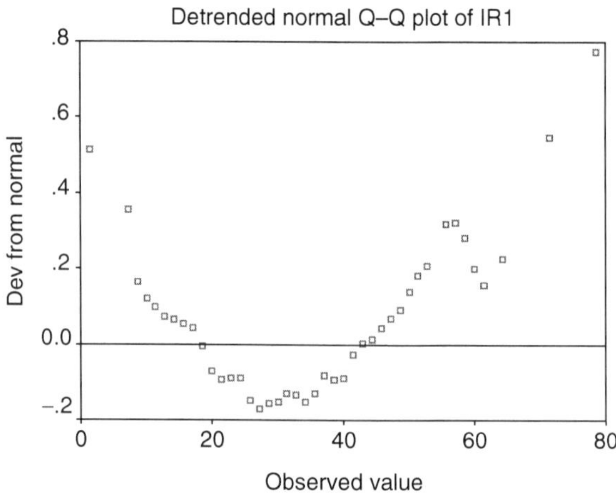

Detrended normal Q–Q plot of IR1
Dev from normal
Observed value

Transform → Compute

(SQRT_IR1=SQRT(IR1))

or COMP SQRT_IR1=SQRT(IR1)

in the syntax mode. Now, to see whether this transformation is sufficient to deal with the problem of positive and marked skewness, we explore the distribution of the so-transformed variable and obtain the following output (presented only using SPSS, since that created by SAS would lead to the same results).

Descriptives

			Statistic	Std. Error
SQRT_IR1	Mean		5.3528	.11178
	95% Confidence Interval for Mean	Lower Bound	5.1319	
		Upper Bound	5.5737	
	5% Trimmed Mean		5.3616	
	Median		5.3460	
	Variance		1.874	
	Std. Deviation		1.36905	
	Minimum		1.20	
	Maximum		8.87	
	Range		7.67	
	Interquartile Range		1.7380	
	Skewness		−.046	.198
	Kurtosis		−.058	.394

As seen by examining this table, the skewness of the transformed variable is no longer significant (like its kurtosis), and the null hypothesis of its distribution being normal is not rejected (see tests of normality in the next table).

Tests of Normality

	Kolmogorov-Smirnov[a]			Shapiro-Wilk		
	Statistic	df	Sig.	Statistic	df	Sig.
SQRT_IR1	.048	150	.200*	.994	150	.840

*. This is a lower bound of the true significance.
a. Lilliefors Significance Correction.

With this in mind, examining the histogram, stem-and-leaf plot, and box plot presented next, given the relatively limited sample size, it is plausible to consider the distribution of the square-rooted inductive reasoning score as much closer to normal than the initial variable. (We should not over-interpret the seemingly heavier left tail in the last histogram, since its appearance is in part due to the default intervals that the software selects internally.) We stress that with samples that are small, some (apparent) deviations from normality may not result from inherent lack of normality of a studied variable in the population of concern, but may be consequences of the sizable sampling error involved. We therefore do not look for nearly "perfect" signs of normality in the graphs to follow, but only for strong and unambiguous deviation patterns (across several of the plots).

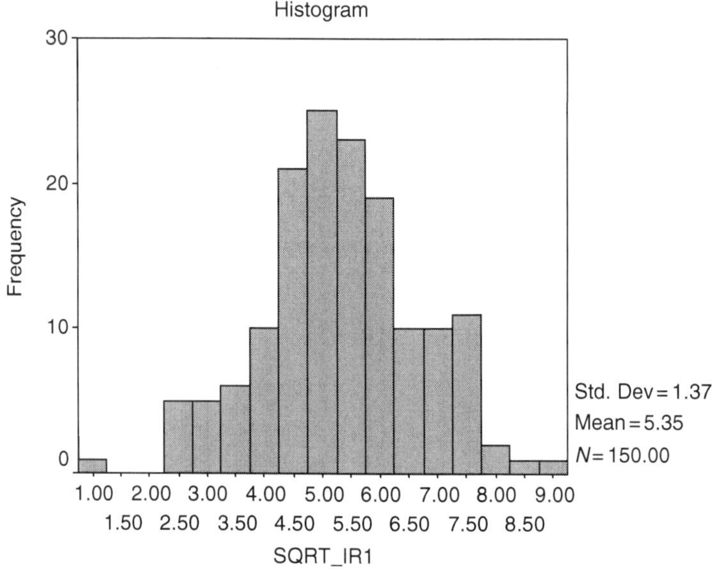

SQRT_IR1 Stem-and-Leaf Plot

Frequency	Stem & Leaf
1.00 Extremes	(=<1.2)
7.00	2. 6666699
5.00	3. 11133
11.00	3. 55557799999
18.00	4. 111333333333444444
17.00	4. 66666677779999999
25.00	5. 0000000000022233333334444
20.00	5. 66666777777788888899
14.00	6. 00022222222344
14.00	6. 55556667788899
7.00	7. 0112244
8.00	7. 55666778
2.00	8. 04
1.00 Extremes	(>=8.9)

Stem width: 1.00
Each leaf: 1 case(s)

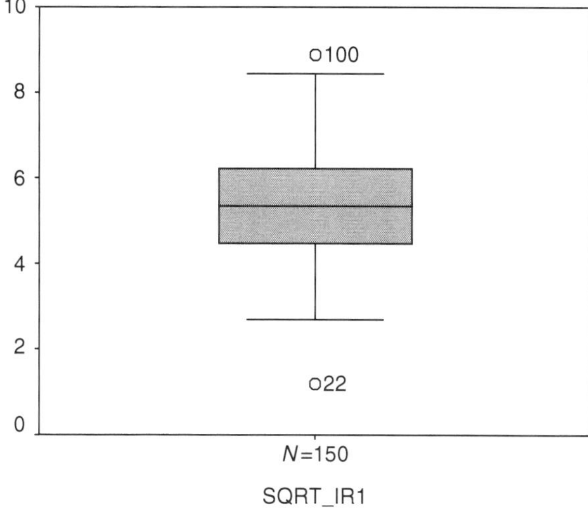

In addition to the last three plots, plausibility of the normality assumption is also suggested from an inspection of the next presented normal probability plots.

As a side note, if we had inadvertently applied the stronger logarithmic transformation instead of the square root, we would have in fact induced negative skewness on the distribution. (As mentioned before, this can happen if too strong a transformation is used.) For illustrative purposes, we present next the relevant part of the data exploration descriptive output that would be obtained then.

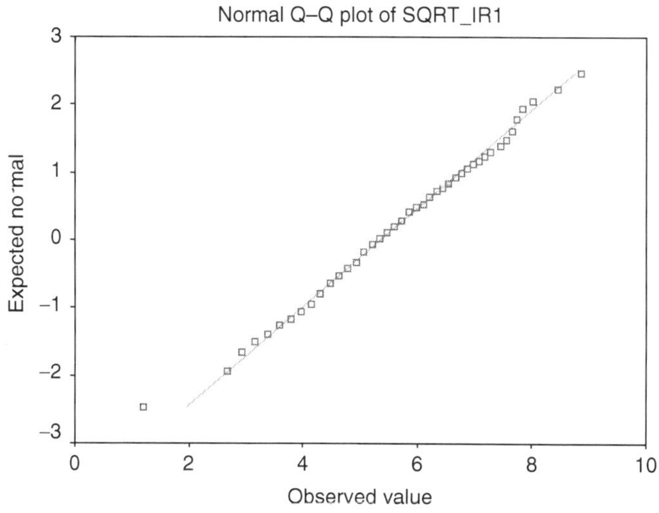

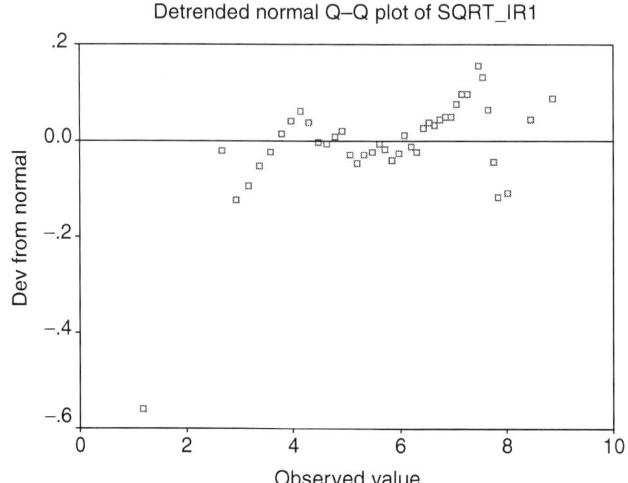

Descriptives

			Statistic	Std. Error
LN_IR1	Mean		3.2809	.04700
	95% Confidence Interval for Mean	Lower Bound	3.1880	
		Upper Bound	3.3738	
	5% Trimmed Mean		3.3149	
	Median		3.3527	
	Variance		.331	
	Std. Deviation		.57565	
	Minimum		.36	
	Maximum		4.36	
	Range		4.01	
	Interquartile Range		.6565	
	Skewness		−1.229	.198
	Kurtosis		3.706	.394

Tests of Normality

	Kolmogorov-Smirnov[a]			Shapiro-Wilk		
	Statistic	df	Sig.	Statistic	df	Sig.
LN_IR1	.091	150	.004	.933	150	.000

a. Lilliefors Significance Correction.

This example demonstrates that considerable caution is advised whenever transformations are used, as one also runs the potential danger of "overdoing" it if an unnecessarily strong transformation is chosen. Although in many cases in empirical research, some of the above-mentioned transformations will often render the resulting variable distribution close to normal, this need not always happen. In the latter cases, it may be recommended that one use the so-called likelihood-based method to determine an appropriate power to which the original measure could be raised in order to achieve closer approximation by the normal distribution. This method yields the most favorable transformation with regard to univariate normality, and does not proceed through examination in a step-by-step manner of possible choices as above. Rather, that transformation is selected based on a procedure considering the likelihood function of the observed data. This procedure is developed within the framework of what is referred to as Box–Cox family of variable transformations, and an instructive discussion of it is provided in the original publication by Box and Cox (1964).

In conclusion, we stress that oftentimes in empirical research, a transformation that renders a variable closer to normality may also lead to comparable variances of the resulting variable across groups in a given study. This variance homogeneity result is then an added bonus of the utilized transformation, and is relevant because many univariate as well as multivariate methods are based on the assumption of such homogeneity (and specifically, as we will see in the next chapter, on the more general assumption of homogeneity of the covariance matrix of the dependent variables).

4

Multivariate Analysis of Group Differences

This chapter is the first in the book to deal with a specific multivariate method. We commence with an example that highlights the general need to use suitable procedures for examining group differences simultaneously on several dependent variables (DVs). This is followed by a formal definition of the multivariate normal distribution. We then introduce methods for testing hypotheses about multidimensional means, discussing initially the case of a known population covariance matrix, followed by that involving an unknown population covariance matrix. Subsequently, we turn to testing hypotheses about multivariate means of two samples or groups, considering first the situation in which the samples are related (dependent) and then that of unrelated (independent) samples. We conclude the chapter with a discussion of an approach for testing hypotheses about means with more than two independent groups, which is widely known as multivariate analysis of variance (MANOVA). The case of more than two related groups is dealt with in Chapter 5.

4.1 A Start-Up Example

We begin by introducing an exemplary empirical setting concerned with studying the outcomes of an instructional program for high school students, where of interest is to examine potential group differences. Suppose that in an effort to address this question, researchers employ two random samples of high school students assessed on their vocabulary acquisition, reading comprehension, verbal and written communication, and their ability to discriminate literary styles (Tatsuoka, 1988). Since these are all measures of main interest, they can be viewed as dependent (response, outcome) variables. Furthermore, let us assume that in this two-group study, one group was designated as control while the students in the other group were exposed to a program whose effectiveness is to be evaluated (experimental group). The researchers are interested in finding out whether there are group differences on the DVs, which cannot be explained only by chance fluctuations in the two samples.

That is, the main concern is to ascertain whether observed differences on the dependent measures can be viewed merely as consequences of sampling error resulting from the fact that one has not studied entire populations of interest, but only portions of them as reflected in the samples at hand.

The first query that one needs to address in this setup is how to conceptualize group differences. A possibility that might perhaps come immediately to mind is individual differences on each of the five outcome variables listed above when considered in turn, which is the well-known univariate approach. At least as interesting, however, would be to evaluate possible group differences on these measures when considered together (i.e., in their "totality"). Obviously the first mentioned approach, the univariate, does not really tell us whether the instructional program is effective overall. The latter issue is addressed by the second method, when the DVs are considered simultaneously. Yet we need to recognize that this is a much more complicated approach to deal with than the univariate. On the other hand, the univariate approach does not seem to be revealing, when interested in ascertaining whether the program was effective as a whole. In particular, what would one have to conclude if the two groups were found to differ only slightly on the DVs, but with no significance in any pertinent univariate test? Should we then conclude that the program was not effective at all? Alternatively, what would one be willing to conclude in the case that significance was found on some of the DVs but not on others? Even more difficult to handle would be the case where significant and positive differences were found on some of the measures, but significant and negative differences on other DVs.

This discussion illustrates that what one needs in order to answer the research question concerning effectiveness of the instructional program, is a method for studying group differences on several (dependent) variables that are considered simultaneously. This method should also be devised in such a way that it possesses the following properties: (a) it controls the overall Type I error; and (b) it allows taking into account these variables' interrelationships, since otherwise one will be wasting information that may be of particular relevance to the research query.

In order to present such a method, we begin with a formal definition of the multivariate normal distribution in Section 4.2. We defined this distribution somewhat informally in Chapter 3, mentioning there that a set of variables was multivariate normal if and only if any linear combination of them was univariate normal. While this is an informative statement, it is hard to capitalize on in empirical research settings, especially those resembling the one described above. To give a more formal definition of multivariate normality (MVN), or simply multinormality, we utilize the analogy to the univariate normal distribution. As indicated in earlier chapters, employing such analogies between multivariate methods and their univariate counterparts will be a great aid when introducing and discussing many of the topics throughout this book. (In fact, we have already used this type of analogy in defining the Mahalanobis distance [MD, also referred to as multivariate distance in this chapter] in Chapter 3, as well as when discussing covariance, correlation, and sum of squares and cross products [SSCP] matrices in Chapters 1 and 2.)

4.2 A Definition of the Multivariate Normal Distribution

Throughout this book, as mentioned earlier, MVN will be a basic assumption for strict validity of significance tests and statistical inference. For this reason, it is important to have a formal definition of MVN, which we provide in this section. To this end, let us first recall the definition of univariate normality. As it is well known, a (unidimensional) random variable X is normally distributed with a mean of μ and variance σ^2 (with $\sigma^2 > 0$), which is usually denoted $X \sim N(\mu, \sigma^2)$, if its probability density function is

$$f(x) = (2\pi\sigma^2)^{-1/2} \exp[-(x-\mu)^2/(2\sigma^2)],$$

which is equivalently rewritten as

$$f(x) = (2\pi)^{-1/2}(\sigma^2)^{-1/2} \exp\left[-\frac{1}{2}(x-\mu)(\sigma^2)^{-1}(x-\mu)\right]. \tag{4.1}$$

Using now the univariate-to-multivariate analogy, which we already employed in previous chapters, we can extend the function in Equation 4.1 to define a p-dimensional random vector $\underline{X}$ (with $p > 1$) as normally distributed with a mean vector $\underline{\mu}$ and covariance matrix Σ (where $\Sigma > 0$), denoted $\underline{X} \sim N_p(\underline{\mu}, \Sigma)$, if its probability density function is

$$f(\underline{x}) = (2\pi)^{-p/2}|\Sigma|^{-1/2} \exp\left[-\frac{1}{2}(\underline{x}-\underline{\mu})'\Sigma^{-1}(\underline{x}-\underline{\mu})\right]. \tag{4.2}$$

We observe that one formally obtains the right-hand side of Equation 4.2 from the right-hand side of Equation 4.1 by exchanging in the latter unidimensional quantities in the exponent with p-dimensional ones, and the variance with the determinant of the covariance matrix (accounting in addition for the fact that now p dimensions are considered simultaneously rather than a single one). We note in this definition the essential requirement of positive definiteness of the covariance matrix Σ (see Chapter 2), because in the alternative case, the right-hand side of Equation 4.2 will not be defined. (In that case, however, which will not be of relevance in the remainder of the book, one could still use the definition of MVN from Chapter 2; Anderson, 1984).

The following two properties, and especially the second one, will be of particular importance later in this chapter. We note that the second property is formally obtained by analogy from the first one.

Property 1: If a sample of size n is drawn from the univariate normal distribution $N(\mu, \sigma^2)$, then the mean $\overline{X}$ of the sample will be distributed as $N(\mu, \sigma^2/n)$, that is, $\overline{X} \sim N(\mu, \sigma^2/n)$.

This property provides the rationale as to why the sample mean is more stable than just a single observation from a studied population—the reason is the smaller variance of the mean.

Property 2: If a sample of size n is drawn from the multivariate normal distribution $N_p(\underline{\mu}, \Sigma)$, then the mean vector $\underline{\overline{X}}$ of the sample will be distributed as $N_p(\underline{\mu}, (1/n)\Sigma)$, that is, $\underline{\overline{X}} \sim N_p(\underline{\mu}, (1/n)\Sigma)$.

We observe that Property 1 is obviously obtained as a special case of Property 2, namely when $p = 1$. We conclude this section by noting that we make the multinormality assumption for the rest of the chapter (see Chapter 3 for its examination).

4.3 Testing Hypotheses About a Multivariate Mean

Suppose for a moment that we were interested in examining whether the means of university freshmen on two distinct intelligence test scores were each equal to 100. Since these are typically interrelated measures of mental ability, it would be wasteful of empirical information not to consider them simultaneously but to test instead each one of them separately for equality of its mean to 100 (e.g., using the single-group t test). For this reason, in lieu of the latter univariate approach, what we would like to do is test whether the mean of the two-dimensional vector of this pair of random variables is the point in the two-dimensional space, which has its both coordinates equal to 100. In more general terms, we would like to test the hypothesis that the two-dimensional mean $\underline{\mu}$ of the random vector consisting of these two interrelated intelligence measures equals $\underline{\mu}_0$, that is, the null hypothesis H_0: $\underline{\mu} = \underline{\mu}_0$, where $\underline{\mu}_0$ is a vector having as elements prespecified numbers. (In the example under consideration, $\underline{\mu}_0 = [100, 100]'$.)

This goal can be accomplished using Property 2 from the preceding section. A straightforward approach to conducting this test is achieved by using the duality principle between confidence interval (confidence region) and hypothesis testing (Hays, 1994), as well as the concept of Mahalanobis distance (MD). According to this principle, a 95%-confidence region for the multivariate mean $\underline{\mu}$ is the area consisting of all points $\underline{\mu}_0$ in the respective multivariate space, for which the null hypothesis H_0: $\underline{\mu} = \underline{\mu}_0$ is not rejected at the $\alpha = .05$ significance level. (With another significance level, the confidence level is correspondingly obtained as the complement to 1 of the former.)

To see this principle at work, let us make use of the univariate-to-multivariate analogy. Recall what is involved in testing the univariate version of the above null hypothesis H_0. Accordingly, testing the hypothesis $\mu = \mu_0$ for a prespecified real number μ_0, is effectively the same as finding out whether μ_0 is a plausible value for the mean μ, given the data. Hence, if μ_0 lies within a plausible range of values for the mean, we do not reject this hypothesis; otherwise we do. This is the same as saying that we do not reject the hypothesis $\mu = \mu_0$ if and only if μ_0 belongs to a plausible range of values for μ, that is, falls in the confidence interval of the mean. Thus, we can carry out hypothesis testing by evaluating a confidence interval and checking whether μ_0 is covered by that interval. If it is, we do not have enough evidence to warrant rejection of the null hypothesis $\mu = \mu_0$, but if it is not, then we can reject that hypothesis. Note that the essence of this principle is a logical one and does not really depend on the dimensionality of μ and μ_0 (whether they are scalars or at least two-dimensional vectors). Therefore, we can use it in the multivariate case as well.

We next note that this testing approach capitalizes on viewing a confidence interval as a set of values for a given parameter, in this case the mean, whose distance from its empirical estimate (here, the sample mean) is sufficiently small. This view is also independent of the dimensionality of the parameter in question. If it is multidimensional, as is the case mostly in this book, the role of distance can obviously be played by the MD that we got familiar with in Chapter 2. That is, returning to our above concern with testing a multivariate null hypothesis, H_0: $\underline{\mu} = \underline{\mu}_0$, we can say that we reject it if the MD of the sample mean vector (point) to the vector (point) $\underline{\mu}_0$ in the p-dimensional space is large enough, while we do not reject this hypothesis if the sample mean is sufficiently close to the hypothetical mean vector $\underline{\mu}_0$. We formalize next these developments further and thereby take into account two possible situations.

4.3.1 The Case of Known Covariance Matrix

We consider again an example, but this time using numerical values. Let us assume that reading and writing scores in a population of third graders follow a two-dimensional normal distribution with a population covariance matrix equal to

$$\Sigma = \begin{bmatrix} 25 & 12 \\ 12 & 27 \end{bmatrix}. \tag{4.3}$$

Suppose also that a sample is taken from this population, which consists of observations on these variables from $n = 20$ students, with an observed mean vector of [57, 44]'. Assume we would like to test whether the population mean is $\underline{\mu}_0 = [55, 46]'$. That is, to test is the null hypothesis H_0: $\underline{\mu} = [55, 46]'$.

As discussed above, we can use the duality principle between hypothesis testing and confidence interval to accomplish this goal. Specifically, for a given significance level α (with $0 < \alpha < 1$), we can check if $\underline{\mu}_0$ falls in the confidence region for the mean vector at confidence level $1 - \alpha$. To this end, we make use of the following result: if a population is studied where a given vector of measures $\underline{X}$ follows a normal distribution, $\underline{X} \sim N_p\ (\underline{\mu}, \Sigma)$, with $\Sigma > 0$, the 95%-confidence region for the mean vector $\underline{\mu}$ is the region enclosed by the ellipsoid with contour defined by the equation ($\underline{Y}$ denoting the running point along the contour)

$$(\underline{Y} - \bar{\underline{X}})'\Sigma^{-1}(\underline{Y} - \bar{\underline{X}}) = \frac{\chi^2_{p,\alpha}}{n}, \tag{4.4}$$

where $\chi^2_{p,\alpha}$ is the pertinent cutoff for the chi-square distribution with p degrees of freedom and n is sample size (as usual in this book; Tatsuoka, 1988).

By way of illustration, in the three-dimensional case (i.e., when $p = 3$) the ellipsoid defined by Equation 4.4 resembles a football, whereas in the two-dimensional case ($p = 2$) it is an ellipse. Just to give an example, a plot of such an ellipse with mean vector (centroid) say at $\underline{\mu} = [55, 46]'$ for the 95%-confidence level, is shown in Figure 4.1 (with the short vertical and horizontal lines assumed to be erected at the scores 55 and 46, respectively, and dispensing for simplicity with representing explicitly the metric of the two coordinate axes and their directions of increase that are implied).

We note that the discussed procedure leading to the confidence region in Equation 4.4 is applicable for any $p \geq 1$, because nowhere is dimensionality mentioned in it (other than in the degrees of freedom, which are

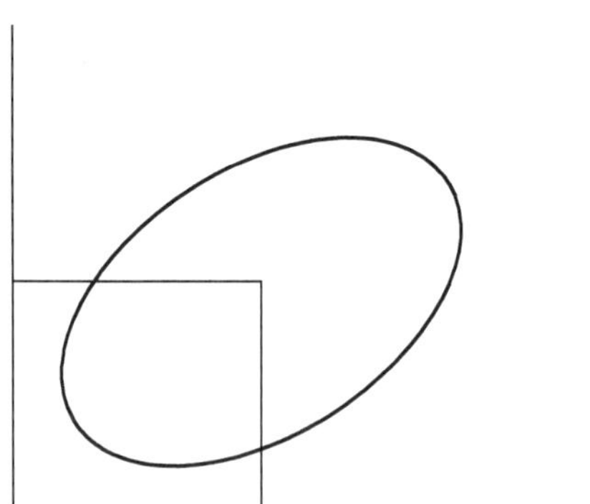

FIGURE 4.1
Two-dimensional ellipsoid with centroid at $\underline{\mu} = [55, 46]'$, representing a 95%-confidence region.

however irrelevant in this regard). In addition, we stress that the left-hand side of Equation 4.4 is precisely the multivariate distance (MD) of the running vector $\underline{Y}$ to the centroid of the data set, $\underline{\bar{X}}$, whereby this distance is taken with regard to the covariance matrix Σ (see Chapter 2). Hence, from Equation 4.4 it follows that all points in the p-dimensional space with an MD from the centroid $\underline{\bar{X}}$, which is less than $\chi^2_{p,\alpha}/n$, constitute the 95%-confidence region for the mean $\underline{\mu}$.

Therefore, using the duality principle mentioned above, the null hypothesis H_0: $\underline{\mu} = \underline{\mu}_0$ may be rejected if the MD of the hypothetical value $\underline{\mu}_0$ to the data centroid $\underline{\bar{X}}$ is large enough, namely larger than $\chi^2_{p,\alpha}/n$, that is, if and only if

$$(\underline{\mu}_0 - \underline{\bar{X}})'\Sigma^{-1}(\underline{\mu}_0 - \underline{\bar{X}}) > \frac{\chi^2_{p,\alpha}}{n}. \tag{4.5}$$

We readily observe the analogy between Equation 4.5 and the well-known test statistic of the univariate z test:

$$z = \frac{(\bar{X} - \mu_0)}{(\sigma/\sqrt{n})}, \tag{4.6}$$

for which we can write its squared value, distributed now as a chi-square variable with 1 degree of freedom, as

$$z^2 = (\bar{X} - \mu_0)/(\sigma^2/n)^{-1}(\bar{X} - \mu_0) \sim \chi^2_1, \tag{4.7}$$

or simply

$$z^2 = (\bar{X} - \mu_0)/(\sigma^2)^{-1}(\bar{X} - \mu_0) \sim \chi^2_1/n. \tag{4.8}$$

(Recall that the square of a random variable following a standard normal distribution, is a chi-square distributed variable with 1 degree of freedom; Hays, 1994.) A comparison of Equations 4.5 and 4.8 reveals that the latter is a special case of the former (leaving aside the reference to significance level in Equation 4.5 that is irrelevant for this comparison).

Returning to our previous example of the reading and writing ability study, since we are interested in two scores, $p = 2$ in it. Hence, given that the sample size was $n = 20$, we easily find that the right-hand side of Equation 4.5 yields here (taking as usual significance level of $\alpha = .05$):

$$\frac{\chi^2_{p,\alpha}}{n} = \frac{\chi^2_{2,.05}}{20} = \frac{5.9948}{20} = .30 \tag{4.9}$$

(This critical value can be obtained from chi-square tables, which are provided in most introductory statistics textbooks, for $\alpha = .05$ and 2 degrees of freedom; Hays, 1994.)

To carry out the corresponding hypothesis testing in an empirical setting, we can use either the following SPSS syntax file or the subsequent SAS PROC IML program file, with explanatory comments correspondingly following each. (For details on using such SPSS and SAS command files, the reader can refer to Chapter 2.)

SPSS command file

```
TITLE 'USING SPSS FOR TWO-GROUP HYPOTHESIS TESTING'.
MATRIX.
COMP X.BAR =    {57, 44}.
COMP MU.0  =    {55, 46}.
COMP SIGMA =    {25, 12;
                 12, 27}.
COMP XBR.DIST= (MU.0-X.BAR)*INV(SIGMA)*T(MU.0-X.BAR).
PRINT XBR.DIST.
END MATRIX.
```

In this command sequence, X.BAR stands for the sample mean ($\underline{\bar{X}}$), MU.0 for the hypothetical mean vector ($\underline{\mu}_0$) with coordinates 55 and 46 (actually, its transpose, for convenience reasons and due to software choice); in addition, SIGMA symbolizes the known population covariance matrix Σ in Equation 4.3, and XBR.DIST is the MD of the sample mean to the hypothetical mean vector. This command file yields the computed value of XBR.DIST = .57. Since it is larger than the above found cutoff value of .30, we conclude that the sample mean is farther away from the hypothetical mean vector than tolerable under the null hypothesis. This finding suggests rejection of H_0.

SAS command file

```
PROC IML;
XBAR={57 44};
MUO={55 46};
SIGMA={25 12,
       12 27}; /* RECALL THE USE OF A COMMA TO SEPARATE ROWS*/
DIFF=MUO - XBAR;
TRANDIFF=T(DIFF);
INVS=INV(SIGMA);
XBRDIST=DIFF*INVS*TRANDIFF;
PRINT XBRDIST;
QUIT;
```

This SAS program (using comparable quantity names to the preceding SPSS input file) generates the same result as that obtained with SPSS.

It is worthwhile noting here that if one were to carry out instead two separate t tests on each of the ability scores from this example, none of the respective univariate null hypotheses would be rejected, that is, both $H_{0,1}$: $\mu_{\text{reading}} = 55$ and $H_{0,2}$: $\mu_{\text{writing}} = 46$ would be retained. The reason can be the lack of power that results from wasting the information about the important interrelationship between these two scores, which one commits in general if univariate rather than multivariate tests are carried out on correlated DVs. In particular, in the present example one can find out from Equation 4.3 that there is a correlation of .46 between the two ability scores (e.g., using Equation 2.39 in Chapter 2). This notable correlation is not accounted for by the univariate testing approach just mentioned. In other words, the multivariate hypothesis was rejected here because the multivariate test accumulated information about violation of H_0 across two interrelated dimensions, while the univariate approach treated these ability scores as unrelated (which they obviously are not). This is a typical example of a difference in results between univariate and multivariate analyses that could be carried out on the same data. The difference usually arises as a result of (a) inflated Type I error when univariate tests are conducted separately on each DV, one at a time; and (b) higher power of the multivariate statistical test, due to more efficient use of available sample information. For this reason, when univariate and multivariate tests disagree in the manner shown in this example, one would tend to trust the multivariate results more.

4.3.2 The Case of Unknown Covariance Matrix

The preceding subsection made a fairly strong assumption that we knew the population covariance matrix for the variables of interest. This assumption is rarely fulfilled in empirical social and behavioral research. In this subsection, we consider the more realistic situation when we do not know that covariance matrix to begin with, yet estimate it in a given sample from a studied population by the empirical covariance matrix S, and want to test the same null hypothesis H_0: $\underline{\mu} = \underline{\mu}_0$.

In this case, one can still employ the same above reasoning but using the MD with regard to the matrix S (because its population counterpart Σ is unknown). Then it can be shown that the 95%-confidence region for the mean consists of all those observed values of $\underline{Y}$ that are close (near) enough to $\underline{\bar{X}}$ in terms of their MD, namely all those $\underline{Y}$ for which

$$(\underline{Y} - \underline{\bar{X}})'S^{-1}(\underline{Y} - \underline{\bar{X}}) < \frac{p(n-1)}{n(n-p)} F_{p,n-p;\alpha}, \tag{4.10}$$

where $F_{p,n-p;\alpha}$ is the pertinent cutoff of the F distribution with p and $n-p$ degrees of freedom (Johnson & Wichern, 2002). As explained above, we stress that this procedure is applicable for any dimensionality $p \geq 1$, that is, with any number of DVs.

From Inequality 4.10 it is seen that the same approach is applied here as in the preceding subsection, with the only differences that (a) the matrix S (i.e., the sample-based estimate of the population covariance matrix) is used in lieu of Σ; and (b) the relevant cutoff is from a different distribution. The latter results from the fact that here we estimate Σ by S rather than use Σ itself as we do not know it. (Recall that in the univariate setup we use a t distribution for tests on means when the population variance is unknown, rather than the normal distribution as we would when that variance were known.)

Hence, by analogy to the last considered testing procedure, and given Inequality 4.10, we reject H_0: $\underline{\mu} = \underline{\mu}_0$ if and only if

$$(\underline{\mu}_0 - \underline{\bar{X}})' S^{-1} (\underline{\mu}_0 - \underline{\bar{X}}) > \frac{p(n-1)}{n(n-p)} F_{p,n-p;\alpha}, \tag{4.11}$$

that is, if the observed mean vector is far enough from the hypothetical mean vector, $\underline{\mu}_0$, in terms of the former's MD.

When the population covariance matrix Σ is unknown, the left-hand side of Inequality 4.11, multiplied by sample size, is called Hotelling's T^2. That is, Hotelling's T^2 is equal to $n(\underline{\mu}_0 - \underline{\bar{X}})' S^{-1} (\underline{\mu}_0 - \underline{\bar{X}})$, and in fact represents the multivariate analog of the univariate t statistic for testing hypotheses about the mean. Specifically, T^2 is a multivariate generalization of the square of the univariate t ratio for testing H_0: $\mu = \mu_0$, that is

$$t = \frac{\bar{X} - \mu_0}{s/\sqrt{n}}. \tag{4.12}$$

Indeed, squaring both sides of Equation 4.12 leads to

$$t^2 = n(\bar{X} - \mu_0)(s^2)^{-1}(\bar{X} - \mu_0), \tag{4.13}$$

which is formally identical to T^2 in case of $p = 1$. To complete this univariate-to-multivariate analogy, recall also that the t-distribution's and particular F-distribution's cutoffs are closely related (at a given sample size n): $t^2_{n-1,\alpha} = F_{1,n-1;\alpha}$; the last relationship should also clarify the use of the F distribution in Equation 4.11.

To illustrate this discussion, let us reconsider the last empirical example, but now asking the same question in the more realistic situation when the population matrix were unknown. Assume that instead of knowing this population matrix, the latter is only estimated by the following sample covariance matrix:

$$S = \begin{bmatrix} 24.22 & 10.98 \\ 10.98 & 27.87 \end{bmatrix}. \tag{4.14}$$

To test the null hypothesis H_0: $\underline{\mu} = [55, 46]'$ using either SPSS or SAS, we employ either of the following two command files. (Note that these are obtained from the last presented, respective command files via a minor modification to accommodate the empirical covariance matrix in Equation 4.14 and the right-hand side of Inequality 4.11.) The utilized below critical value of $F = 3.55$ is obtained from F-distribution tables, which can be found in appendices to most introductory statistics textbooks, based on $\alpha = .05$, $p = 2$, and $n - p = 18$ degrees of freedom (Hays, 1994).

SPSS command file

```
TITLE 'USING SPSS FOR HOTELLING'S TEST'.
MATRIX.
COMP X.BAR      = {57, 44}.
COMP MU.0       = {55, 46}.
COMP S          = {24.22, 10.98;
                   10.98, 27.87}.
COMP XBR.DIST   = (MU.0-X.BAR)*INV(S)*T(MU.0-X.BAR).
PRINT XBR.DIST.
COMP CUTOFF     = 2*19/(20*18)*3.55.
PRINT CUTOFF.
END MATRIX.
```

SAS command file

```
PROC IML;
XBAR={57 44};
MUO={55 46};
S={24.22 10.98,
   10.98 27.87};
DIFF=MUO - XBAR;
TRANDIFF =T(DIFF);
INVS=INV(S);
XBRDIST=DIFF*INVS*TRANDIFF;
PRINT XBRDIST;
CUTOFF=2*19/(20*18)*3.55;
PRINT CUTOFF;
QUIT;
```

Submitting either of these SPSS or SAS programs yields the values of XBR.DIST = .53 and CUTOFF = .37 (recalling of course from Equation 4.11 that CUTOFF $= p(n-1)/[n(n-p)]F = 2(19)/[20(18)]3.55 = 0.37$). Because the value of XBR.DIST is larger than the value of the relevant cutoff, we reject H_0 and conclude that there is evidence in the analyzed data to warrant rejection of the null hypothesis stating that the reading and writing score means were equal to 55 and 46, respectively.

4.4 Testing Hypotheses About Multivariate Means of Two Groups

Many times in empirical research, we do not have precise enough information to come up with meaningful hypothetical values for the means of analyzed variables. Furthermore, situations may often occur in which there is interest in comparing means across two groups. This section deals with methods that can be used in such circumstances.

4.4.1 Two Related or Matched Samples (Change Over Time)

Suppose $\underline{Y} = (Y_1, Y_2, \ldots, Y_p)'$ is a set of p multinormal measures that have been administered to n subjects on two occasions, with resulting scores $(y_{11}, y_{12}, \ldots, y_{1p})$ for the first occasion and $(y_{21}, y_{22}, \ldots, y_{2p})$ for the second occasion (with $p > 1$). We are interested in testing whether there are mean differences across time, that is, whether there is mean change over time. In other words, we are concerned with testing the null hypothesis H_0: $\underline{\mu}_1 = \underline{\mu}_2$, where $\underline{\mu}_1$ and $\underline{\mu}_2$ are the population mean vectors at first and second assessment occasions, respectively. The following approach is also directly applicable when these measurements result from two related, dependent, or so-called matched samples.

In order to proceed, we can reduce the problem to an already handled case—that of testing the hypothesis $\underline{\mu}_1 - \underline{\mu}_2 = \underline{0}$. Indeed, we note that the mean difference, $\underline{\mu}_1 - \underline{\mu}_2$, is equivalent to the mean of the difference score, that is, $\underline{\mu}_1 - \underline{\mu}_2 = \underline{\mu}_{\underline{D}}$, where $\underline{D} = \underline{Y}_1 - \underline{Y}_2$ is the vector of differences on all $\underline{Y}$ components across the two assessments. Thus the hypothesis of interest, H_0: $\underline{\mu}_1 = \underline{\mu}_2$, being the same as $\underline{\mu}_1 - \underline{\mu}_2 = \underline{0}$, is also equivalent to $\underline{\mu}_{\underline{D}} = \underline{0}$. The latter hypothesis, however, is a special case of the one we have already dealt with in Section 4.3, and is obtained from that hypothesis when $\underline{\mu}_0 = \underline{0}$.

Therefore, using Inequality 4.11 with $\underline{\mu}_0 = \underline{0}$, we reject the null hypothesis H_0 under consideration if and only if

$$(\underline{0} - \underline{\bar{D}})' S^{-1} (\underline{0} - \underline{\bar{D}}) > \frac{p(n-1)}{n(n-p)} F_{p,n-p;\alpha}, \tag{4.15}$$

or simply, if and only if

$$\underline{\bar{D}}' S^{-1} \underline{\bar{D}} > \frac{p(n-1)}{n(n-p)} F_{p,n-p;\alpha}. \qquad (4.16)$$

In other words, we reject the hypothesis of equal means when and only when $\underline{\bar{D}}$ is far enough from $\underline{0}$ (in terms of the former's MD from the origin; throughout this section, n stands for number of pairs of subject scores in the analyzed data set; see Section 4.3.2).

The test statistic in the right-hand side of Equation 4.16 is also readily seen to be the multivariate analog of the univariate t statistic for testing differences in two related means. Indeed, that univariate statistic is $t = \frac{\bar{d}}{s_d/\sqrt{n}}$, where s_d is the standard deviation of the difference score and $\bar{d}$ is its mean in the sample. We emphasize that here, as well as in the rest of this section, n stands for the number of pairs of recorded observations (e.g., studied subjects in a pretest/posttest design, or pairs in a matched-pairs design) rather than for the total number of all available measurements that is obviously $2n$.

To illustrate this discussion, consider the following research study setting. Two intelligence tests, referred to as test 1 and test 2, are administered to a sample of 160 high school students at the beginning and at the end of their 11th-grade year. A researcher is interested in finding out whether there is any change in intelligence, as evaluated by these two measures, across the academic year. To answer this question, let us first denote test 1 by Y_1 and test 2 by Y_2, and for their two administrations let us add as a second subscript 1 and 2, respectively. Next, with SPSS or SAS we can correspondingly calculate the difference scores for each of the two tests, denoted D_1 and D_2, using, for example, the following syntax commands (where Y_{11} corresponds to test 1 on occasion 1, and the other symbols are defined correspondingly):

```
COMP D1 = Y11 - Y12.
COMP D2 = Y21 - Y22.
```

or

```
D1 = Y11 - Y12;
D2 = Y21 - Y22;
```

Then we compute, as usual, the means $\bar{D}_1$ and $\bar{D}_2$, and the covariance matrix S of the two difference scores from the available sample. With these estimates, to evaluate the left-hand side of Inequality 4.16 we proceed as follows with either SPSS or SAS. (Clarifying comments are inserted immediately after command lines where needed; for generality of the following two programs, we refer to means, variances, and covariances by symbols/names.)

SPSS command file

```
TITLE 'USING SPSS FOR HOTELLING'S RELATED SAMPLES TEST'.
MATRIX.
COMP D.BAR={D1.BAR, D2.BAR}.
* ENTER HERE THE 2 DIFFERENCE SCORE MEANS FROM THE SAMPLE.
COMP MU.0 =  {0, 0}.
COMP S    =  {S11, S12;
              S21, S22}.
* ENTER HERE COVARIANCE MATRIX OBTAINED FROM SAMPLE.
COMP DBR.DIST=(MU.0-D.BAR)*INV(S)*T(MU.0-D.BAR).
PRINT DBR.DIST.
COMP CUTOFF=2*159/(160*158)*F.CUTOFF.
* F.CUTOFF IS THE CUTOFF OF F WITH 2 AND 158 DF'S, AT ALPHA=.05.
* WHICH IS FOUND FROM APPENDICES IN INTRO STATS BOOKS.
PRINT CUTOFF.
END MATRIX.
```

SAS command file

```
PROC IML;
DBAR={D1BAR D2BAR}; /* ENTER HERE THE TWO DIFFERENCE SCORE MEANS */;
MUO={0 0 0};
S={S11 S12,
    S21 S22}; /* ENTER HERE COVARIANCE MATRIX OBTAINED FROM SAMPLE*/;
DIFF=MUO - DBAR;
TRANDIFF =T(DIFF);
INVS=INV(S);
DBRDIST=DIFF*INVS*TRANDIFF;
PRINT DBRDIST;
CUTOFF=2*(n−1)/(n*(n−2))*F;
PRINT CUTOFF;
QUIT;
```

As indicated before, if the computed value of DBR.DIST exceeds CUTOFF in a given data set, we reject the null hypothesis of no mean difference (in the case of no change over time); otherwise we retain it. We stress that both above SPSS and SAS command files can be readily modified in case a different number of measures are taken at both assessments (or observed in related samples); they produce numerical results as soon as one inserts into the appropriate places empirical statistics (viz., corresponding sample means, variances, and covariances, as well as cutoff value for the respective F distribution).

As an example, assume that in the last considered empirical study the following results were obtained for the means $\bar{D}_1$ and $\bar{D}_2$: $\bar{D}_1 = 0.7$ and

$\bar{D}_2 = 2.6$; and that the relevant covariance matrix was $S = \begin{bmatrix} 22.1 & 2.8 \\ 2.8 & 90.4 \end{bmatrix}$. Using either of the above SPSS or SAS command files, we readily obtain DBR.DIST = 0.092, which is larger than the CUTOFF value of 0.056 furnished thereby (with 2 and 158 degrees of freedom, $F_{\alpha=.05} = 4.46$). Hence, we can reject H_0 and conclude that there is a significant change in intelligence, as evaluated by these two measures, across the academic year.

4.4.2 Two Unrelated (Independent) Samples

We begin this subsection with a motivating example in which two methods of teaching algebra topics are used in a study with an experimental and a control group. Suppose three achievement tests are administered to $n_1 = 45$ subjects designated as the experimental group and $n_2 = 110$ subjects designated as the control group. Assume that a researcher is interested in finding out whether there is a differential effect upon average performance of the two teaching methods.

To proceed here, we must make an important assumption, namely that the covariance matrices of the observed variables are the same in the populations from which these samples were drawn (we denote that common covariance matrix by Σ). That is, designating by $\underline{X}$ and $\underline{Y}$ the vectors of three achievement measures in the experimental and control groups, respectively, we assume that $\underline{X} \sim N_3(\underline{\mu}_1, \Sigma)$ and $\underline{Y} \sim N_3(\underline{\mu}_2, \Sigma)$. Note that this assumption is the multivariate analog of that of equal variances in a corresponding univariate analysis, which underlies the conventional t test for mean differences across two unrelated groups.

Since the common population covariance matrix is unknown, it is desirable to estimate it using the sample data in order to proceed. To this end, we first note that as can be shown under the above distributional assumptions, the difference between the two group means is normally distributed, and specifically

$$\bar{\underline{X}} - \bar{\underline{Y}} \sim N_p\left[\underline{\mu}_1 - \underline{\mu}_2, \left(\frac{1}{n_1} + \frac{1}{n_2}\right)\Sigma\right], \tag{4.17}$$

that is, follows the multinormal distribution with mean $\underline{\mu}_1 - \underline{\mu}_2$ and covariance matrix equal to $(1/n_1 + 1/n_2)\Sigma$ (Johnson & Wichern, 2002).

Consequently, based on the previous discussion, we can use the following estimator of the common covariance matrix (see Chapter 2):

$$S^* = \frac{\left(\frac{1}{n_1} + \frac{1}{n_2}\right)(\text{SSCP}_X + \text{SSCP}_Y)}{\text{df}}, \tag{4.18}$$

where $SSCP_X$ and $SSCP_Y$ are the SSCP matrices for the X and Y measures in each of the groups, respectively, and $df = n_1 + n_2 - 2$ are the underlying degrees of freedom. As an aside at this moment, let us note the identity of the degrees of freedom here and in the corresponding univariate setup when estimating a common population variance in the two-group study (we remark on this issue later again). In a univariate setup, Equation 4.18 would correspond to the well-known relationship $s_p^2 = (1/n_1 + 1/n_2)(\Sigma x_1^2 + \Sigma y_2^2)/(n_1 + n_2 - 2)$, where Σx_1^2 and Σy_2^2 are the sum of squared mean deviations in each of the two groups. We refer to the covariance matrix S^* in Equation 4.18 as the pooled covariance matrix (similarly to pooled variance in the univariate case).

Returning now to our initial concern in this subsection, which is to test the null hypothesis H_0: $\underline{\mu}_1 = \underline{\mu}_2$, or equivalently the null hypothesis $\underline{\mu}_1 - \underline{\mu}_2 = \underline{0}$, we use the same reasoning as earlier in this chapter. Accordingly, if $\underline{0}$ is close enough to $\underline{\mu}_1 - \underline{\mu}_2$, in terms of the latter's MD (to the origin), we do not reject H_0; otherwise we reject H_0. How do we find the MD between $\underline{\mu}_1 - \underline{\mu}_2$ and the origin $\underline{0}$ in the present situation? According to its definition (see Chapter 2, and notice the analogy to the case of two related samples), this MD is

$$T_2^2 = (\underline{\bar{X}} - \underline{\bar{Y}})' S^{*-1} (\underline{\bar{X}} - \underline{\bar{Y}}), \tag{4.19}$$

which is similarly called Hotelling's T^2 for the two-sample problem.

We further note that the right-hand side of Equation 4.19 is a multivariate analog of the t-test statistic in the univariate two-sample setup, in which our interest lies in testing the hypothesis of equality of two independent group means (Hays, 1994). That is,

$$t = \frac{(\bar{X} - \bar{Y})}{\left[s_p\sqrt{\left(\frac{1}{n_1} + \frac{1}{n_2}\right)}\right]}, \tag{4.20}$$

from which it also follows by squaring

$$t^2 = (\bar{X} - \bar{Y})\left[s_p^2\left(\frac{1}{n_1} + \frac{1}{n_2}\right)\right]^{-1}(\bar{X} - \bar{Y}), \tag{4.21}$$

where s_p is the pooled estimate of the common variance in both samples and can be obtained as a special case from Equation 4.18 when there is a single studied variable. We note that the expression in the right-hand side of Equation 4.21 is identical to the right-hand side of Equation 4.19 in that special case.

Considering again the empirical example regarding the two methods of teaching algebra topics, let us assume that we have first computed the necessary mean vectors and SSCP matrices using the procedures described in Chapters 1 and 2, based on the common covariance matrix assumption. Now we can use either of the following SPSS or SAS command files to test for group differences. (The raw data are found in the file named ch4ex1.dat available from www.psypress.com/applied-multivariate-analysis, where TEST.1 through TEST.3 denote the three achievement test scores, in percentage correct metric, and GROUP stands for the experimental vs. control group dichotomy.)

SPSS command file

```
TITLE 'HOTELLING'S INDEPENDENT SAMPLES TEST'.
MATRIX.
COMP XBAR= {27.469, 46.244, 34.868}.
COMP YBAR= {31.698, 50.434, 37.856}.
COMP DIFF= XBAR - YBAR.
COMP SSCPX={9838.100, 7702.388, 10132.607;
            7702.388, 11559.293, 8820.509;
            10132.607, 8820.509, 13295.321}.

COMP SSCPY={23463.794, 15979.138, 23614.533;
            15979.138, 29152.887, 22586.075;
            23614.533, 22586.075, 34080.625}.
* THE LAST ARE THE SSCP MATRICES OBTAINED FROM BOTH GROUPS.
COMP SUMS=SSCPX+SSCPY.
COMP S= ((1/45+1/110)*(SUMS))/153.
COMP DBR.DIST= (DIFF)*INV(S)*T(DIFF).
PRINT DBR.DIST.
COMP CUTOFF= (45+110-2)*2/(45+110-3)*2.99.
* 2.99 IS THE F VALUE WITH 2 AND 158 DF'S, AT ALPHA= .05, FROM BOOKS.
PRINT CUTOFF.
END MATRIX.
```

SAS command file

```
PROC IML;
XBAR={27.469 46.244 34.868};
YBAR={31.698 50.434 37.856};
SSCPX={9838.100 7702.388 10132.607,
       7702.388 11559.293 8820.509,
       10132.607 8820.509 13295.321};
SSCPY={23463.794 15979.138 23614.533,
       15979.138 29152.887 22586.075,
       23614.533 22586.075 34080.625};
```

```
DIFF=XBAR - YBAR;
TRANDIFF =T(DIFF);
SUMS=SSCPX+SSCPY;
S=((1/45+1/110)*(SUMS))/153;
INVS=INV(S);
DBRDIST=DIFF*INVS*TRANDIFF;
PRINT DBRDIST;
CUTOFF=(45+110-2)*2/(45+110-3)*2.99;
PRINT CUTOFF;
QUIT;
```

Submitting either of these two program files to the software, with the observed data, yields XBAR.DIST = 4.154 (i.e., equal to the MD of the difference between the two sample means from the origin) and CUTOFF = 6.02. Because XBAR.DIST is not larger than CUTOFF, we do not reject H_0, and can conclude that there is no evidence in the analyzed samples that the two methods of teaching algebra topics differ in their effectiveness.

4.5 Testing Hypotheses About Multivariate Means in One-Way and Higher Order Designs (Multivariate Analysis of Variance, MANOVA)

Many empirical studies in the social and behavioral sciences are concerned with designs that have more than two groups. The statistical approach followed so far in this chapter can be generalized to such cases, and the resulting extension is referred to as multivariate analysis of variance (MANOVA). A main concern of MANOVA is the examination of mean differences across several groups when more than one DVs are considered simultaneously. That is, a MANOVA is essentially an analysis of variance (ANOVA) with $p > 1$ response (dependent) variables; conversely, ANOVA is a special case of MANOVA with $p = 1$ outcome variable.

In analogy to ANOVA, a major question in MANOVA is whether there is evidence in an analyzed data set for an "effect" (i.e., a main effect or an interaction), when all p DVs are considered together ($p \geq 1$). As could be expected, there is a helpful analogy between the statistical procedure behind MANOVA and the one on which its univariate counterpart, ANOVA, is based. We use this analogy in our discussion next.

We first recall from UVS that in a one-way ANOVA with say g groups ($g \geq 2$), a hypothesis of main interest is that of equality of their means on the single DV, that is,

$$H_0\colon \mu_1 = \mu_2 = \ldots = \mu_g. \qquad (4.22)$$

The testing approach to address this question then is based on the well-known sum of squares partitioning,

$$SS_T = SS_B + SS_W, \tag{4.23}$$

where
SS_T is the total sum of squares
SS_B is the sum of squares between (for) the means of the groups
SS_W is the sum of squares within the groups

The statistic for testing the null hypothesis in Equation 4.22 is then the following *F* ratio:

$$F = \frac{\text{Mean SS}_B}{\text{Mean SS}_W} = \frac{[SS_B/(g-1)]}{[SS_W/(n-g)]} \sim F_{g-1, n-g} \tag{4.24}$$

(Hays, 1994).

Turning to the case of a one-way MANOVA design with p DVs ($p \geq 2$), a null hypothesis of main interest is

$$H_0: \underline{\mu}_1 = \underline{\mu}_2 = \ldots = \underline{\mu}_g, \tag{4.25}$$

whereby we stress that here equality of vectors is involved. That is, the hypothesis in Equation 4.25 states that the g population centroids, in the p-dimensional space of concern now, are identical. Note that this hypothesis is the multivariate analog of the above ANOVA null hypothesis H_0: $\mu_1 = \mu_2 = \ldots = \mu_g$, and may actually be obtained from the latter by formally exchanging scalars (the μs in Equation 4.22) with vectors.

We next recall our earlier remark (see Chapter 1) that in the multivariate case the SSCP matrix is the analog of the univariate concept of sum of squares. In fact, testing the multivariate null hypothesis in Equation 4.25 can be seen by analogy as being based on the partitioning of the SSCP matrix of observations on the p response variables, which proceeds in essentially the same manner as the one in Equation 4.23. Specifically, for the multivariate case this breakdown is

$$SSCP_T = SSCP_B + SSCP_W, \tag{4.26}$$

where the subindexes T, B, and W stand for total, between, and within (SSCP), and

$$SSCP_W = SSCP_1 + SSCP_2 + \cdots + SSCP_g, \tag{4.27}$$

which represents the sum of the group-specific SSCP matrices across all g groups considered. We stress that each SSCP appearing in the right-hand side of Equation 4.27 not only combines all dependent variability indices (sum of squares, along its main diagonal) but also takes the response variables' interrelationships into account (reflected in its off-diagonal elements, the cross products). We also note that Equation 4.27 helps clarify the reason why we want to assume homogeneity of group-specific covariance matrices. (We present a detailed discussion of this matter later.)

Now, in order to test the null hypothesis H_0 in Equation 4.25, we can make use of the so-called likelihood ratio theory (LRT), which has some very attractive properties. This theory is closely related to the method of maximum likelihood (ML), which follows a major principle in statistics and is one of the most widely used estimation approaches in its current applications. Accordingly, as estimates of unknown parameters—here, a mean vector and covariance matrix per group—one takes those values for them, which maximize the "probability" of observing the data at hand. (Strictly speaking, maximized is the likelihood of the data as a function of the model parameters; Roussas, 1997.) These estimates are commonly called ML estimates.

The LRT is used to test hypotheses in the form of various parameter restrictions, whenever the ML estimation method is employed. The LRT utilizes instrumentally the ratio of (a) the probability (actually, likelihood in the continuous variable case) of observing the data at hand if a hypothesis under consideration were to be true, to (b) the probability (likelihood) of observing the data without assuming validity of the hypothesis, whereby both (a) and (b) are evaluated at the values for the unknown parameters that make each of these probabilities (likelihoods) maximal. This ratio is called the likelihood ratio (LR), and it is the basic quantity involved in hypothesis testing within this framework. If the LR is close to 1, its numerator and denominator are fairly similar, and thus the two probabilities (likelihoods) involved are nearly the same. For this reason, such a finding is interpreted as suggesting that there is lack of evidence against the tested null hypothesis. If alternatively the LR is close to 0, then its numerator is much smaller than its denominator. Hence, the probability (likelihood) of the data if the null hypothesis were true is quite different from the probability (likelihood) of the data without assuming validity of the null hypothesis; this difference represents evidence against the null hypothesis (Wilks, 1932). We note that this application of the LRT considers, for the first time in this book, evidence against H_0 as being provided not by a large test statistic value but by a small value of a test statistic (i.e., a value that is close to 0, as opposed to close to 1). We emphasize that the LR cannot be larger than 1 since its numerator cannot exceed its denominator, due to the fact that in the former maximization occurs across a subspace of the multivariate space that is of relevance for the denominator; similarly, the LR cannot be negative as it is a ratio of two probabilities.

In the current case of one-way MANOVA, it can be shown that the LR test statistic is a monotone function of what is called the Wilks' lambda (Λ, capital Greek letter lambda):

$$\Lambda = \frac{|\mathrm{SSCP_W}|}{|\mathrm{SSCP_T}|} = \frac{|\mathrm{SSCP_W}|}{|\mathrm{SSCP_B} + \mathrm{SSCP_W}|}, \tag{4.28}$$

which represents a ratio of the determinants of the within-group and total sample SSCP matrices (Johnson & Wichern, 2002). More specifically, if we denote

$$S_\mathrm{W} = \frac{\mathrm{SSCP_W}}{(n-g)} = \frac{(\mathrm{SSCP}_1 + \cdots + \mathrm{SSCP}_g)}{(n-g)},$$

it can also be shown that Wilks' Λ criterion can be written as

$$\Lambda = \frac{|S_\mathrm{W}|}{|S_\mathrm{T}|} \frac{(n-g)^p}{(n-1)^p}, \tag{4.29}$$

where S_T is the total covariance matrix (of the entire sample, disregarding group membership).

The Λ criterion in Equations 4.28 and 4.29 can be viewed as a multivariate generalization of the univariate ANOVA F ratio. Since the Λ criterion is defined in terms of determinants of appropriate matrices, it takes into account involved variable variances as well as interrelationships (reflected in the off-diagonal elements of these matrices). Also, it can be shown that in the univariate case Λ is inversely proportional to the ANOVA F ratio, denoted next F:

$$\Lambda_{(p=1)} = \frac{1}{1 + [(k-1)/(n-k)]F}. \tag{4.30}$$

Because the right-hand side of Equation 4.30 is a monotone function of the F ratio, it demonstrates that testing the null hypothesis equation (Equation 4.25) effectively reduces in the univariate case to the familiar ANOVA F test (Tatsuoka, 1988).

In the general case of $p \geq 1$ DVs, as indicated earlier the smaller Λ the more evidence there is in the data against H_0; that is, Λ is an inverse measure of disparity between groups. In other words, the logic of Wilks' Λ in relation to the null hypothesis of interest is the "reverse" to that of the F ratio. Also, from Equation 4.28 follows that Wilks' Λ can be presented as

$$\frac{1}{\Lambda} = |\mathrm{SSCP_W^{-1}}(\mathrm{SSCP_B} + \mathrm{SSCP_W})| = |\mathrm{SSCP_W^{-1}}\ \mathrm{SSCP_B} + I|, \tag{4.31}$$

where I is the correspondingly sized identity matrix. Another interesting fact is that Λ and Hotelling's T^2 are related in the two-group case, for any number of DVs ($p \geq 1$), as follows:

$$\frac{1}{\Lambda} = \left(\frac{T^2}{n-2}\right) - 1 \qquad (4.32)$$

(Johnson & Wichern, 2002). Equation 4.32 indicates that the bigger the group disparity as determined by Hotelling's T^2, the smaller the value of Λ. Consequently, an alternative way of testing mean differences for two groups is to use, in lieu of Hotelling's T^2, Wilks' Λ. Although there are also a number of other test criteria that can be used to examine group differences, for the moment we restrict our attention to Wilks' Λ criterion; we discuss those test criteria in Section 4.5.3.

The sampling distribution of Wilks' Λ has been worked out a long time ago, under the assumption of homogeneity of the covariance matrix across groups: $\Sigma_1 = \Sigma_2 = \ldots = \Sigma_g$, where the Σs denote the covariance matrices of the DVs in the studied populations. This assumption is testable using the so-called Box's M test, or Bartlett's homogeneity test (Morrison, 1976). As explained later in this section, SPSS provides Box's M test and SAS yields Bartlett's test (see also Chapter 10). These tests represent multivariate generalizations of the test of variance homogeneity in an ANOVA setup (and in particular of the t test for mean differences in a two-group setting). Considerable robustness, however, is in effect against violations of this assumption when large samples of equal size are used (this robustness applies also to the earlier considered case of $g = 2$ groups, as do all developments in this section unless stated otherwise explicitly).

Box's M test of homogeneity compares the determinant—that is, the generalized variance (see discussion in Chapter 1)—of the pooled covariance matrix estimate with those of the group covariance matrices. Its test statistic is computed as an approximate F-statistic proportional to the difference

$$U = (n-p)ln|S^*| - \sum (n_k - 1)ln|S_k|, \qquad (4.33)$$

where n is total sample size (disregarding group membership), p is number of variables, n_k the size of kth sample ($k = 1, \ldots, g$), and the summation in the right-hand side of Equation 4.33 runs across all groups. The proportionality constant, not shown in Equation 4.33, simply renders the value of U to follow a known distribution, and is of no intrinsic interest here. Consequently, an evaluation of Box's M test relative to an F distribution with appropriate degrees of freedom at a prespecified significance level can be used to statistically test the assumption of covariance matrix homogeneity.

Bartlett's test of homogeneity operates in much the same way as Box's M test except that the test statistic of the former is computed as an approximately χ^2-distributed statistic. For this reason, an evaluation of Bartlett's statistic relative to a value from the χ^2 distribution with $p(p+1)(g-1)/2$ degrees of freedom is used to test the assumption of covariance matrix homogeneity. (As indicated before, the test statistic is provided in a pertinent SAS output section.) We note in passing that both Box's M and Bartlett's tests are notoriously sensitive to nonnormality, a fact that we remark on again later in this chapter. As an alternative, one can use structural equation modeling techniques, and specifically robust methods within that framework, to test this homogeneity assumption (Raykov, 2001). As another possibility, with either Box's M or Bartlett's test one may consider utilizing a more conservative significance level of $\alpha = .01$ say, that is, proclaim the assumption violated if the associated p-value is less than .01.

To demonstrate the discussion in this section, let us consider the following example study. In it, three measures of motivation are obtained on three socioeconomic status (SES) groups—lower, medium, and high—with $n_1 = 47$, $n_2 = 50$, and $n_3 = 48$ children, respectively. (The three measures of motivation are denoted MOTIV.1, MOTIV.2, and MOTIV.3 in the file ch4ex2.dat available from www.psypress.com/applied-multivariate-analysis, whereas SES runs from 1 through 3 and designates their respective group membership there). Let us assume that a researcher is concerned with the question of whether there are any SES group differences in motivation.

Such a question can easily be handled via the MANOVA approach in which $p = 3$ responses are to be compared across $g = 3$ groups. This MANOVA can be readily carried out with SPSS using the following menu options/sequence:

Analyze → General Linear Model → Multivariate (choose DVs, SES as "fixed factor"; Options: homogeneity test and means for SES).

With SAS, two procedures can be used to conduct a MANOVA: PROC GLM or PROC ANOVA. The main difference between them, for the purposes of this chapter, has to do with the manner in which the two procedures handle the number of observations in each group. PROC ANOVA was designed to deal with situations where there were an equal number of observations in each group, in which case the data are commonly referred to as being balanced. For the example under consideration, the number of observations in the study groups is not the same; under these circumstances, the data are referred to as unbalanced. PROC GLM was developed to handle such unbalanced data situations. When the data are balanced, both procedures produce identical results. Due to the fact that PROC ANOVA does not check to determine whether the data are balanced or not, if used with unbalanced data it can produce incorrect results. Therefore, only with balanced data could one use PROC ANOVA; otherwise one proceeds with PROC GLM.

Because the data in the example under consideration are not balanced, we use PROC GLM to conduct the analysis with the following program setup:

```
DATA motivation;
INFILE 'ch4ex2.dat';
INPUT MOTIV1 MOTIV2 MOTIV3 SES;
PROC GLM;
  CLASS SES;
  MODEL MOTIV1 MOTIV2 MOTIV3 = SES;
  MEANS SES/HOVTEST = LEVENE;
  MANOVA H = SES;
RUN;
PROC DISCRIM pool = test;
  CLASS SES;
RUN;
```

A number of new statements are used in this command file and require some clarification. In particular, we need to comment on the statements CLASS, MODEL, MEANS, MANOVA, and PROC DISCRIM. The CLASS statement is used to define the independent variables (IVs) (factors) in the study—in this case SES. The MODEL statement specifies the DVs (the three motivation measures) and IVs—to the left of the equality sign, one states the dependent measures, and to the right of it the independent ones. The MEANS statement produces the mean values for the IVs specified on the CLASS statement, while HOVTEST = LEVENE requests the Levene's univariate test of homogeneity of variance for each DV. (One may look at this test as a special case of Box's M test for the univariate case.) The MANOVA statement requests the multivariate tests of significance on the IV and also provides univariate ANOVA results for each DV. Finally, in order to request Bartlett's test of homogeneity, the "pool = test" option within PROC DISCRIM can be used, which procedure conducts in general discriminant function analysis. We note that our only interest at this time in employing PROC DISCRIM is to obtain Bartlett's test of covariance matrix homogeneity, and we consider the use of this procedure in much more detail in Chapter 10.

The resulting outputs produced by the above SPSS and SAS command sequences, follow next. For ease of presentation, the outputs are reorganized into sections and clarifying comments are inserted at their end.

SPSS output

Box's Test of Equality of Covariance Matrices[a]

Box's M	13.939
F	1.126
df1	12
df2	97138.282
Sig.	.333

Tests the null hypothesis that the observed covariance matrices of the dependent variables are equal across groups.
a. Design: Intercept+SES

Levene's Test of Equality of Error Variances[a]

	F	df1	df2	Sig.
MOTIV.1	.672	2	142	.512
MOTIV.2	.527	2	142	.591
MOTIV.3	1.048	2	142	.353

Tests the null hypothesis that the error variance of the dependent variable is equal across groups.
a. Design: Intercept+SES

SAS output

The GLM Procedure

Levene's Test for Homogeneity of MOTIV1 Variance

ANOVA of Squared Deviations from Group Means

Source	DF	Sum of Squares	Mean Square	F Value	Pr > F
SES	2	155078	77538.9	0.80	0.4503
Error	142	13724135	96640.8		

Levene's Test for Homogeneity of MOTIV2 Variance

ANOVA of Squared Deviations from Group Means

Source	DF	Sum of Squares	Mean Square	F Value	Pr > F
SES	2	26282.7	13141.3	0.19	0.8270
Error	142	9810464	69087.8		

```
          Levene's Test for Homogeneity of MOTIV3 Variance
             ANOVA of Squared Deviations from Group Means

                                               Mean
Source        DF        Sum of Squares        Square       F Value      Pr > F

 SES            2              209125         104563          0.95      0.3874
 Error        142            15554125         109536
```

```
                          The DISCRIM Procedure

            Test of Homogeneity of Within Covariance Matrices

 Notation: K     = Number of Groups
           P     = Number of Variables
           N     = Total Number of Observations - Number of Groups
           N(i)  = Number of Observations in the i'th Group - 1

                          _                      N(i)/2
                          || |Within SS Matrix(i)|
           V     = -----------------------------------

                                             N/2
                             |Pooled SS Matrix|

                          _               _          2
                         |        1        1 |  2P + 3P - 1
           RHO   = 1.0 - | SUM -----  -  --- |  -----------
                         |_       N(i)     N _|  6(P+1)(K-1)

           DF    = .5(K-1)P(P+1)
                                                 _                    _
                                                |        PN/2          |
                                                |      N          V    |
 Under the null hypothesis:      -2 RHO ln      | ---------------------|
                                                |   _         PN(i)/2  |
                                                |_  ||  N(i)          _|

 is distributed approximately as Chi-Square(DF).

                    Chi-Square        DF       Pr > ChiSq
                     13.512944        12         0.3329

 Reference: Morrison, D.F. (1976) Multivariate Statistical
 Methods p252.
```

The displayed results indicate that according to Box's M (and Bartlett's) test, there is not sufficient evidence in the data to warrant rejection of the assumption of equal covariance matrices for the motivation measures considered across the three groups. In particular, given that SPSS yields a value for Box's M of 13.939 with associated p-value of 0.333, and SAS furnished a Bartlett's test statistic of 13.513 with associated $p = 0.333$, we conclude that the dependent measure covariance matrices do not differ significantly across the groups. We note in passing that Levene's univariate tests of homogeneity of variance also suggests that each motivation measure has equal variances across groups, but these results are not of interest because the counterpart multivariate test is not significant in this example. When covariance matrix homogeneity is rejected, however, Levene's test can help identify which measure may be contributing singly to such a finding.

SPSS output

Multivariate Tests[c]

Effect		Value	F	Hypothesis df	Error df	Sig.
Intercept	Pillai's Trace	.902	430.301[a]	3.000	140.000	.000
	Wilks' Lambda	.098	430.301[a]	3.000	140.000	.000
	Hotelling's Trace	9.221	430.301[a]	3.000	140.000	.000
	Roy's Largest Root	9.221	430.301[a]	3.000	140.000	.000
SES	Pillai's Trace	.015	.351	6.000	282.000	.909
	Wilks' Lambda	.985	.350[a]	6.000	280.000	.910
	Hotelling's Trace	.015	.349	6.000	278.000	.910
	Roy's Largest Root	.015	.686[b]	3.000	141.000	.562

a. Exact statistic
b. The statistic is an upper bound on F that yields a lower bound on the significance level.
c. Design: Intercept+SES

In this SPSS output section with multivariate results, we are interested only in the SES part of the table since the one titled "intercept" pertains to the means of the three measures, and whether they are significant or not is really of no particular interest. (Typically, measures used in the social and behavioral sciences yield positive scores and thus it cannot be unexpected that their means are significant—in fact, the latter could be viewed as a trivial finding.) For the moment, in that SES part, we look only at the row for Wilks' Λ; we discuss in more detail the other test criteria in Section 4.5.3. An examination of Wilks' Λ and its statistical significance, based on the F distribution with 6 and 280 degrees of freedom, shows no evidence of SES group differences in the means on the three motivation measures when considered together. We note again that the degrees of freedom for Wilks' test are identical to those in the univariate case. The reason is that the uni and multivariate designs are the same for this study—it is a three-group investigation regardless of number of outcome variables—with the only difference that more dependent measures have been used here, a fact that does not affect degrees of freedom.

SAS output

```
                         The GLM Procedure

                  Multivariate Analysis of Variance

  MANOVA Test Criteria and F Approximations for the Hypothesis of No
                        Overall SES Effect

                  H = Type III SSCP Matrix for SES

                       E = Error SSCP Matrix

                         S = 2  M = 0  N = 69

Statistic                        Value  F Value  Num DF  Den DF   Pr > F

Wilks' Lambda               0.98516192     0.35       6     280   0.9095
Pillai's Trace              0.01484454     0.35       6     282   0.9088
Hotelling-Lawley Trace      0.01505500     0.35       6   184.9   0.9092
Roy's Greatest Root         0.01460592     0.69       3     141   0.5617

     NOTE: F Statistic for Roy's Greatest Root is an upper bound.

            NOTE: F Statistic for Wilks' Lambda is exact.
```

As can be readily seen by examining this SAS output part, identical results are found with respect to Wilks' Λ. The next set of output sections provided by each program corresponds to the means of the SES groups on the three measures of motivation and their univariate ANOVA tests. Given that the research question of interest was conceptualized as a multivariate one in the first instance, and that as already found there was no evidence of SES group differences, the univariate results contained in the subsequent tables are of no particular interest for the moment. We provide them here only for the sake of completeness, so that the reader can get a more comprehensive picture of the output generated by the SPSS and SAS command files under consideration.

SPSS output

Tests of Between-Subjects Effects

Source	Dependent Variable	Type III Sum of Squares	df	Mean Square	F	Sig.
Corrected Model	MOTIV.1	4.639[a]	2	2.320	.011	.989
	MOTIV.2	87.561[b]	2	43.781	.164	.849
	MOTIV.3	67.277[c]	2	33.639	.110	.896
Intercept	MOTIV.1	134031.059	1	134031.059	614.101	.000
	MOTIV.2	346645.340	1	346645.340	1298.135	.000
	MOTIV.3	196571.511	1	196571.511	642.188	.000
SES	MOTIV.1	4.639	2	2.320	.011	.989
	MOTIV.2	87.561	2	43.781	.164	.849
	MOTIV.3	67.277	2	33.639	.110	.896
Error	MOTIV.1	30992.332	142	218.256		
	MOTIV.2	37918.729	142	267.033		
	MOTIV.3	43465.713	142	306.097		
Total	MOTIV.1	165115.351	145			
	MOTIV.2	384743.280	145			
	MOTIV.3	240415.613	145			
Corrected Total	MOTIV.1	30996.971	144			
	MOTIV.2	38006.290	144			
	MOTIV.3	43532.991	144			

a. R Squared = .000 (Adjusted R Squared = −.014)
b. R Squared = .002 (Adjusted R Squared = −.012)
c. R Squared = .002 (Adjusted R Squared = −.013)

Estimated Marginal Means

SES

Dependent Variable	SES	Mean	Std. Error	95% Confidence Interval	
				Lower Bound	Upper Bound
MOTIV.1	1.00	30.252	2.155	25.992	34.512
	2.00	30.323	2.089	26.193	34.454
	3.00	30.664	2.132	26.449	34.879
MOTIV.2	1.00	48.684	2.384	43.972	53.396
	2.00	48.096	2.311	43.528	52.665
	3.00	49.951	2.359	45.289	54.614
MOTIV.3	1.00	36.272	2.552	31.227	41.317
	2.00	37.783	2.474	32.892	42.674
	3.00	36.440	2.525	31.448	41.431

SAS output

The GLM Procedure

Dependent Variable: MOTIV1

Source	DF	Sum of Squares	Mean Square	F Value	Pr > F
Model	2	4.63938	2.31969	0.01	0.9894
Error	142	30992.33186	218.25586		
Corrected Total	144	30996.97124			

R-Square	Coeff Var	Root MSE	MOTIV1 Mean
0.000150	48.57612	14.77348	30.41306

Source	DF	Type I SS	Mean Square	F Value	Pr > F
SES	2	4.63937832	2.31968916	0.01	0.9894

Source	DF	Type III SS	Mean Square	F Value	Pr > F
SES	2	4.63937832	2.31968916	0.01	0.9894

The GLM Procedure

Dependent Variable: MOTIV2

Source	DF	Sum of Squares	Mean Square	F Value	Pr > F
Model	2	87.56128	43.78064	0.16	0.8489
Error	142	37918.72865	267.03330		
Corrected Total	144	38006.28993			

R-Square	Coeff Var	Root MSE	MOTIV2 Mean
0.002304	33.41694	16.34115	48.90081

Source	DF	Type I SS	Mean Square	F Value	Pr > F
SES	2	87.56127654	43.78063827	0.16	0.8489

Source	DF	Type III SS	Mean Square	F Value	Pr > F
SES	2	87.56127654	43.78063827	0.16	0.8489

The GLM Procedure

Dependent Variable: MOTIV3

Source	DF	Sum of Squares	Mean Square	F Value	Pr > F
Model	2	67.27715	33.63857	0.11	0.8960
Error	142	43465.71338	306.09657		
Corrected Total	144	43532.99052			

R-Square	Coeff Var	Root MSE	MOTIV3 Mean
0.001545	47.47987	17.49562	36.84849

Source	DF	Type I SS	Mean Square	F Value	Pr > F
SES	2	67.27714571	33.63857286	0.11	0.8960

Source	DF	Type III SS	Mean Square	F Value	Pr > F
SES	2	67.27714571	33.63857286	0.11	0.8960

The GLM Procedure

Level of SES		------MOTIV1------		------MOTIV2------		------MOTIV3------	
	N	Mean	Std Dev	Mean	Std Dev	Mean	Std Dev
1	47	30.2522340	15.7199013	48.6838511	16.0554160	36.2722766	18.9356935
2	50	30.3233800	13.1297030	48.0962000	16.9047920	37.7827600	16.2551049
3	48	30.6639583	15.4217050	49.9513958	16.0174023	36.4395000	17.2742199

4.5.1 Statistical Significance Versus Practical Importance

As is the case when examining results obtained in a univariate analysis, finding differences among sample mean vectors in a MANOVA to be statistically significant does not necessarily imply that they are important in a practical sense. With large enough samples, the null hypothesis of no mean differences will be rejected anyway, even if only violated to a substantively irrelevant degree. As is well known, one measure that can be used to address the practical relevance of group differences in an ANOVA design is the correlation ratio (also referred to as "eta squared"—sometimes also considered an effect size index):

$$\eta^2 = 1 - \left(\frac{SS_W}{SS_T}\right). \tag{4.35}$$

In the multivariate case, this equation can be generalized to the so-called multivariate correlation ratio:

$$\eta^2_{\text{mult}} = 1 - \Lambda. \tag{4.36}$$

The ratio in Equation 4.36 is interpretable as the proportion of generalized variance of the set of DVs that is attributable to group membership. In other words, the right-hand side of Equation 4.36 describes the strength of association between IVs and DVs in a sample at hand. It seems to be still one of the most widely used measures of practical significance from a set of related measures, for which we refer the reader to Rencher (1998) and references therein.

4.5.2 Higher Order MANOVA Designs

So far in this section we have considered MANOVA designs with only one factor. (In the example used in the preceding subsection, this was SES group membership.) Oftentimes in social and behavioral research, however, it is necessary to include additional factors, with several levels each. These circumstances lead to two-way, three-way, or higher order multivariate designs. In such designs, once again an analogy to the univariate two-way and higher order ANOVA settings will turn out to be quite helpful. In particular, recall the two-way ANOVA design partitioning of sum of squares on the response measure:

$$SS_T = SS_A + SS_B + SS_{AB} + SS_W, \tag{4.37}$$

where the factors are denoted using the letters A and B, SS_W is the within-group sum of squares, and SS_{AB} is the sum of squares due to their interaction, that is, represents DV variation in design cells over and above what could be attributed to the factors A and B. In other words, the interaction component is viewed as a unique effect that cannot be explained from knowledge of the effects of the other two factors.

In a two-way MANOVA design, using the earlier indicated uni-to-multivariate analogy, the same partitioning principle is upheld but with respect to the SSCP matrices:

$$SSCP_T = SSCP_A + SSCP_B + SSCP_{AB} + SSCP_W, \tag{4.38}$$

because as mentioned earlier the SSCP matrix is the multivariate generalization of the univariate sum of squares notion (see Chapter 1), where $SSCP_W$ is the earlier defined sum of within-group SSCP matrices across all cells of the design. (Like in a univariate two-way ANOVA design, the number of cells equals the number of combinations of all factor levels.) Because in this multivariate case the design is not changed but only the number of DVs is, as indicated before the degrees of freedom remain equal to those in the univariate case, for each hypothesis test discussed next.

The null hypotheses that can be tested in a two-way MANOVA are as follows, noting the complete design-related analogy with univariate ANOVA tests (observe also that in part due to the interaction effect being viewed as a unique effect that cannot be predicted from knowledge of the effects of the other two factors, interaction is commonly tested first):

$H_{0,1}$: Factors A and B do not interact,
$H_{0,2}$: Factor A does not have an effect (no main effect of A), and
$H_{0,3}$: Factor B does not have an effect (no main effect of B).

This analogy can be carried out further to the test statistics used. Specifically, employing Wilks' Λ as a test criterion, it can be written here in its more general form as

$$\Lambda_H = \frac{|\mathrm{SSCP_E}|}{|\mathrm{SSCP_E} + \mathrm{SSCP_H}|}, \tag{4.39}$$

or equivalently its reciprocal as

$$\frac{1}{\Lambda_\mathrm{H}} = |\mathrm{SSCP}_\mathrm{E}^{-1}(\mathrm{SSCP_E} + \mathrm{SSCP_H})| = |\mathrm{SSCP}_\mathrm{E}^{-1}\mathrm{SSCP_H} + \mathrm{I}|, \tag{4.40}$$

where H denotes each of the above effect hypotheses taken in turn, E stands for error, and I is the appropriately sized identity matrix. Thereby, the effect hypothesis SSCP is chosen as follows (Johnson & Wichern, 2002):

1. For $H_{0,1}$, $\mathrm{SSCP_H} = \mathrm{SSCP_{AB}}$ (i.e., formally substitute "H" for "AB");
2. For $H_{0,2}$: $\mathrm{SSCP_H} = \mathrm{SSCP_A}$ (substitute "H" for "A"); and
3. For $H_{0,3}$: $\mathrm{SSCP_H} = \mathrm{SSCP_B}$ (substitute "H" for "B").

For all three hypotheses, $\mathrm{SSCP_E} = \mathrm{SSCP_W}$ (i.e., formally, substitute "E" = "W"). The SSCP matrices in Equations 4.39 or 4.40 are readily calculated using software like SPSS or SAS once the design is specified, as are the pertinent test statistics.

For higher order MANOVA designs or more complicated ones, the basic principle in multivariate hypothesis testing follows from Equation 4.39:

1. Compute the SSCP matrix corresponding to each sum of squares in the respective ANOVA design (sum of square partitioning);
2. Select the appropriate error SSCP matrix (with fixed effects, as in this chapter, it is the $\mathrm{SSCP_W}$ from Equation 4.38); and
3. Calculate Λ_H from Equation 4.39.

All these steps are carried out by the used software once the needed detail about the underlying design and data are provided. In cases where the effects in the design are not fixed, but random or mixed (i.e., some are fixed and others are random), the choice of the appropriate error SSCP matrix will depend on the particular type of model examined. Since SPSS and SAS use fixed effects as default design when computing the necessary test statistics, whenever the effects being considered are not fixed the programs must be provided with this information in order to correctly compute the test statistic (e.g., see Marcoulides & Hershberger, 1997, for a table of error SSCP matrices in a two-way MANOVA then).

4.5.3 Other Test Criteria

In case of $g = 2$ groups, and regardless of number of DVs, a single test statistic is obtained in MANOVA, as indicated before, which we can treat for this reason just as Wilks' Λ. As soon as we have more than two groups, however, there are three additional test criteria that reflect the complexity of handling multivariate questions and are discussed in this subsection. Before we proceed, we note that it is not infrequent in empirical research for all four test statistics to suggest the same substantive conclusion.

The three other test statistics that are commonly computed are (a) Hotelling's Trace, (b) Pillai's Trace, and (c) Roy's Largest Root. Because each of these tests can be reasonably approximated using an F distribution (Rao, 1952; Schatzoff, 1964), it is also common practice to assess their significance based upon that F distribution rather than looking at the exact value of each of these test statistics. Only when $p = 1$ (i.e., in the univariate case) will all four test statistics provide identical F ratios as a special case of them.

In order to define these three test statistics, we need to introduce the notion of an eigenvalue, also sometimes referred to as a characteristic root or latent root. We spend a considerable amount of time on this concept again in Chapter 7, including some detailed numerical and graphical illustrations as well as use of software, but for the purposes of this section we briefly touch upon it here. To this end, suppose that $\underline{x}$ is a $p \times 1$ data vector of nonzero length, that is, $\underline{x} \neq \underline{0}$. As mentioned in Chapter 1, each data vector (row in the data matrix) can be represented in the multivariate space by a p-dimensional point, and hence also by the vector—or one-way arrow—that extends from the origin and ends into that point.

Consider now the vector $\underline{y}$ defined as $\underline{y} = \mathrm{A}\,\underline{x}$, where A is a $p \times p$ matrix (for example a SSCP matrix for a set of variables). This definition implies that we can obviously look at $\underline{y}$ as the result of a transformation working on $\underline{x}$. In general, $\underline{y}$ can be in any position and have any direction in the p-dimensional space, as a result of this transformation. However, the particular case when $\underline{y}$ is collinear with $\underline{x}$ is of special interest. In this case, both $\underline{x}$ and $\underline{y}$ share a common direction, but need not have the same length.

Under such a circumstance, there exists a number that we label λ, which ensures that $\underline{y} = \lambda \underline{x}$. This number λ is called eigenvalue of A while $\underline{x}$ is called eigenvector of A that pertains to λ. The matrices that are of interest and pursued in this book will have as many eigenvalues as their size, denoted say $\lambda_1, \lambda_2, \ldots, \lambda_p$ from largest to smallest, that is, $\lambda_1 \geq \lambda_2 \geq \ldots \geq \lambda_p$, whereby the case of some of them being equal is in general not excluded.

For example, if A is the following correlation matrix with $p = 2$ variables, $A = \begin{bmatrix} 1 & 0.4 \\ 0.4 & 1 \end{bmatrix}$, it can be readily found with software that it will have two eigenvalues, which will be equal to $\lambda_1 = 1.4$ and $\lambda_2 = 0.6$. (Specific details for using software to accomplish this aim are postponed to Chapter 7.) Similarly can be found that if the matrix were $A = \begin{bmatrix} 1 & 0 \\ 0 & 1 \end{bmatrix}$, which implies that the two variables would be unrelated, the two eigenvalues would be $\lambda_1 = 1$ and $\lambda_2 = 1$. Also, if the two variables were perfectly correlated, so that $A = \begin{bmatrix} 1 & 1 \\ 1 & 1 \end{bmatrix}$, then $\lambda_1 = 2$ and $\lambda_2 = 0$.

We discuss further eigenvalues and eigenvectors in Chapter 7 when we will be concerned also with how they are computed. For the moment, let us emphasize that the eigenvalues carry information characterizing the matrix A. In particular, it can be shown that the product of all eigenvalues for a symmetric matrix equals its determinant. Indeed, recalling from Chapter 2 that the determinant of $A = \begin{bmatrix} 1 & 0.4 \\ 0.4 & 1 \end{bmatrix}$ equals $|A| = [(1)(1) - (0.4)(0.4)] = 0.84$, we readily see that it is the product of its eigenvalues $\lambda_1 = 1.4$ and $\lambda_2 = 0.6$. For this reason, if a symmetric matrix has one or more eigenvalues equal to zero, it will be singular and thus cannot be inverted (i.e., its inverse does not exist; see Chapter 2). Such a problematic case will occur when two variables are perfectly correlated and A is their covariance or correlation matrix. For instance, in the above case where the two eigenvalues of a matrix were equal to $\lambda_1 = 2$ and $\lambda_2 = 0$, it can be implied that the variables contain redundant information (which also follows from the observation that they are perfectly correlated). Last but not least, as can be directly shown from the definition of eigenvalue and positive definiteness of a matrix, when all eigenvalues of a matrix are positive, then that matrix is positive definite. (The inverse statement is also true—a positive definite matrix has only positive eigenvalues.)

With this discussion of eigenvalues, we can now move on to the additional multivariate test statistics available generally in MANOVA, which are defined as follows:

Hotelling's trace criterion. This statistic, denoted τ, is the trace of a particular matrix product, specifically

$$\tau = \text{trace}\,(\text{SSCP}_E^{-1}\text{SSCP}_H). \tag{4.41}$$

It can be shown that this definition is equivalent to stating

$$\tau = \text{sum}\,(\text{eigenvalues of } \mathrm{SSCP}_\mathrm{E}^{-1}\mathrm{SSCP}_\mathrm{H}) \\ = \lambda_1 + \lambda_2 + \cdots + \lambda_p,$$

where the λs now denote the eigenvalues of the matrix product $\mathrm{SSCP}_\mathrm{E}^{-1}\mathrm{SSCP}_\mathrm{H}$ (recalling Equation 4.39 where this matrix product originated). In this definition, a particular feature of the trace operator is used, that is, for a symmetric positive definite matrix it equals the sum of its eigenvalues.

Roy's largest root criterion. This statistic, denoted θ, is defined as the following nonlinear function of the first eigenvalue of a related matrix:

$$\theta = \frac{\lambda_1}{(1 + \lambda_1)}, \tag{4.42}$$

where λ_1 denotes the largest eigenvalue of the matrix product

$$(\mathrm{SSCP}_\mathrm{E} + \mathrm{SSCP}_\mathrm{H})^{-1}\mathrm{SSCP}_\mathrm{H}.$$

Pillai–Bartlett trace criterion. This test statistic, denoted V, utilizes all eigenvalues of the last matrix product, $(\mathrm{SSCP}_\mathrm{E} + \mathrm{SSCP}_\mathrm{H})^{-1}\mathrm{SSCP}_\mathrm{H}$:

$$V = \frac{\lambda_1}{(1 + \lambda_1)} + \frac{\lambda_2}{(1 + \lambda_2)} + \cdots + \frac{\lambda_p}{(1 + \lambda_p)}, \tag{4.43}$$

where the λs are the same as for Roy's criterion.

We observe that all three criteria are increasing functions of the eigenvalues involved in them. That is, large eigenvalues indicate strong (and possibly significant) effect that is being tested with the statistics. It can also be shown that Wilks' Λ is indirectly related to these three test criteria, since

$$\Lambda = \frac{1}{[(1 + \lambda_1)(1 + \lambda_2) \cdots (1 + \lambda_p)]}, \tag{4.44}$$

where the λs are the eigenvalues participating in Equation 4.43. Notice that, contrary to the other three test statistics, Λ is a decreasing function of the eigenvalues involved in it. That is, larger eigenvalues lead to a smaller test statistic value Λ (i.e., possibly significant), while entailing higher alternative test statistics.

There is to date insufficient research, and lack of strong consensus in the literature, regarding the issue of which statistic is best used when. Usually, with pronounced (or alternatively only weak) effects being tested, all four would be expected to lead to the same substantive conclusions. Typically, if groups differ mainly along just one of the outcome variables, or along a single direction in the multivariate space (i.e., the group means are collinear), Roy's criterion θ will likely be the most powerful test statistic. With relatively small samples, Pillai–Bartlett's V statistic may be most robust to violation of the covariance matrix homogeneity assumption. Some simulation studies have found that this statistic V can also be more powerful in more general situations, in particular when the group means are not collinear (Olson, 1976; Schatzoff, 1966). In all cases, Wilks' Λ compares reasonably well with the other test statistics, however, and this is part of the reason why Λ is so popular, in addition to resulting from the framework of the LR theory (ML estimation).

To illustrate a MANOVA using the above test criteria, consider the following example study in which $n = 240$ juniors were examined with respect to their educational aspiration. As part of the study, 155 of them were randomly assigned to an experimental group with an instructional program aimed at enhancing their motivation for further education, and the remaining 85 students were assigned to a control group. At the end of the program administration procedure, $p = 3$ educational aspiration measures were given to both groups, and in addition data on their SES were collected. The research question is whether there are any effects of the program or SES on students' aspiration. (The data are available in the file ch4ex3.dat available from www.psypress.com/applied-multivariate-analysis; in that file, the motivation measures are denoted ASPIRE.1 through ASPIRE.3, with self-explanatory names for the remaining two variables; for SES, 1 denotes low, 2 middle, and 3 high SES; and for GROUP, 0 stands for control and 1 for experimental subjects.)

To conduct a MANOVA of this two-way design using SPSS, the following menu options/sequence would be used:

Analyze → General linear model → Multivariate (ASPIRE.1 through ASPIRE.3 as DVs, SES and GROUP as fixed factors; Options: homogeneity test, GROUP * SES means)

With SAS, the following command file could be employed with PROC GLM (detailed descriptions of each command were given in an earlier section of this chapter):

```
DATA ASPIRATION;
INFILE 'ch4ex3.dat';
INPUT MOTIV1 MOTIV2 MOTIV3 GROUP SES;
PROC GLM;
  CLASS GROUP SES;
  MEANS GROUP SES;
  MANOVA h = GROUP SES GROUP*SES;
RUN;
```

The outputs produced by SPSS and SAS are displayed next. For ease of presentation, the outputs are organized into sections, with clarifying comments inserted at appropriate places.

SPSS output

Box's Test of Equality of Covariance Matrices[a]

Box's M	37.354
F	1.200
df1	30
df2	68197.905
Sig.	.208

Tests the null hypothesis that the observed covariance matrices of the dependent variables are equal across groups.

a. Design: Intercept+GROUP+SES+GROUP* SES

The lack of significance of Box's M test (and of Bartlett's test with SAS, thus not presented below) suggests that the covariance matrix homogeneity assumption for the aspiration measures could be considered plausible in this two-way study.

Multivariate Tests[c]

Effect		Value	F	Hypothesis df	Error df	Sig.
Intercept	Pillai's Trace	.954	1603.345[a]	3.000	232.000	.000
	Wilks' Lambda	.046	1603.345[a]	3.000	232.000	.000
	Hotelling's Trace	20.733	1603.345[a]	3.000	232.000	.000
	Roy's Largest Root	20.733	1603.345[a]	3.000	232.000	.000
GROUP	Pillai's Trace	.084	7.052[a]	3.000	232.000	.000
	Wilks' Lambda	.916	7.052[a]	3.000	232.000	.000
	Hotelling's Trace	.091	7.052[a]	3.000	232.000	.000
	Roy's Largest Root	.091	7.052[a]	3.000	232.000	.000
SES	Pillai's Trace	.050	1.990	6.000	466.000	.066
	Wilks' Lambda	.951	1.986[a]	6.000	464.000	.066
	Hotelling's Trace	.052	1.983	6.000	462.000	.067
	Roy's Largest Root	.037	2.901[b]	3.000	233.000	.036
GROUP * SES	Pillai's Trace	.089	3.612	6.000	466.000	.002
	Wilks' Lambda	.912	3.646[a]	6.000	464.000	.002
	Hotelling's Trace	.096	3.680	6.000	462.000	.001
	Roy's Largest Root	.084	6.559[b]	3.000	233.000	.000

a. Exact statistic

b. The statistic is an upper bound on F that yields a lower bound on the significance level.

c. Design: Intercept+GROUP+SES+GROUP * SES

SAS output

The GLM Procedure
Multivariate Analysis of Variance

MANOVA Test Criteria and Exact F Statistics for the Hypothesis of No Overall GROUP Effect
H = Type III SSCP Matrix for GROUP
E = Error SSCP Matrix
S = 1 M = 0.5 N = 115

Statistic	Value	F Value	Num DF	Den DF	Pr > F
Wilks' Lambda	0.91642568	7.05	3	232	0.0001
Pillai's Trace	0.08357432	7.05	3	232	0.0001
Hotelling-Lawley Trace	0.09119596	7.05	3	232	0.0001
Roy's Greatest Root	0.09119596	7.05	3	232	0.0001

NOTE: F Statistic for Roy's Greatest Root is an upper bound.

NOTE: F Statistic for Wilks' Lambda is exact.

MANOVA Test Criteria and F Approximations for the Hypothesis of No Overall SES Effect
H = Type III SSCP Matrix for SES
E = Error SSCP Matrix
S = 2 M = 0 N = 115

Statistic	Value	F Value	Num DF	Den DF	Pr > F
Wilks' Lambda	0.95054327	1.99	6	464	0.0662
Pillai's Trace	0.04995901	1.99	6	466	0.0657
Hotelling-Lawley Trace	0.05150154	1.99	6	307.56	0.0672
Roy's Greatest Root	0.03735603	2.90	3	233	0.0357

NOTE: F Statistic for Roy's Greatest Root is an upper bound.

NOTE: F Statistic for Wilks' Lambda is exact.

MANOVA Test Criteria and F for the Hypothesis of No Overall GROUP*SES Effect
H = Type III SSCP Matrix for GROUP*SES
E = Error SSCP Matrix
S = 2 M = 0 N = 115

Statistic	Value	F Value	Num DF	Den DF	Pr > F
Wilks' Lambda	0.91197238	3.65	6	464	0.0015
Pillai's Trace	0.08888483	3.61	6	466	0.0016
Hotelling-Lawley Trace	0.09558449	3.69	6	307.56	0.0015
Roy's Greatest Root	0.08445494	6.56	3	233	0.0003

NOTE: F Statistic for Roy's Greatest Root is an upper bound.

NOTE: F Statistic for Wilks' Lambda is exact.

As in most applications of ANOVA, when examining the output provided by either SPSS or SAS, we usually look first at the interaction term. All four test statistics indicate that it is significant, so we can interpret this finding as evidence that warrants rejection of the null hypothesis of no interaction between GROUP and SES. Hence, the effect of the instructional program on students' aspiration is related to their SES. Since our main research concern here is with whether (and, possibly, when) the program has an effect, we move on to carrying out so-called "simple effect" tests, as we would in a univariate setup in the face of a finding of factorial interaction. To this end, we select the cases from each of the three SES groups in turn, and test within them whether the program has an effect on the aspiration measures. We note that this latter analysis is equivalent to conducting a one-way MANOVA within each of these three groups (in fact, testing then for multivariate mean differences across two independent samples per SES group), and thus can be readily carried out with the corresponding methods discussed earlier in this chapter. In particular, for the low SES group (with value of 1 on this variable), the so-obtained results are as follows (presented only using SPSS, to save space, since the SAS output would lead to identical conclusions).

These results indicate the presence of a significant program effect for low SES students, as far as the multivariate mean differences are concerned. In order to look at them more closely, we decide to examine the mean group differences for each aspiration measure separately, as provided by the univariate analysis results next.

SPSS output

Multivariate Tests[b]

Effect		Value	F	Hypothesis df	Error df	Sig.
Intercept	Pillai's Trace	.953	517.569[a]	3.000	76.000	.000
	Wilks' Lambda	.047	517.569[a]	3.000	76.000	.000
	Hotelling's Trace	20.430	517.569[a]	3.000	76.000	.000
	Roy's Largest Root	20.430	517.569[a]	3.000	76.000	.000
GROUP	Pillai's Trace	.189	5.898[a]	3.000	76.000	.001
	Wilks' Lambda	.811	5.898[a]	3.000	76.000	.001
	Hotelling's Trace	.233	5.898[a]	3.000	76.000	.001
	Roy's Largest Root	.233	5.898[a]	3.000	76.000	.001

a. Exact statistic

b. Design: Intercept+GROUP

Tests of Between-Subjects Effects

Source	Dependent Variable	Type III Sum of Squares	df	Mean Square	F	Sig.
Corrected Model	ASPIRE.1	3888.114[a]	1	3888.114	11.559	.001
	ASPIRE.2	4028.128[b]	1	4028.128	17.940	.000
	ASPIRE.3	4681.383[c]	1	4681.383	12.010	.001
Intercept	ASPIRE.1	167353.700	1	167353.700	497.548	.000
	ASPIRE.2	334775.484	1	334775.484	1490.945	.000
	ASPIRE.3	196890.228	1	196890.228	505.098	.000
GROUP	ASPIRE.1	3888.114	1	3888.114	11.559	.001
	ASPIRE.2	4028.128	1	4028.128	17.940	.000
	ASPIRE.3	4681.383	1	4681.383	12.010	.001
Error	ASPIRE.1	26235.845	78	336.357		
	ASPIRE.2	17514.046	78	224.539		
	ASPIRE.3	30404.863	78	389.806		
Total	ASPIRE.1	215241.332	80			
	ASPIRE.2	385735.384	80			
	ASPIRE.3	253025.216	80			
Corrected Total	ASPIRE.1	30123.959	79			
	ASPIRE.2	21542.174	79			
	ASPIRE.3	35086.246	79			

a. R Squared = .129 (Adjusted R Squared = .118)
b. R Squared = .187 (Adjusted R Squared = .177)
c. R Squared = .133 (Adjusted R Squared = .122)

Accordingly, each of the three aspiration measures shows statistically significant experimental versus control group differences. This finding, plus an inspection of the group means next, suggest that for low SES students the instructional program had an effect on each aspiration measure. For more specific details regarding these effects, we take a close look at the group means (typically referred to as marginal means). Alternatively, we can also examine a graph of these means to aid in visualizing the group differences that are present at each level of the SES factor.

GROUP

Dependent Variable	GROUP	Mean	Std. Error	95% Confidence Interval	
				Lower Bound	Upper Bound
ASPIRE.1	.00	39.565	3.242	33.111	46.020
	1.00	53.796	2.647	48.526	59.066
ASPIRE.2	.00	58.781	2.649	53.507	64.055
	1.00	73.265	2.163	68.959	77.571
ASPIRE.3	.00	42.825	3.490	35.877	49.774
	1.00	58.440	2.850	52.767	64.114

From this table is seen that each of the measures is on average higher in the experimental group, suggesting that the program may have led to some enhancement of aspiration for low SES students.

Next, in order to examine the results for students with middle SES, we repeat the same analyses after selecting their group. The following results are obtained then:

Multivariate Tests[b]

Effect		Value	F	Hypothesis df	Error df	Sig.
Intercept	Pillai's Trace	.966	715.311[a]	3.000	76.000	.000
	Wilks' Lambda	.034	715.311[a]	3.000	76.000	.000
	Hotelling's Trace	28.236	715.311[a]	3.000	76.000	.000
	Roy's Largest Root	28.236	715.311[a]	3.000	76.000	.000
GROUP	Pillai's Trace	.128	3.715[a]	3.000	76.000	.015
	Wilks' Lambda	.872	3.715[a]	3.000	76.000	.015
	Hotelling's Trace	.147	3.715[a]	3.000	76.000	.015
	Roy's Largest Root	.147	3.715[a]	3.000	76.000	.015

a. Exact statistic
b. Design: Intercept + GROUP

Considered together, the aspiration measures differ on average across the experimental and control groups also in the middle SES group, as evinced by the significant p-value of .015 for the group effect in the last table. To examine these differences more closely, we decide again to look at the univariate ANOVA results.

None of the three aspiration measures considered separately from the other two appears to have been affected notably by the program for middle SES students. It would seem that the multivariate effect here has capitalized on the relatively limited mean differences on each of the aspiration measures as well as on their interdependency. This phenomenon is consistent with the fact that the practical importance measure of the multivariate effect is quite negligible (as $\eta^2 = 1 - \Lambda = 1 - .872 = .128$). This is also noticed by looking at the cell means below and seeing small differences across groups that are with opposite directions, keeping in mind that they are actually not significant per measure. (Note also the substantial overlap of the confidence intervals for the means of the same measure across groups.)

Tests of Between-Subjects Effects

Source	Dependent Variable	Type III Sum of Squares	df	Mean Square	F	Sig.
Corrected Model	ASPIRE.1	88.228[a]	1	88.228	.312	.578
	ASPIRE.2	134.494[b]	1	134.494	.818	.369
	ASPIRE.3	743.237[c]	1	743.237	2.437	.123
Intercept	ASPIRE.1	148239.017	1	148239.017	524.093	.000
	ASPIRE.2	332086.264	1	332086.264	2018.959	.000
	ASPIRE.3	185905.713	1	185905.713	609.600	.000
GROUP	ASPIRE.1	88.228	1	88.228	.312	.578
	ASPIRE.2	134.494	1	134.494	.818	.369
	ASPIRE.3	743.237	1	743.237	2.437	.123
Error	ASPIRE.1	22062.200	78	282.849		
	ASPIRE.2	12829.743	78	164.484		
	ASPIRE.3	23787.163	78	304.964		
Total	ASPIRE.1	185278.464	80			
	ASPIRE.2	389142.549	80			
	ASPIRE.3	223936.342	80			
Corrected Total	ASPIRE.1	22150.427	79			
	ASPIRE.2	12964.237	79			
	ASPIRE.3	24530.400	79			

a. R Squared = .004 (Adjusted R Squared = −.009)
b. R Squared = .010 (Adjusted R Squared = −.002)
c. R Squared = .030 (Adjusted R Squared = .018)

GROUP

Dependent Variable	GROUP	Mean	Std. Error	95% Confidence Interval	
				Lower Bound	Upper Bound
ASPIRE.1	.00	46.628	3.237	40.184	53.071
	1.00	44.407	2.310	39.808	49.006
ASPIRE.2	.00	66.756	2.468	61.842	71.670
	1.00	69.498	1.762	65.991	73.005
ASPIRE.3	.00	54.196	3.361	47.505	60.887
	1.00	47.750	2.399	42.975	52.526

Finally, when examining the results for the high SES students, we observe the following findings:

Multivariate Tests[b]

Effect		Value	F	Hypothesis df	Error df	Sig.
Intercept	Pillai's Trace	.945	436.512[a]	3.000	76.000	.000
	Wilks' Lambda	.055	436.512[a]	3.000	76.000	.000
	Hotelling's Trace	17.231	436.512[a]	3.000	76.000	.000
	Roy's Largest Root	17.231	436.512[a]	3.000	76.000	.000
GROUP	Pillai's Trace	.138	4.040[a]	3.000	76.000	.010
	Wilks' Lambda	.862	4.040[a]	3.000	76.000	.010
	Hotelling's Trace	.159	4.040[a]	3.000	76.000	.010
	Roy's Largest Root	.159	4.040[a]	3.000	76.000	.010

a. Exact statistic
b. Design: Intercept + GROUP

As can be seen also here, there is indication of program effect when the aspiration measures are analyzed simultaneously and their interrelationship is taken into account. In difference to students of middle SES, however, there are notable program effects also when each measure is considered on its own, as we see next.

Tests of Between-Subjects Effects

Source	Dependent Variable	Type III Sum of Squares	df	Mean Square	F	Sig.
Corrected Model	ASPIRE.1	1810.066[a]	1	1810.066	5.510	.021
	ASPIRE.2	1849.893[b]	1	1849.893	7.020	.010
	ASPIRE.3	4048.195[c]	1	4048.195	9.917	.002
Intercept	ASPIRE.1	134832.701	1	134832.701	410.442	.000
	ASPIRE.2	306088.252	1	306088.252	1161.537	.000
	ASPIRE.3	145442.177	1	145442.177	356.303	.000
GROUP	ASPIRE.1	1810.066	1	1810.066	5.510	.021
	ASPIRE.2	1849.893	1	1849.893	7.020	.010
	ASPIRE.3	4048.195	1	4048.195	9.917	.002
Error	ASPIRE.1	25623.504	78	328.506		
	ASPIRE.2	20554.561	78	263.520		
	ASPIRE.3	31839.425	78	408.198		
Total	ASPIRE.1	193803.985	80			
	ASPIRE.2	390463.464	80			
	ASPIRE.3	221555.322	80			
Corrected Total	ASPIRE.1	27433.570	79			
	ASPIRE.2	22404.453	79			
	ASPIRE.3	35887.620	79			

a. R Squared = .066 (Adjusted R Squared = .054)
b. R Squared = .083 (Adjusted R Squared = .071)
c. R Squared = .113 (Adjusted R Squared = .101)

To explore these group differences more closely, we take a look at the corresponding means.

GROUP

Dependent Variable	GROUP	Mean	Std. Error	95% Confidence Interval Lower Bound	Upper Bound
ASPIRE.1	.00	38.748	3.555	31.671	45.824
	1.00	48.904	2.466	43.993	53.814
ASPIRE.2	.00	60.899	3.184	54.561	67.237
	1.00	71.165	2.209	66.768	75.563
ASPIRE.3	.00	37.923	3.962	30.035	45.812
	1.00	53.111	2.749	47.638	58.585

As is evident from the last table, the experimental group has a higher mean than the control group on each of the aspiration measures, indicating also for high SES students enhanced levels of aspiration after program administration.

We conclude our analyses for this empirical example by stating that there is evidence suggesting program effect for students with low and students with high SES, for whom they are found on each aspiration measure as well as overall. For middle SES students, although there is some indication of overall instructed versus uninstructed group mean differences, they are relatively small and inconsistent in direction when particular motivation measures are considered. In this SES group, there is no evidence for a uniformly enhanced aspiration levels across measures that would be due to the program administered.

4.6 MANOVA Follow-Up Analyses

When no significant differences are observed in a MANOVA across $g \geq 2$ studied groups, one usually does not follow it up with univariate analyses. In particular, if the initial research question is multivariate in nature, no subsequent univariate analyses would be necessary or appropriate. However, it is possible—and in empirical research perhaps not infrequently the case—that there is not sufficient information available to begin with, which would allow unambiguous formulation of the substantive research question as a multivariate one. Under such circumstances, it may be helpful to examine which DV shows significant group differences and which not. This latter question may also be of interest when a MANOVA group difference is found to be significant.

The most straightforward approach to the last query would be to examine the univariate test results. Typically, if the latter were not of concern prior to a decision to carry out a MANOVA, they should better be conducted at a lower significance level. To be precise, if one is dealing with p DVs ($p \geq 1$), then examining each one of them at a significance level of α/p would be recommendable. That is, any of these variables would be proclaimed as significantly differing across groups only if the pertinent p-value is less than α/p. In other words, if $\alpha = .05$ was the initial choice for a significance level, the tests in question would be carried out each at a significance level $\alpha' = .05/p$. For the fixed effects of concern so far in the book, these tests are readily conducted using the output obtained with SPSS, by inspecting the pertinent entries in the output section entitled "Tests of Between-Subject Effects" and correspondingly checking if the p-value for an effect of interest with regard to a given DV is smaller than α'. In the corresponding SAS output, one would simply look at the univariate ANOVA results provided separately for each DV and check their associated p-value accordingly.

The general significance level adjustment procedure underlying this approach is often referred to as Bonferroni correction. It turns out to be somewhat conservative, but if one finds univariate differences with it one can definitely have confidence in that statistical result. The correction is frequently recommended, especially with a relatively small number of DVs, p (up to say half a dozen), instead of counterparts of what has become known as simultaneous multiple testing procedures (Hays, 1994). The latter turn out to yield too wide confidence intervals for mean differences on individual DVs, and for this reason they are not generally recommendable unless p is quite large (such as say $p > 10$; Johnson & Wichern, 2002). These procedures capitalize on the interrelationship between the DVs and aim at securing a prespecified significance level to all possible comparisons between means. This is the reason why their resulting confidence intervals can be excessively wide. These procedures may be viewed as multivariate analogs of the so-called Scheffe's multiple-comparison procedure in the context of univariate ANOVA, which is also known to be rather conservative but is well suited for "data-snooping" or a posteriori mean comparisons.

As an alternative to Bonferroni's correction, one may wish to carry out conventional unidimensional F tests (i.e., t tests in cases of $g = 2$ groups), which are the univariate ANOVA tests for group differences on each DV considered separately, at the conventional α level (say .05). By opting for such a strategy only in the case of a significant MANOVA group difference, this approach would be nearly optimal in terms of controlling the overall Type I error and not being overly conservative (Rencher, 1998). Regardless which of these follow-up analysis approaches is chosen, however, one should keep in mind that their results are not independent of one another across response variables, since the latter are typically interrelated. Another important alternative that can be used to follow up significant MANOVA tests is discriminant function analysis, which is the subject of Chapter 10.

4.7 Limitations and Assumptions of MANOVA

MANOVA is a fairly popular procedure among social, behavioral, and educational scientists. It is therefore important when applying it to be aware of its assumptions and limitations.

First of all, one cannot base any causality conclusions only on the results of statistical tests carried out within the MANOVA framework. That is, a finding of a significant group mean difference does not justify a statement that an IV in a study under consideration has produced differences observed in the DVs. Second, in order for the underlying parameter estimation procedure to be computationally feasible, each cell of the design should have more cases than DVs. If this is not the case, some levels of pertinent factors need to be combined to insure fulfillment of this condition (see also Chapter 7 for an alternative approach). Third, excessively highly correlated DVs present a problem akin to multicollinearity in regression analysis. Thus, when such redundant variables are initially planned as outcome variables in a study, one should either drop one or more of them or use possibly a linear combination of an appropriate subset of them instead, which has clear substantive meaning.

Furthermore, as mentioned earlier, throughout this chapter we have assumed that the MVN assumption holds. Although the methods discussed have some degree of robustness against violations of this assumption, it is recommended that consideration be given to use of appropriate variable transformations prior to analysis, or of alternative methods (such as path analysis or structural equation modeling; Raykov & Marcoulides, 2006) where corrections for some nonnormality can be carried out on respective fit statistics and parameter standard errors. It is also noted that depending on outcome variable relationships, a multivariate test may turn out to be nonsignificant while there are considerable or even significant group differences on individual response measures, a circumstance sometimes referred to as "washing out" of univariate differences. Hence, it is of special importance that a researcher conceptualizes carefully, and before the analysis is commenced, whether they need to use a multivariate test or one or more univariate ones.

Another important MANOVA assumption that we mentioned repeatedly in this chapter is that of covariance matrix homogeneity. While it can be tested using Box's M test, or Bartlett's test, they are notorious for being quite sensitive to nonnormality. In such cases, our note in the preceding paragraph applies, that is, consideration should better be given to appropriate variable transformations before analysis is begun. Alternatively, covariance matrix homogeneity can be tested using structural equation modeling (Raykov, 2001). The latter approach, when used with corresponding fit statistics' and standard error corrections, provides another means for these purposes in cases of up to mild nonnormality (and absence of piling of cases at an end of any measurement scale used, e.g., with absence of ceiling and floor effects).

5

Repeated Measure Analysis of Variance

In this chapter, we will be concerned with the analysis of data obtained from studies with more than two related groups. We first introduce the notion of within-subject design (WSD) that underlies investigations where repeated assessments are carried out on a given sample(s) of subjects, and parallel it with that of between-subject design (BSD). Subsequently, we discuss the necessity to have special methods for analyzing repeated observations collected in such studies. Then we deal with the univariate approach to repeated measures analysis (RMA), which is developed within the general framework of analysis of variance (ANOVA). Following that discussion, we will be concerned with the multivariate approach to RMA, which is presented within the context of multivariate analysis of variance (MANOVA). The settings where the methods discussed in this chapter will be applicable are those where several dependent variables result from multiple assessments of a single outcome measure across several occasions, regardless of whether there are any independent variables; when there are, typically a grouping factor will be present (e.g., control vs. experimental group). (Studies with more than one repeatedly administered measure are considered in Chapter 13.) Statistical approaches of relevance for these settings are also sometimes called profile analysis methods, and empirical investigations with such designs are frequently referred to as longitudinal or panel studies in the social and behavioral sciences.

Repeated measure designs (RMDs) have some definitive advantages relative to cross-sectional designs (CSDs) that have been widely used for a large part of the past century. In particular, CSDs where say a number of different age groups are measured at one point in time on an outcome or a set of outcomes have as a main limitation the fact that they are not well-suited for studying developmental processes that are of special relevance in these disciplines. The reason is that they confound the effects of time and age (cohort). As a viable alternative, RMDs have become popular since the late 1960s and early 1970s. To give an example of such designs, suppose one were interested in studying intellectual development of high school students in grades 10 through 12. A longitudinal design utilizing repeated assessments will then call for measuring say once a year a given sample of students in grade 10, then in grade 11, and finally in grade 12.

Alternatively, a cross-sectional design (CSD) would proceed by assessing at only one time point a sample from each of the 10th, 11th, and 12th grades. As it turns out, there are numerous threats to the validity of a CSD that limit its utility for studying processes of temporal development. Central amongst these threats is the potential bias that results from the above mentioned confounding. A thorough discussion of the validity threats is beyond the scope of this text, and we refer the reader to some excellent available sources (Campbell & Stanley, 1963; Nesselroade & Baltes, 1979).

5.1 Between-Subject and Within-Subject Factors and Designs

In a repeated measure study, it is particularly important to distinguish between two types of factors: (a) between-subject factors (BSFs; if any), and (b) within-subject factors (WSFs). As indicated in Chapter 4, a BSF is a factor whose levels are represented by independent subjects that thus provide unrelated measurements (observations). For example, the distinction between an experimental and control group represents a BSF since each person can be (usually) either in an experimental or a control group; for this reason, each subject gives rise only to one score on this factor (say "0" if belonging to the control group and "1" if from the experimental group). Similarly, gender (male vs. female) is another BSF since persons come to a study in question with only a single value on gender. Further examples of BSFs would include factors like ethnicity, political party affiliation, or religious affiliation (at least at a given point in time), which may have more than a pair of levels. Hence, in a design with a BSF, the measurements collected from studied subjects are independent from one another across that factor's levels.

By way of contrast, a within-subject factor (WSF) has levels that represent repeated measurements conducted on the same persons or in combination with others related to them (e.g., in studies with matched samples). In a typical longitudinal design, any given assessment occasion is a level of a WSF (frequently referred to as "time factor"). In a design with a WSF, measurements on this factor are not independent of one another across its levels. For example, consider a cognitive development study of boys and girls in grades 7 through 9, in which mental ability is assessed by a test once in each of these three grades on a given sample of students. Since these are repeated measurements of the same subjects, the assessment (or time) factor is a WSF with three levels, while the gender factor is a BSF with two levels.

For convenience, designs having only BSFs are commonly called between-subject designs (BSDs), while those with only WSFs are referred to as within-subject designs (WSDs). In the social and behavioral sciences, designs that include both BSF(s) and WSF(s) are rather frequently used and are generally called mixed designs. In fact, it may be even fair to say that mixed designs of this kind are at present perhaps typical for repeated measure studies in these disciplines.

So far in this text, we have mostly been concerned with designs that include BSFs. For example, one may recall that the designs considered in Chapter 4 were predominantly based on BSFs. Specifically, when testing for mean differences across independent groups, and the entire discussion of MANOVA in that chapter, utilized BSDs. However, we also used a WSD there without explicitly referring to it in this way. Indeed, when testing for mean differences with related samples (or two repeated assessments), a WSF with two levels was involved. Given that we have already devoted considerable attention to BSDs and procedures for analysis of data resulting from such studies, it may be intriguing to know why it would be important to have special methods for handling WSD.

To motivate our interest in such methods, let us consider the following example that utilizes a WSD. Assume we are interested in motivation, which is considered by many researchers a critically important variable for academic learning and achievement across childhood and through adolescence. To highlight better the idea behind the necessity to use special methods in RMDs, we use only a small portion of the data resulting from that study, which is presented in Table 5.1 (cf. Howell, 2002).

Two observations can be readily made when looking at the data in Table 5.1. First, we note that although the average score (across students) is not impressively changing over time, there is in fact some systematic growth within each subject of about the same magnitude as the overall mean change. Second, the variability across the three assessment means is considerably smaller than subject score variability within each measurement occasion. Recalling that what contributes toward a significant main effect in a univariate ANOVA is the relationship of variability in the group means to that within groups (Hays, 1994), these two observations suggest that if one were to analyze data from a WSD—like the presently considered one—using methods appropriate for between-subject designs, one is

TABLE 5.1

Scores From Four Students in a Motivation Study Across Three Measurement Occasions (Denoted by Time_1 Through Time_3)

Student	Time_1	Time_2	Time_3
1	3	4	5
2	16	17	19
3	4	5	7
4	22	24	25
Average total score:	11.25	12.50	14

likely to face a nonsignificant time effect finding even if all subjects unambiguously demonstrate temporal change in the same direction (e.g., growth or decline).

The reason for such an incorrect conclusion would be the fact that unlike BSFs, as mentioned, measurements across levels of WSFs are not independent. This is because repeatedly assessed subjects give rise to correlated measurements across the levels of any WSF (which relationship is often referred to as serial correlation). This is clearly contrary to the case with a BSF. Thus, an application of a BSD method to repeated measure data like that in Table 5.1 effectively disregards this correlation—that is, wastes empirical information stemming from the interrelationships among repeated assessments for the same subjects—and therefore is likely to lead to an incorrect conclusion. Hence, different methods from those applicable with BSDs are needed for studies having WSFs. These different methods should utilize that empirical information, i.e., take into account the resulting serial correlation—in a sense "partial it out"—before testing hypotheses of interest. Univariate and multivariate approaches to repeated measure analysis provide such methods for dealing with studies containing within-subject factors, and we turn to them next.

5.2 Univariate Approach to Repeated Measure Analysis

This approach, also sometimes referred to as ANOVA method of RMA, can be viewed as following the classical ANOVA procedure of variance partitioning. For example, suppose that a sample of subjects from a studied population is repeatedly observed on m occasions ($m \geq 2$). The method is then based on the following decomposition (cf. Howell, 2002):

$$SS_T = SS_{BS} + SS_{WS} = SS_{BS} + SS_{BO} + SS_E, \tag{5.1}$$

where the subscript notation used is as follows: T for total, BS for between-subjects, WS for within-subjects, BO for between-occasions, and E for error. According to Equation 5.1, if two observations were randomly picked from the set of all observations for all subjects at all measurement points, then their dependent variable values could differ from one another as a result of these values belonging to two different persons and/or stemming from two distinct occasions, or being due to unexplained reasons.

The model underlying the decomposition in Equation 5.1 can be written as

$$X_{ij} = \mu + \pi_i + \tau_j + e_{ij}, \tag{5.2}$$

where the score X_{ij} of the ith subject at jth assessment is decomposed into an overall mean (μ), subject effect (π_i), assessment occasion effect (τ_j), and error (e_{ij}) ($i = 1, \ldots, n$, with n denoting as usual sample size; $j = 1, \ldots, m$). As seen from Equation 5.2, no interaction between subject and occasion is assumed in the model. This implies that any possible interaction between subject and occasion is relegated to the error term. In other words, in this model we assume that there is no occasion effect that is differential by subject. Such a model is said to be additive. In case this assumption is invalid, an interaction term between subject and occasion would need to be included in the model that will then be referred to as nonadditive (for an excellent discussion of nonadditive models, see Kirk, 1994). Although a number of tests have been developed for examining whether a model is additive or nonadditive, for the goals of our discussion in this chapter we will assume an additive model.

The test for change over time, which is of particular interest here, is based on the following F ratio that compares the between-occasion mean sum of squares to that associated with error:

$$F = \frac{\text{Mean(SSBO)}}{\text{Mean(SSE)}} = \frac{\text{SSBO}/(m-1)}{\text{SSE}/[(n-1)(m-1)]} \sim F_{m-1,(n-1)(m-1)}. \tag{5.3}$$

The validity of this test, that is, the distribution of its F ratio stated in Equation 5.3, is based on a special assumption called sphericity (Maxwell & Delaney, 2004). This condition is fulfilled when the covariance matrix of the repeated assessments, denoted by Ψ, has the same quantity along its main diagonal (say σ^2), and off its diagonal an identical quantity (say θ) appears. That is, sphericity holds for that matrix when

$$\Psi = \begin{bmatrix} \sigma^2 & & & & \\ \theta & \sigma^2 & & & \\ \theta & \theta & \sigma^2 & & \\ . & . & . & . & . \\ \theta & \theta & \theta & \ldots & \sigma^2 \end{bmatrix}, \tag{5.4}$$

for two generally distinct numbers σ^2 and θ (such that Ψ is positive definite; see Chapter 2).

The sphericity condition implies that any two repeated assessments correlate to the same extent, as can be readily seen by obtaining the correlation of two occasions from the right-hand side of Equation 5.4. For example, if the following covariance matrix involving three repeated assessments were to hold in a studied population, the assumption of sphericity would be fulfilled:

$$\Psi = \begin{bmatrix} 25 & & \\ 10 & 25 & \\ 10 & 10 & 25 \end{bmatrix}.$$

This assumption is, however, not very likely to be valid in most longitudinal behavioral and social studies, since repeated measurements in these disciplines tend to correlate more highly when taken closely in time than when they are further apart. When the sphericity assumption is tenable, however, the univariate approach provides a powerful method for RMA, also with relatively small samples. We will return to this issue in more detail in a later section in the chapter.

We note in passing that the sphericity condition could be viewed, loosely speaking, as an "extension" to the RMD context of the independent-group ANOVA assumption of variance homogeneity, which states that the variances of all groups on a single dependent variable are the same. (Recall that the latter assumption underlies conventional applications of the t test as well, in cases with $g = 2$ groups.) Specifically, the variance homogeneity assumption might be seen as a special case of the sphericity assumption (cf. Howell, 2002) because when dealing with independent groups the corresponding formal covariances are all 0, and thus the value θ (see Equation 5.4) equals 0 in a respective counterpart of the above matrix Ψ (while all its main diagonal entries are the same, when variance homogeneity holds across groups). That is, as an aside, as soon as variance homogeneity is fulfilled in a multiple-population setting with a single outcome variable, in the above sense also the sphericity assumption is fulfilled. The much more important concern with sphericity is, however, in the repeated measure context of interest in this chapter, when variance homogeneity alone is obviously not sufficient for sphericity because the latter also requires covariance coefficient homogeneity (see Equation 5.4).

Since the univariate approach to RMA hinges on the sphericity assumption being tenable, it is important to note that it represents a testable condition. That is, in an empirical repeated measure setting, one can use special statistical tests to ascertain whether sphericity is plausible. A widely available test for this purpose is the so-called Mauchly's sphericity test, whose result is automatically provided in most statistical software output (see discussion of SAS and SPSS RMA procedures below). An alternative test is also readily obtained within the latent variable modeling framework (Raykov, 2001). When the sphericity assumption is violated, however, the F test in Equation 5.3 is liberal. That is, it tends to reject more often than it should the null hypothesis when the latter is true; in other words, its Type I error rate is higher than the nominal level (usually set at .05). For this reason, when the F test in Equation 5.3 is not significant, one might have considerable confidence in its result and a suggestion based on it not to reject the pertinent null hypothesis.

When the sphericity condition is not plausible for a given data set, one possibility is to employ a different analytic approach, viz. multivariate repeated measure analysis of variance that is the subject of the next section. (As another option, one could use with large samples unidimensional versions of the models discussed in Chapter 13; see that chapter for further discussion on their utility.) The multivariate approach to RMA is particularly attractive with large samples. Alternatively, and especially with small samples (when the multivariate approach tends to lack power, as indicated in the next section), one could still use the F test in Equation 5.3 but after correcting its degrees of freedom when working out the pertinent cut-off value for significance testing. This results in the so-called *ε-correction* procedure for the univariate RMA F test. Specifically, the new degrees of freedom to be used with that F test are $df_1 = \varepsilon(m-1)$ and $df_2 = \varepsilon(m-1)(n-1)$, rather than those stated immediately after Equation 5.3 above. We note that the value of the ε factor is 1 when sphericity is not violated; in all other cases, ε will be smaller than 1 (see further discussion below), thus ensuring that smaller degrees of freedom values are utilized. This correction effectively increases the cut-off value of the pertinent F- distribution of reference for significance testing, thus counteracting the earlier mentioned "liberalism" feature of the F test in Equation 5.3 when sphericity is violated.

There are three procedures for accomplishing this degrees of freedom correction: (a) Box's lower bound, (b) Greenhouse–Geisser's adjustment, and (c) Huynh–Feldt's correction. Each of them produces a different value for ε with which the degrees of freedom need to be modified by multiplication as indicated above. Hence, each of these three procedures leads to different degrees of freedom for the reference F- distribution, against which the test statistic in the right-hand side of Equation 5.3 is to be judged for significance.

The effect of correction procedures (a) through (c) is that they make that F test in Equation 5.3 more conservative, in order to compensate for its being liberal when sphericity is violated. However, each of the three corrections actually "overshoots" the target—the resulting, corrected F test becomes then conservative. That is, it tends to reject the null hypothesis, when true, less often than it should; therefore, the associated Type I error rate is then lower than the nominal level (commonly taken as .05). The three corrections actually lead to tests showing to a different extent this "conservatism" feature. Box's procedure is the most conservative, Greenhouse–Geisser's is less so, and the least conservative of the three is Huynh–Feldt's adjustment (cf. Timm, 2002). Hence, if Huynh–Feldt's procedure rejects the null hypothesis—that is, the F test in Equation 5.3 is significant when its value is referred to the F- distribution with degrees of freedom being $\varepsilon(m-1)$ and $\varepsilon(m-1)(k-1)$, where ε results from Huynh–Feldt's adjustment—then we can have trust in its result; otherwise interpretation is ambiguous (within this univariate approach to RMA).

To demonstrate this discussion, consider the following example. One hundred and fifty juniors are measured with an intelligence test at a pretest and twice again after a 2 week training aimed at improving their test-relevant cognitive skills. The research question is whether there is evidence for change over time. Here, as typical in single-group longitudinal studies, of main interest is the null hypothesis H_0: $\mu_1 = \mu_2 = \mu_3$, where μ_t stands for the observed mean at tth occasion ($t = 1, 2, 3$). This hypothesis of no change over time is also sometimes called the flatness hypothesis. (The data for this example are found in the file ch5ex1.dat available from www.psypress.com/applied-multivariate-analysis, where the variables at the three repeated assessments are named Test.1, Test.2, and Test.3.)

To answer this question, with SPSS we use the following menu option/sequence:

General Linear Model → Repeated Measures (WSF name = "time," 3 levels, click "Add"; define levels as the successive measures; Options: mean for the time factor).

To accomplish this analysis with SAS, the following command statements within the PROC GLM procedure can be used:

```
DATA RMA;
INFILE 'ch4ex1.dat';
INPUT t1-t3;
PROC GLM;
MODEL t1 - t3 = /nouni;
REPEATED time POLYNOMIAL /summary mean printe;
RUN;
```

As one may recall from the previous chapter, the MODEL statement names the variables containing the data on the three assessments in the study under consideration. There are, however, a number of new statements in this program setup that require clarification. In particular, we need to comment on the use of the statements "nouni," "REPEATED," "POLYNOMIAL," and "summary mean printe." The "nouni" statement suppresses the display of output relevant to certain univariate statistics that are in general not of interest with this type of repeated measures design. (In fact, not including the "nouni" subcommand would make SAS compute and print F tests for each of the three assessments, which results are not usually of much interest in within-subjects designs.) When the variables in the MODEL statement represent repeated measures on each subject, the REPEATED statement enables one to test hypotheses about the WSF (in this case the time factor). For example, in the present study the three levels of the time factor used are "t1," "t2," and "t3." One can also choose to provide a number that reflects exactly the number of assessments considered. If one used this option, the pertinent statement above could also be specified as "REPEATED 3." As a result, both univariate and multivariate tests are generated as well as hypothesis tests for various patterns of change over time (also called contrasts; see next section). To specifically request examination of linear, quadratic, and higher-order patterns of change, if applicable, the POLYNOMIAL statement is utilized (see *SAS User's Guide* for a complete listing of the various contrasts that can be obtained). Finally, the options "summary mean printe" produce ANOVA tables for each pattern of change under consideration, the means for each of the within-subject variables, and Mauchly's test of sphericity.

The outputs produced by the above SPSS instructions and SAS command file follow next. For ease of presentation, they are re-organized into sections, and clarifying comments are inserted after each of them.

SPSS output

General Linear Model

Within-Subjects Factors

Measure: MEASURE_1

TIME	Dependent Variable
1	TEST.1
2	TEST.2
3	TEST.3

In this table, the levels are explicitly named after the measures comprising the WSF, whose effect we are interested in testing. Next, these three repeated measures are examined with respect to the sphericity assumption.

Mauchly's Test of Sphericity[b]

Measure: MEASURE_1

Within Subjects Effect	Mauchly's W	Approx. Chi-Square	df	Sig.	Epsilon[a]		
					Greenhouse-Geisser	Huynh-Feldt	Lower-bound
TIME	.851	23.888	2	.000	.870	.880	.500

Tests the null hypothesis that the error covariance matrix of the orthonormalized transformed dependent variables is proportional to an identity matrix.

a. May be used to adjust the degrees of freedom for the averaged tests of significance. Corrected tests are displayed in the Tests of Within-Subjects Effects table.

b. Design: Intercept
Within Subjects Design: TIME

From the column titled "Sig." in this table, Mauchly's test of sphericity is found to be significant. This means that the data contain evidence warranting rejection of the null hypothesis of sphericity for the repeated measure covariance matrix (i.e., the covariance matrix resulting when considering the repeated measures as three interrelated variables). Consequently, in order to deal with the lack of sphericity, epsilon corrections may be used in determining the more appropriate F test. (As mentioned earlier, an alternative approach is discussed in the next section, and another one in Chapter 13, both best applied with large samples.) As indicated before, Huynh–Feldt's adjustment is least affecting the degrees of freedom for the F test in Equation 5.3, because its pertinent ε factor is closest to 1, while Box's modification provides a lower bound (labeled "Lower-bound" in the output) of this multiplier. Since degrees of freedom are inversely related to the extent an F test is conservative (when it is so), the last presented output table also demonstrates our earlier comment that Box's correction leads to the most conservative test statistic whereas Huynh–Feldt's entails the least conservative F test version that is therefore preferable.

SAS output

The GLM Procedure

Repeated Measures Analysis of Variance

Repeated Measures Level Information

Dependent Variable	t1	t2	t3
Level of time	1	2	3

Similar to the preceding SPSS output, this table explicitly names each of the measures representing the WSF levels. Next, the sphericity assumption is evaluated.

The GLM Procedure

Repeated Measures Analysis of Variance

Sphericity Tests

Variables	DF	Mauchly's Criterion	Chi-Square	Pr > ChiSq
Transformed Variates	2	0.8509465	23.888092	<.0001
Orthogonal Components	2	0.8509465	23.888092	<.0001

As observed previously, there is evidence that the assumption of sphericity is not tenable and that ε-corrections may be used in determining the more appropriate use of the F test statistics. Keeping this in mind, we move on to the next displayed output sections provided by each program.

SPSS output

Tests of Within-Subjects Effects

Measure: MEASURE_1

Source		Type III Sum of Squares	df	Mean Square	F	Sig.
TIME	Sphericity Assumed	26982.714	2	13491.357	189.835	.000
	Greenhouse-Geisser	26982.714	1.741	15502.291	189.835	.000
	Huynh-Feldt	26982.714	1.759	15336.541	189.835	.000
	Lower-bound	26982.714	1.000	26982.714	189.835	.000
Error(TIME)	Sphericity Assumed	21178.493	298	71.069		
	Greenhouse-Geisser	21178.493	259.344	81.662		
	Huynh-Feldt	21178.493	262.147	80.789		
	Lower-bound	21178.493	149.000	142.138		

SAS output

```
                         The GLM Procedure

                Repeated Measures Analysis of Variance

        Univariate Tests of Hypotheses for Within Subject Effects

                                                              Adj Pr>F
Source          DF  Type III SS  Mean Square  F Value  Pr>F   G - G    H - F

time             2  26982.71434  13491.35717  189.84  <.0001  <.0001  <.0001
Error(time)    298  21178.49291     71.06877

                 Greenhouse-Geisser Epsilon    0.8703
                 Huynh-Feldt Epsilon           0.8797
```

Given our earlier note on the conservatism feature of the *F* test after its degrees of freedom are corrected (as well as the fact that the uncorrected *F* test is significant—see row labeled "sphericity assumed"), it is prudent that we only look at the test related information based on Huynh–Feldt's adjustment. (In the SPSS output, it is found in the row labeled Huynh–Feldt, whereas in the SAS output it is referred to both by name and its initials, H-F.) Accordingly, we reject the null hypothesis of no change over time and conclude that significant differences across the three assessments are present. We also observe that the *F* ratio in this example (see columns titled "F" and "F Value") is the same in all four *F* test statistics considered—with sphericity assumed, and with the three corrections in case of sphericity violation. The reason for this invariance of the *F* ratio is that what is being changed by any of the three ε-correction procedures is not that ratio itself, but the cut-off to which its value is compared when judged for significance. That cut-off, for each of the three corrections, is reflected in the p-values provided by the programs (labeled "Sig." and "Pr > F" in SPSS and SAS, respectively), which are all smaller than .001 in the present example.

SPSS output

Tests of Within-Subjects Contrasts

Measure: MEASURE_1

Source	TIME	Type III Sum of Squares	df	Mean Square	F	Sig.
TIME	Linear	3231.332	1	3231.332	73.863	.000
	Quadratic	23751.382	1	23751.382	241.401	.000
Error(TIME)	Linear	6518.392	149	43.748		
	Quadratic	14660.101	149	98.390		

SAS output

```
                        The GLM Procedure
           Repeated Measures Analysis of Variance

          Analysis of Variance of Contrast Variables

time_N represents the nth degree polynomial contrast for time

Contrast Variable: time_1

Source      DF      Type III SS     Mean Square    F Value    Pr > F

Mean         1      3231.332485     3231.332485      73.86    <.0001
Error      149      6518.392269       43.747599

Contrast Variable: time_2

Source      DF      Type III SS     Mean Square    F Value    Pr > F

Mean         1     23751.38185     23751.38185      241.40    <.0001
Error      149     14660.10064        98.38994
```

These output sections provide the results pertaining to the query what types of patterns of change are exhibited by the test scores across the three measurements. In case there is evidence in the data for change over time, as in this example, it is of interest to examine whether this evidence points to linear growth or decline over time, or perhaps even to a quadratic pattern of change. Such a question can be addressed by testing corresponding null hypotheses of no linear, quadratic, or higher-order (if applicable) pattern of change. While SPSS names these patterns explicitly, in SAS they are referred to as "Contrast Variable: time_1" for the linear pattern, and "Contrast Variable: time_2" for the quadratic pattern (and correspondingly for higher-order patterns if applicable). Note that with only three occasions, as in this example, no trend pattern higher than second degree (quadratic trend) can be tested for, since through any three points in the plane there could be only up to a quadratic curve going through them that would be uniquely determined. As can be seen by examining the displayed output for the present example, there is evidence of a linear pattern of change as well as deviations from it that could be described as a quadratic pattern. We will return to this issue further below, when we explore in more detail the relationships between empirical means over time.

SPSS output

Tests of Between-Subjects Effects

Measure: MEASURE_1

Transformed Variable: Average

Source	Type III Sum of Squares	df	Mean Square	F	Sig.
Intercept	682100.709	1	682100.709	1032.659	.000
Error	98418.715	149	660.528		

SAS output

```
                        The GLM Procedure

Dependent Variable: Average

                                  Sum of
Source                   DF      Squares    Mean Square   F Value    Pr > F

Model                     1  2046302.128    2046302.128   1032.66    <.0001
Error                   149   295256.145       1981.585

Uncorrected Total       150  2341558.273

Source                   DF  Type III SS    Mean Square   F Value    Pr > F

Intercept                 1  2046302.128    2046302.128   1032.66    <.0001
```

We stress that in this example only one group of subjects are repeatedly measured over time, and hence there is no BSF. Consequently, the *F* test of the intercept mentioned in the last pair of SPSS and SAS output sections is of no particular interest to us since it only shows that on average the subject scores are nonzero, which is a trivial finding.

SPSS output

Estimated Marginal Means

Time

Measure: MEASURE_1

TIME	Mean	Std. Error	95% Confidence Interval	
			Lower Bound	Upper Bound
1	30.514	1.208	28.127	32.901
2	49.207	1.350	46.539	51.875
3	37.078	1.438	34.236	39.920

SAS output

```
                The GLM Procedure
    Repeated Measures Analysis of Variance

        Means of Within Subjects Effects

Level of
time          N          Mean           Std Dev

1           150     30.51391333     14.79710773
2           150     49.20733333     16.53674977
3           150     37.07778667     17.61383930
```

These results suggest that after an initial average improvement in test scores, there was some drop at the end of the study, but even then the average score was considerably above that at first occasion (i.e., pretest). This may be most easily seen by comparing the confidence intervals of the assessment means (see last two columns of the marginal means section of the SPSS output)—in particular, this interval for pretest is completely to the left of that for second assessment, and the latter interval is so with regard to the one for the last occasion, while the interval for 1st assessment is entirely below that for last occasion. (These are informal observations rather than resulting from statistical tests.) This drop in test scores at final measurement could be explained with the possibility that forgetting of training modules and rules taught may have taken place, which is not unlikely to have occurred by the third assessment carried out a number of months after administration of these modules.

Repeated Measure Analysis With Between-Subject Factors

The example discussed so far involved only a within-subject factor. If a BSF was also included in the design, as would frequently be the case in social and behavioral research, the same logic and testing approach implemented above could be used with regard to the WSF(s). An important test to examine initially is again that of the sphericity assumption; if it is not fulfilled, one may invoke a degrees of freedom correction for the F test in Equation 5.3, with the best option being Huynh–Feldt's adjustment procedure. (As indicated earlier, one could alternatively use the multivariate approach to RMA discussed in the next section, or unidimensional versions of the modeling procedures in Chapter 13, both best applied with large samples.)

To exemplify a situation in which a BSF is included in a design, consider the last study we dealt with but now assuming that we also had information from each subject on their socioeconomic status (SES, treated in this example as a categorical variable with values of 1 for low, 2 for middle, and 3 for high; the data are included in the file ch5ex2.dat available from www.psypress.com/applied-multivariate-analysis). The main question of interest here is whether the three SES groups exhibit the same pattern of change over time. In this mixed design context, the so-called flatness hypothesis considered earlier remains of interest, but it formally states here that the occasion means stay the same over time when disregarding group membership. In addition, two further hypotheses become of particular interest in this setting: (a) the hypothesis of no time-by-group interaction, also sometimes referred to as parallelism hypothesis; and (b) the hypothesis of no overall group differences, also called the level hypothesis. That is, formally hypotheses (a) and (b) are stated, respectively, as

$$H_{0,a}: \mu_{k,Gi} - \mu_{k,Gj} = \mu_{k+1,Gi} - \mu_{k+1,Gj},$$

where $k = 1, \ldots, m-1$; $i \neq j$; $i, j = 1, \ldots, g$ (g being the number of levels of the between-subject factor, here 3) and the individual groups are symbolized by G_i; and

$$H_{0,b}: \mu_1 = \mu_2 = \ldots = \mu_g,$$

where $\mu_1, \ldots, \mu_g$ are the means of the g groups after pooling over assessment occasions (i.e., averaging over all assessments).

Geometrically, the parallelism hypothesis states that if one connects graphically the dependent variable group means across time, then all resulting group-specific profiles will be parallel. The level hypothesis states that disregarding time, the group means are the same. The parallelism hypothesis is often substantively the most interesting one, since it reflects differential by group effect of time (i.e., treatment, intervention, therapy, program administered) on the response variable, and for this reason is typically addressed first. If this hypothesis is found not to be plausible, both the level and flatness hypotheses lose much of their attraction if not meaningfulness. Hence, as in earlier chapters, also here we will attend first to the no-interaction hypothesis (that is, of parallelism), and only if it is found not to be significant will we proceed to testing the flatness and level hypotheses.

Returning now to the empirical example of interest, in order to respond to the research question of concern we use with SPSS the same menu option sequence employed earlier, but add SES as a between-subject factor (also, in Options we ask for all effect means and for the homogeneity test). For completeness, we restate this sequence next:

General Linear Model → Repeated Measures (WSF name = "time," 3 levels, click "Add"; define levels as the successive measures; Between-subject factor: SES; Options: mean for all factors and interactions, homogeneity test).

To accomplish this analysis with SAS, the following command statements within PROC GLM can be used:

```
DATA RMA;
INFILE 'ch4ex2.dat';
INPUT t1-t3 ses;
PROC GLM;
CLASS ses;
MODEL t1 - t3 = ses /nouni;
REPEATED time / summary printe;
RUN;
```

The resulting outputs are as follows (provided again in segments, to simplify the discussion).

SPSS output

General Linear Model

Within-Subjects Factors

Measure: MEASURE_1

TIME	Dependent Variable
1	TEST.1
2	TEST.2
3	TEST.3

Between-Subjects Factors

		N
SES	1.00	50
	2.00	50
	3.00	50

In these tables, both the within-subject and between-subject factors are explicitly named. Note now the added sample split by the BSF, socioeconomic status (SES).

Box's Test of Equality of Covariance Matrices[a]

Box's M	15.797
F	1.277
df1	12
df2	104720.5
Sig.	.224

Tests the null hypothesis that the observed covariance matrices of the dependent variables are equal across groups.
a. Design: Intercept+SES
Within Subjects Design: TIME

When a BSF is present in a repeated measure study, covariance homogeneity is again an assumption for the validity of its tests. In the present case, it is not violated, as judged by Box's test.

Levene's Test of Equality of Error Variances[a]

	F	df1	df2	Sig.
TEST.1	1.083	2	147	.341
TEST.2	.628	2	147	.535
TEST.3	1.701	2	147	.186

Tests the null hypothesis that the error variance of the dependent variable is equal across groups.
a. Design: Intercept+SES
Within Subjects Design: TIME

We are not particularly interested in the results of Levene's test, since we already know that the covariance matrix homogeneity assumption is plausible (based on Box's M test results stated above). Levene's test would be of interest if that assumption was violated, particularly in an attempt to find out possible assessment occasions that contribute to lack of variance homogeneity.

Mauchly's Test of Sphericity[b]

Measure: MEASURE_1

Within Subjects Effect	Mauchly's W	Approx. Chi-Square	df	Sig.	Epsilon[a]		
					Greenhouse-Geisser	Huynh-Feldt	Lower-bound
TIME	.853	23.201	2	.000	.872	.893	.500

Tests the null hypothesis that the error covariance matrix of the orthonormalized transformed dependent variables is proportional to an identity matrix.

a. May be used to adjust the degrees of freedom for the averaged tests of significance. Corrected tests are displayed in the Tests of Within-Subjects Effects table.
b. Design: Intercept+SES
Within Subjects Design: TIME

The last results indicate that the sphericity assumption is violated. Thus, we would like to invoke the previously discussed degree-of-freedom correction(s) (since for the purposes of this section we are willing to remain within the framework of the univariate approach to RMA; see next section and Chapter 13 for alternative approaches). Notice that, as mentioned earlier, the Huynh–Feldt correction is again associated with the largest ε, and thus leads to the least conservative F test correction. In the remaining output sections where corrected F test results are reported, we will therefore focus only on the Huynh–Feldt's row when examining effect significance.

SAS output

```
                    The GLM Procedure

                Class Level Information

               Class         Levels   Values
               ses                3    1 2 3

          Repeated Measures Analysis of Variance

            Repeated Measures Level Information

Dependent Variable           t1        t2        t3
Level of time                 1         2         3

          Number of Observations       150
```

Similar to the SPSS output, the last table explicitly names each of the within-subject and between-subject factors. We note that since the earlier reported Box's M test was not significant, we do not provide the SAS results for the test of covariance matrix homogeneity; we also dispense with reporting the results for Levene's test. Instead, we turn next to the sphericity assumption. The following output section indicates, as observed previously, evidence that this assumption is violated, and that ε-corrections would be used with the pertinent F tests (notice also that the uncorrected F test, in the row labeled "sphericity assumed," is significant).

```
                          The GLM Procedure

               Repeated Measures Analysis of Variance

                          Sphericity Tests

                                Mauchly's
Variables                 DF    Criterion    Chi-Square    Pr > ChiSq

Transformed Variates       2    0.8108592     30.610491        <.0001
Orthogonal Components      2     0.853071     23.201217        <.0001
```

SPSS output

Tests of Within-Subjects Effects

Measure: MEASURE_1

Source		Type III Sum of Squares	df	Mean Square	F	Sig.
TIME	Sphericity Assumed	26982.714	2	13491.357	188.517	.000
	Greenhouse-Geisser	26982.714	1.744	15473.628	188.517	.000
	Huynh-Feldt	26982.714	1.787	15099.726	188.517	.000
	Lower-bound	26982.714	1.000	26982.714	188.517	.000
TIME * SES	Sphericity Assumed	138.184	4	34.546	.483	.748
	Greenhouse-Geisser	138.184	3.488	39.622	.483	.723
	Huynh-Feldt	138.184	3.574	38.664	.483	.727
	Lower-bound	138.184	2.000	69.092	.483	.618
Error(TIME)	Sphericity Assumed	21040.309	294	71.566		
	Greenhouse-Geisser	21040.309	256.337	82.081		
	Huynh-Feldt	21040.309	262.684	80.097		
	Lower-bound	21040.309	147.000	143.131		

SAS output

```
                         The GLM Procedure

                 Repeated Measures Analysis of Variance

        Univariate Tests of Hypotheses for Within Subject Effects

                                                            Adj Pr > F
Source         DF  Type III SS   Mean Square  F Value   Pr > F   G - G     H - F

time            2  26982.71434   13491.35717   188.52   <.0001  <.0001    <.0001
time*ses        4    138.18435      34.54609     0.48   0.7484  0.7229    0.7275
Error(time)   294  21040.30857      71.56568

                  Greenhouse-Geisser Epsilon        0.8719
                  Huynh-Feldt Epsilon               0.8935
```

We first note from these results that there is no indication of a significant interaction effect (see the variance source row labeled "time*ses"). Hence, the parallelism hypothesis may be retained, which is interpreted as lack of evidence warranting a conclusion of differential by assessment group effect (or differential by group time effect). Thus, we can proceed to evaluation of the within-subject and between-subject effects. In fact, with regard to the within-subject effect, there is evidence in the same output section (see line above that for the time-by-group interaction), which allows us to reject the flatness hypothesis. We thus conclude that there is significant change over time, after pooling over the levels of the BSF (examined in the next tables).

SPSS output

Tests of Between-Subjects Effects

Measure: MEASURE_1
Transformed Variable: Average

Source	Type III Sum of Squares	df	Mean Square	F	Sig.
Intercept	682100.709	1	682100.709	1020.113	.000
SES	126.865	2	63.433	.095	.910
Error	98291.850	147	668.652		

SAS output

The GLM Procedure

Repeated Measures Analysis of Variance

Tests of Hypotheses for Between Subjects Effects

Source	DF	Type III SS	Mean Square	F Value	Pr > F
ses	2	126.86545	63.43273	0.09	0.9095
Error	147	98291.84952	668.65204		

The last results indicate that there is no evidence for SES differences; that is, pooling over time, the means of the three SES groups do not differ significantly from on another.

SPSS output

Estimated Marginal Means

1. TIME

Measure: MEASURE_1

TIME	Mean	Std. Error	95% Confidence Interval Lower Bound	95% Confidence Interval Upper Bound
1	30.514	1.216	28.111	32.917
2	49.207	1.355	46.529	51.886
3	37.078	1.448	34.216	39.939

SAS output

```
                 The GLM Procedure

      Repeated Measures Analysis of Variance

         Means of Within Subjects Effects

Level of
time          N         Mean            Std Dev

1           150    30.51391333    14.79710773
2           150    49.20733333    16.53674977
3           150    37.07778667    17.61383930
```

As expected from the earlier discussed findings in the previous output sections, the results presented in the last tables show overall mean change over time, which follows a linear and quadratic trend, with a drop after an initial improvement in test scores.

Because the descriptive statistics for the time-by-group interaction created by SPSS and SAS would obviously lead to identical results, to save space we only provide next the remaining output generated by SPSS.

SPSS output

2. SES*TIME

Measure: MEASURE_1

				95% Confidence Interval	
SES	**TIME**	**Mean**	**Std. Error**	**Lower Bound**	**Upper Bound**
1.00	1	30.895	2.106	26.733	35.057
	2	49.477	2.347	44.838	54.116
	3	36.983	2.508	32.026	41.939
2.00	1	30.038	2.106	25.875	34.200
	2	47.525	2.347	42.886	52.164
	3	37.068	2.508	32.112	42.024
3.00	1	30.609	2.106	26.447	34.772
	2	50.620	2.347	45.981	55.259
	3	37.183	2.508	32.227	42.139

As can be seen from the last table, within each SES group a very similar trend is evinced over time as that found overall. This result is fully compatible with the earlier mentioned finding of no significant interaction of SES and time, which could also be rephrased as no differential effect of SES upon change over time.

3. SES

Measure: MEASURE_1

SES	Mean	Std. Error	95% Confidence Interval	
			Lower Bound	Upper Bound
1.00	39.118	2.111	34.946	43.291
2.00	38.210	2.111	34.038	42.383
3.00	39.471	2.111	35.298	43.643

From the last output, there is little overall difference across SES groups, which is not unexpected given the earlier finding of no main effect for SES. We thus conclude that the three SES groups seem to follow very similar patterns of change over time. Accordingly, there is initial improvement in performance, which is followed by some drop at the end of the study; thereby, the three SES group mean score profiles across time are quite alike.

5.3 Multivariate Approach to Repeated Measure Analysis

As mentioned at the beginning of the preceding section, the sphericity assumption is likely to be violated considerably in behavioral, social, and educational research dealing with repeated assessments. While the degrees of freedom corrections attempt to accommodate these violations, their effect is not completely satisfactory in many cases. In fact, each correction leads to a conservative F test, which therefore does not preserve the Type I error rate. For this reason, other avenues for RMA need to be explored. One viable alternative to the univariate approach is the so-called multivariate approach to RMA, which is the subject of this section. A third general possibility can be developed within the framework of latent variable modeling and is discussed in Chapter 13.

To set the context of following developments, we begin by first recalling the use of the simple t test for related samples, which is applicable for instance in what may be referred to as a pretest/posttest design. Denote the measures taken at two assessments of interest by X_1 and X_2, with means μ_1 and μ_2. The null hypothesis of concern then is that of no mean change over time,

$$H_0\colon \mu_1 = \mu_2 \text{ (e.g., } \mu_{\text{pretest}} = \mu_{\text{posttest}}).$$

To test this hypothesis, one can take the difference score, $D = X_2 - X_1$, and examine instead the hypothesis

$$H_0\colon \mu_D = 0,$$

where μ_D is its mean, i.e., test an equivalent hypothesis on a derived measure, viz. the difference score.

The multivariate approach to RMA, also called MANOVA approach to RMA, proceeds in a similar manner and is based on two steps. In the first, from the repeated measures, say $X_1, X_2, \ldots, X_m$, one obtains $m-1$ new derived ones—referred to as contrasts and denoted by $D_1, D_2, \ldots, D_{m-1}$—that are (linearly) independent of one another ($m > 1$; see below for pertinent details). In the second step, a MANOVA is conducted on $D_1, D_2, \ldots, D_{m-1}$ treating them as dependent variables. Carrying out these two steps in succession represents the multivariate approach to RMA.

This discussion also suggests that an assumption of the multivariate approach to RMA is that of covariance matrix homogeneity across groups, when a BSF(s) is present, and involves two new concepts—contrast and linear independency. A contrast for a given set of variables, say $X_1, X_2, \ldots, X_m$, is any linear combination of them, $c_1X_1 + c_2X_2 + \ldots + c_mX_m$, such that $c_1 + c_2 + \ldots + c_m = 0$. In other words, a contrast is a weighted combination of variables with such weights (values of the c's) that sum up to zero. For example, any of the consecutive differences between the above variables $X_1, X_2, \ldots, X_m$,

$$U_1 = X_2 - X_1, U_2 = X_3 - X_2, \ldots, U_{m-1} = X_m - X_{m-1},$$

is a contrast. Further, a set of k variables $Y_1, Y_2, \ldots, Y_k$ ($k > 1$) are called linearly independent if there are no such constant coefficients, say $a_1, a_2, \ldots, a_k$, not all of them being 0, for which $a_1Y_1 + a_2Y_2 + \ldots + a_kY_k = 0$; that is, the only choice for $a_1, a_2, \ldots, a_k$ so that the last stated equality to 0 is fulfilled, is when they all are selected as 0.

The reason we are concerned here with the notions of "contrast" and "linear independence" is that as mentioned the MANOVA approach to RMA is based on derived measures, $D_1, D_2, \ldots, D_{m-1}$, which are constructed as linearly independent contrasts. Thereby, this approach is not bound to any particular set of contrasts, but can instead use any set of linearly independent contrasts chosen by the researcher. In fact, for the test of the BSF(s), it is irrelevant which set is chosen. However, different sets of contrasts reveal different aspects of the patterns of change over time, and thus reflect different ways in which the WSF is handled for analytic purposes.

For the following developments, it would be informative to provide two additional illustrations of the concept of linearly independent contrasts for a given set of m repeated measures. For example, the differences-to-preceding-average,

$$D_1 = X_2 - X_1, D_2 = X_3 - (X_1 + X_2)/2, \ldots,$$
$$D_{m-1} = X_m - (X_1 + \ldots + X_{m-1})/(m - 1),$$

represent such a set of contrasts. Moreover, the differences-to-last,

$$D_1 = X_1 - X_m, D_2 = X_2 - X_m, \ldots, D_{m-1} = X_{m-1} - X_m,$$

is also another possible choice. Also, the polynomial set

$$D_1 = \text{linear trend}, D_2 = \text{quadratic trend}, D_3 = \text{cubic trend}, \ldots,$$
$$D_{m-1} = m - 1\text{st order trend}$$

is yet another possibility for a group of linearly independent contrasts. We note that the actual coefficients $a_1, \ldots, a_m$ that participate in each of the last given $m-1$ contrasts (polynomial set) are constants that depend on the number m of repeated assessments. The particular choices for these coefficients have been worked out a long time ago and are readily available in many different sources (Kirk, 1994). They are also implemented in and automatically invoked by most statistical software (see illustration of SPSS and SAS utilization for these purposes below). Regardless of which set of linearly independent contrasts is elected by the researcher, we emphasize that once the choice of transformed variables D_1, $D_2, \ldots,$ and D_{m-1} is done (for a given number $m > 1$ of repeated assessments), the multivariate approach to RMA treats the new variables D_1, $D_2, \ldots, D_{m-1}$ just like MANOVA treats multiple dependent variables.

To demonstrate this discussion, consider the following example. One hundred and sixty one male and female high school sophomore students are assessed four consecutive times using an induction reasoning test at the beginning of each quarter in a school year. (The data for this example are found in the file ch5ex3.dat available from www.psypress.com/applied-multivariate-analysis, where the variables containing the four successive assessments are named ir.1, ir.2, ir.3, and ir.4.) The research question is whether there are gender differences in the pattern of change over time.

To respond to this query, with SPSS we follow the same steps as in the univariate approach when a BSF was involved but focus on those parts of the output produced thereby, which deal with the multivariate approach, viz. the multivariate tests. For completeness of this discussion, we include this menu options sequence:

General Linear Model → Repeated Measures (WSF name = "time," 3 levels, click "Add"; define levels as the successive measures; Between-subject factor: Gender; Options: mean for all factors and interactions, homogeneity test).

To accomplish the same analysis with SAS, the following command statements within PROC GLM can be used. We note that with exception of the added statement "printm," this command sequence is quite similar to that used in the previous section for the analysis of measures with a BSF. Including the new statement "printm" ensures that the used linearly independent contrasts are displayed in the output.

```
DATA RMA;
INFILE 'ch4ex2.dat';
INPUT ir1-ir4 gender;
PROC GLM;
CLASS gender;
MODEL ir1 - ir4 = gender/nouni;
REPEATED time POLYNOMIAL/summary printe printm;
run;
```

The results produced are as follows (provided again in segments, to simplify the discussion).

SPSS output

General Linear Model

Within-Subjects Factors

Measure: MEASURE_1

TIME	Dependent Variable
1	IR.1
2	IR.2
3	IR.3
4	IR.4

Between-Subjects Factors

		N
GENDER	.00	50
	1.00	111

Box's Test of Equality of Covariance Matrices[a]

Box's M	4.173
F	.403
df1	10
df2	44112.470
Sig.	.946

Tests the null hypothesis that the observed covariance matrices of the dependent variables are equal across groups.

a. Design: Intercept+GENDER
Within Subjects Design: TIME

SAS output

```
                  The GLM Procedure

              Class Level Information

              Class    Levels    Values

              gender      2        0 1

        Repeated Measures Level Information

Dependent Variable     ir1        ir2       ir3    ir4
      Level of time      1          2         3      4

         Number of Observations            161
```

The preceding output tables identify the variables associated with the four levels of the within subject factor, referred to as "time," and indicate the gender (BSF) split. In addition, we note the results of Box's M test (displayed only in the SPSS output, to save space), which show no evidence for violation of the covariance matrix homogeneity assumption.

SPSS output

Multivariate Tests[b]

Effect		Value	F	Hypothesis df	Error df	Sig.
TIME	Pillai's Trace	.764	169.284[a]	3.000	157.000	.000
	Wilks' Lambda	.236	169.284[a]	3.000	157.000	.000
	Hotelling's Trace	3.235	169.284[a]	3.000	157.000	.000
	Roy's Largest Root	3.235	169.284[a]	3.000	157.000	.000
TIME * GENDER	Pillai's Trace	.023	1.233[a]	3.000	157.000	.300
	Wilks' Lambda	.977	1.233[a]	3.000	157.000	.300
	Hotelling's Trace	.024	1.233[a]	3.000	157.000	.300
	Roy's Largest Root	.024	1.233[a]	3.000	157.000	.300

a. Exact statistic
b. Design: Intercept+GENDER
Within Subjects Design: TIME

Mauchly's Test of Sphericity[b]

Measure: MEASURE_1

Within Subjects Effect	Mauchly's W	Approx. Chi-Square	df	Sig.	Epsilon[a] Greenhouse-Geisser	Huynh-Feldt	Lower-bound
TIME	.693	57.829	5	.000	.798	.816	.333

Tests the null hypothesis that the error covariance matrix of the orthonormalized transformed dependent variables is proportional to an identity matrix.
a. May be used to adjust the degrees of freedom for the averaged tests of significance. Corrected tests are displayed in the Tests of Within-Subjects Effects table.
b. Design: Intercept+GENDER
Within Subjects Design: TIME

SAS output

The GLM Procedure

Repeated Measures Analysis of Variance

Sphericity Tests

Variables	DF	Mauchly's Criterion	Chi-Square	Pr > ChiSq
Transformed Variates	5	0.6930509	57.829138	<.0001
Orthogonal Components	5	0.6930509	57.829138	<.0001

The GLM Procedure

Repeated Measures Analysis of Variance

MANOVA Test Criteria and Exact F Statistics for the Hypothesis of no time Effect

H = Type III SSCP Matrix for time
E = Error SSCP Matrix
S = 1 M = 0.5 N = 77.5

Statistic	Value	F Value	Num DF	Den DF	Pr > F
Wilks' Lambda	0.23614229	169.28	3	157	<.0001
Pillai's Trace	0.76385771	169.28	3	157	<.0001
Hotelling-Lawley Trace	3.23473484	169.28	3	157	<.0001
Roy's Greatest Root	3.23473484	169.28	3	157	<.0001

MANOVA Test Criteria and Exact F Statistics for the Hypothesis of no time*gender Effect

H = Type III SSCP Matrix for time*gender
E = Error SSCP Matrix
S = 1 M = 0.5 N = 77.5

Statistic	Value	F Value	Num DF	Den DF	Pr > F
Wilks' Lambda	0.97698592	1.23	3	157	0.2997
Pillai's Trace	0.02301408	1.23	3	157	0.2997
Hotelling-Lawley Trace	0.02355620	1.23	3	157	0.2997
Roy's Greatest Root	0.02355620	1.23	3	157	0.2997

The first issue to mention with respect to these output sections is that, since we are not dealing here with the univariate approach to RAM, its sphericity assumption and test of it are of no relevance (but we present them here for completeness reasons). Hence, we move on to examination only of the multivariate results. An inspection of the pertinent MANOVA findings reveals that there is no evidence for group-by-time interaction, and a clear indication of change over time.

SPSS output

Tests of Within-Subjects Effects

Measure: MEASURE_1

Source		Type III Sum of Squares	df	Mean Square	F	Sig.
TIME	Sphericity Assumed	36725.163	3	12241.721	289.446	.000
	Greenhouse-Geisser	36725.163	2.395	15337.153	289.446	.000
	Huynh-Feldt	36725.163	2.449	14996.321	289.446	.000
	Lower-bound	36725.163	1.000	36725.163	289.446	.000
TIME * GENDER	Sphericity Assumed	143.242	3	47.747	1.129	.337
	Greenhouse-Geisser	143.242	2.395	59.821	1.129	.331
	Huynh-Feldt	143.242	2.449	58.491	1.129	.332
	Lower-bound	143.242	1.000	143.242	1.129	.290
Error(TIME)	Sphericity Assumed	20174.090	477	42.294		
	Greenhouse-Geisser	20174.090	380.729	52.988		
	Huynh-Feldt	20174.090	389.382	51.811		
	Lower-bound	20174.090	159.000	126.881		

SAS output

```
                          The GLM Procedure

               Repeated Measures Analysis of Variance

        Univariate Tests of Hypotheses for Within Subject Effects

                                                              Adj Pr > F
Source          DF  Type III SS   Mean Square  F Value   Pr > F   G - G    H - F

time             3  36725.16331   12241.72110   289.45  <.0001  <.0001   <.0001
time*gender      3    143.24178      47.74726     1.13  0.3369  0.3309   0.3316
Error(time)    477  20174.08954      42.29369

                  Greenhouse-Geisser Epsilon     0.7982

                  Huynh-Feldt Epsilon            0.8163
```

For completeness of output reasons, we display the above tables but note that these are the results of testing within-subject effects with the univariate RMA approach, and as mentioned above we are not interested in them here.

SPSS output

Tests of Between-Subjects Effects

Measure: MEASURE_1
Transformed Variable: Average

Source	Type III Sum of Squares	df	Mean Square	F	Sig.
Intercept	1079408.511	1	1079408.511	985.998	.000
GENDER	2442.392	1	2442.392	2.231	.137
Error	174063.147	159	1094.737		

SAS output

The GLM Procedure

Repeated Measures Analysis of Variance

Tests of Hypotheses for Between Subjects Effects

Source	DF	Type III SS	Mean Square	F Value	Pr > F
gender	1	2442.3916	2442.3916	2.23	0.1372
Error	159	174063.1472	1094.7368		

As can be observed from these tables, the data do not provide evidence for an overall Gender effect, that is, there is no Gender effect present after pooling over the levels of the time factor.

SPSS output

Tests of Within-Subjects Contrasts

Measure: MEASURE_1

Source	TIME	Type III Sum of Squares	df	Mean Square	F	Sig.
TIME	Linear	20223.797	1	20223.797	366.602	.000
	Quadratic	15745.604	1	15745.604	328.783	.000
	Cubic	755.762	1	755.762	31.721	.000
TIME * GENDER	Linear	133.642	1	133.642	2.423	.122
	Quadratic	.289	1	.289	.006	.938
	Cubic	9.311	1	9.311	.391	.533
Error(TIME)	Linear	8771.316	159	55.166		
	Quadratic	7614.609	159	47.891		
	Cubic	3788.165	159	23.825		

SAS output

```
                     The GLM Procedure

         Repeated Measures Analysis of Variance

        Analysis of Variance of Contrast Variables

time_N represents the nth degree polynomial contrast for time

Contrast Variable: time_1

  Source     DF     Type III SS    Mean Square    F Value    Pr > F
  Mean        1     20223.79673    20223.79673     366.60    <.0001
  gender      1       133.64183      133.64183       2.42    0.1216
  Error     159      8771.31593       55.16551


Contrast Variable: time_2

  Source     DF     Type III SS    Mean Square    F Value    Pr > F
  Mean        1     15745.60449    15745.60449     328.78    <.0001
  gender      1         0.28882        0.28882       0.01    0.9382
  Error     159      7614.60891       47.89062


Contrast Variable: time_3

  Source     DF     Type III SS    Mean Square    F Value    Pr > F
  Mean        1      755.762097     755.762097      31.72    <.0001
  gender      1        9.311132       9.311132       0.39    0.5328
  Error     159     3788.164706      23.824935
```

The above tables contain the default contrasts used by the programs, that is, the polynomial contrasts. SPSS labels them Linear (D_1), Quadratic (D_2), and Cubic (D_3). These are the transformed, or new/derived, variables utilized here, which represent the linear, quadratic, and cubic trends of change over time. (Note that in this study with four repeated assessments, one could also examine a cubic trend, i.e., a third-order polynomial trend, which is, however, in general difficult to interpret substantively.) In contrast, SAS labels these as "time_N represents the nth degree polynomial contrast for time," where "time_1" corresponds to the linear contrast, "time_2" to the quadratic, and "time_3" to the cubic contrast. The above results reveal that there is indication for all three patterns of change over time (see also means considered below), but no effect of time-by-gender interaction on any of them.

SPSS output

Transformation Coefficients (M Matrix)

TIME[a]

Measure: MEASURE_1

Dependent Variable	TIME		
	Linear	Quadratic	Cubic
IR.1	−.671	.500	−.224
IR.2	−.224	−.500	.671
IR.3	.224	−.500	−.671
IR.4	.671	.500	.224

a. The contrasts for the within subjects factors are:
TIME: Polynomial contrast

SAS output

```
                         The GLM Procedure
              Repeated Measures Analysis of Variance

    time_N represents the nth degree polynomial contrast for time

              M Matrix Describing Transformed Variables

               ir1            ir2            ir3            ir4

time_1    -.6708203932   -.2236067977   0.2236067977   0.6708203932
time_2    0.5000000000   -.5000000000   -.5000000000   0.5000000000
time_3    -.2236067977   0.6708203932   -.6708203932   0.2236067977
```

As indicated before, with the earlier specifications used, SPSS and SAS automatically generate and utilize the linearly independent contrasts we wanted to use for our purposes. For clarity, we provide the underlying transformation tables so that the reader can view the precise definition of the derived variables D_1, D_2, and D_3 for this example, which are obtained from the original four repeated measures (cf., e.g., Kirk, 1994).

In conclusion of this section, we display the descriptive statistics created by SPSS and dispense with the duplicating results by SAS.

SPSS output

Estimated Marginal Means

1. GENDER

Estimates

Measure: MEASURE_1

GENDER	Mean	Std. Error	95% Confidence Interval	
			Lower Bound	Upper Bound
.00	42.134	2.340	37.513	46.755
1.00	46.343	1.570	43.242	49.444

As would be expected from the finding of no Gender effect mentioned above, the overall means for males and females are very close (see also their confidence intervals that overlap to a considerable degree; this is an informal observation, while the Gender effect test discussed above is a formal statistical finding).

2. TIME

Estimates

Measure: MEASURE_1

TIME	Mean	Std. Error	95% Confidence Interval	
			Lower Bound	Upper Bound
1	30.248	1.256	27.768	32.728
2	48.444	1.542	45.399	51.489
3	50.719	1.558	47.641	53.797
4	47.543	1.574	44.434	50.652

From this table, it appears that after some considerable increase in test performance a small drop is observed at the end of the study, perhaps due to measurement being obtained at the end of the last quarter of the academic year.

3. GENDER * TIME

Estimates

Measure: MEASURE_1

GENDER	TIME	Mean	Std. Error	95% Confidence Interval	
				Lower Bound	Upper Bound
.00	1	28.723	2.085	24.605	32.841
	2	46.757	2.560	41.701	51.813
	3	48.243	2.588	43.132	53.354
	4	44.813	2.614	39.651	49.976
1.00	1	31.773	1.400	29.009	34.537
	2	50.131	1.718	46.737	53.524
	3	53.195	1.737	49.765	56.625
	4	50.272	1.754	46.807	53.738

Finally, within each gender the same pattern of change over time appears—initial increase up until the third assessment, and then a slight drop at the end of the study, as noticed also overall and mentioned above. Our conclusion is that both genders have shown quite a similar pattern of change over time: improvement until third assessment and then some drop. Thereby, there is no indication of differential by gender time effect upon the induction reasoning intelligence test scores (that is, their group means follow essentially parallel profiles), and no overall gender effect.

5.4 Comparison of Univariate and Multivariate Approaches to Repeated Measure Analysis

After having discussed two different approaches to repeated measure analysis of variance, the natural question is when to use which one of them. Here we offer several guidelines to help answer this question in empirical research, which we have already partly indicated earlier in the chapter. First of all, if the sphericity assumption is fulfilled, the univariate approach may well be more powerful than the multivariate one, particularly when relatively small groups are being used in a study. For this reason, as long as the sphericity assumption is fulfilled, the univariate approach can be recommended. However, if this assumption is violated—as it is likely to be more often than not the case in empirical social and behavioral research—it is reasonable to use the multivariate approach unless with small samples. The latter approach does not depend on sphericity and can proceed regardless of whether or not this assumption is plausible. If then sample sizes are small, it may be best to still use the univariate approach, with the Huynh–Feldt correction to the degrees of freedom. One should keep in mind, though, that it is conservative; that is, if one rejects a hypothesis with it, one can have trust in this result but not otherwise. Last but not least, although the multivariate approach does not have sphericity as an assumption, it still requires covariance matrix homogeneity. One might suggest that it may be less often violated than the sphericity assumption in empirical research in the social, behavioral, and educational sciences. If covariance matrix homogeneity is seriously violated (and there are no transformations that render at least the individual dependent variables having homogeneous variances—see the discussion on variable transformation in Chapter 3), yet another alternative to repeated measure analysis with large samples is the one provided within the latent variable modeling framework that is the subject of Chapter 13.

6

Analysis of Covariance

A primary goal of statistical analyses in the behavioral, social, and educational sciences is to develop models that are useful representations of the studied phenomena. In the two previous chapters dealing with multivariate analysis of variance (MANOVA) and related repeated measure analysis (RMA) approaches, we used as explanatory variables only group membership (e.g., experimental vs. control group, gender, and SES status). However, a frequent finding in empirical research is that membership in groups under consideration often only accounts for a small part of total variance in dependent variables. Moreover, a commonly made informal observation in many behavioral and social studies is that preexisting differences among subjects at the start of an investigation are at least as important predictors of response variables as are many treatments, programs, or interventions used. These preexisting differences per se are not typically of main substantive interest in an empirical study, yet possibly relate to the outcome measures of interest and therefore contain information about the latter, which needs to be incorporated into the analyses in order to make their results more accurate and trustworthy. Oftentimes, these differences are on variables that are similar and notably correlated to the outcome measures, although they do not have to be commensurate with the latter.

In this chapter, we discuss ways in which preexisting differences on variables markedly correlated with outcome measures can be profitably incorporated into a modeling process. We refer to the latter as analysis of covariance (abbreviated to ANCOVA) in the univariate case, and multivariate analysis of covariance (abbreviated to MANCOVA) when two or more dependent variables are considered simultaneously. Initially, for didactic reasons only, we presume that we are dealing with just a single dependent variable, and will discuss ANCOVA. Subsequently, we will generalize the developments to the case with a larger number of dependent measures or responses and introduce MANCOVA.

6.1 Logic of Analysis of Covariance

Let us begin with a simple illustrative example study. In it, two different methods of teaching number division in two respective samples of elementary school students were compared using a test of division ability, which was administered at the end of the study (with resulting scores denoted Y). In addition, the children were also measured at the beginning of the study on an intelligence test yielding for each child an intelligence score (denoted by X). For the purposes of this section, a small portion of the resulting data is presented in Table 6.1.

Examining the means for the ability and intelligence measures indicates that there are considerable group differences on both the test score X and the variable Y that is of main interest in the study. Furthermore, we note that if the overall correlation (disregarding group) between division score and intelligence were to be computed, its value would be found to be considerable, that is, 48. Similarly it can be found that the relationship between X and Y is of comparable magnitude within each of the two groups.

These observations naturally give rise to the following question: Are the observed mean differences on the division test due to method differences (i.e., different effectiveness of the two methods), or are they alternatively a consequence of preexisting intelligence differences, or do they result from both these sources? This query asks whether the observed mean differences on the outcome variable (division test score) are due to chance effects only, and if not, what their origin is.

These types of questions can be addressed using the statistical method of ANCOVA. In the example under consideration, there is an inherent query about group differences, in which sense it relates to the typical ANOVA question of mean group differences in a multigroup setup (see Chapter 4). However, in addition, there is now also a continuous

TABLE 6.1

Data From a Division Ability Study (Variable Means per Group in Last Row)

	Group 1 (Method 1)		Group 2 (Method 2)	
	Y	X	Y	X
	18	10	28	16
	17	20	31	17
	23	15	38	30
	19	12	40	31
	40	22	41	18
	22	31	40	22
Means	23.17	18.33	36.33	22.33

variable, X, called a covariate because it covaries with the dependent variable Y. Further, the covariate is typically considerably correlated with the response variable Y (i.e., contains information about the latter and therefore possibly also about group differences on that response measure). In general, empirical settings where ANCOVA is appropriate have at least one categorical factor(s) and one or more continuous covariate(s).

The present discussion allows us to define ANCOVA as an extension of ANOVA, where main effects and interactions of independent variables (factors) are assessed in the presence of covariate(s) that are markedly related to the outcome measures of main interest. More specifically, evaluation of effects occurs in ANCOVA after the dependent variable is adjusted for potential preexisting group differences associated with the covariate(s). The covariate is assumed to be at least approximately continuous, and is commonly measured at the beginning of the study. If the study includes a treatment, training, or therapy (intervention), then the covariate is presumed to be measured before its administration. Obviously, a minimal requirement when one would consider using ANCOVA is the availability of $g \geq 2$ groups of subjects that are evaluated on at least one additional characteristic, the covariate(s), beyond a response variable to which the covariate is related.

In such settings, the typical ANCOVA question is whether the groups would differ on the dependent variable if they were equivalent on the covariate(s) to begin with. That is, the query is whether there would be any group differences on the dependent variable if all groups started out the same on the covariate(s), that is, in case there were no group differences on the latter. In those circumstances, an application of ANCOVA (under its specific assumptions; see below) aims to increase sensitivity, that is, enhance power of tests of main effects and interactions, by reducing the error term of pertinent F test statistics (F ratios). In addition, ANCOVA permits the researcher to adjust the group means on a response variable to what they would have been if all subjects had scored the same on the covariate(s) in the first instance. Moreover, ANCOVA may also be applied after MANOVA for the purpose of what is referred to as step-down analysis (SDA), a topic discussed in Section 6.3.

The most common use of ANCOVA is to employ the additional information available in a variable considerably related to an outcome measure, leading to a decrease in the error variance. As a consequence of this reduction, the power pertaining to the mean difference test is increased, because predictable variance is removed by the covariate(s) from the error term associated with the F ratio for a test of main effect or interaction. This variance part would otherwise be integrated into the error and thus increase it. Hence, when there is appreciable relationship between covariate(s) and dependent variables within each group, error terms in ANCOVA tend to be smaller relative to the corresponding ones in ANOVA. The amount by which these terms are reduced may be thought of being proportional to the correlation between covariate and outcome measure. Then tests of main effects and interactions would have higher power in ANCOVA since the pertinent F ratios for these tests tend to be larger due to that reduction of their denominators.

Another common application of ANCOVA is that it is often utilized in nonexperimental situations where subjects cannot be completely randomly assigned to groups. For example, one cannot randomly assign children to poverty conditions or to abusive environments to see how these circumstances affect their learning. In such situations, use of the ANCOVA method is primarily made with the goal of statistically making the studied groups somewhat more comparable. However, we stress that this method does not fully account for the lack of random assignment of subjects to groups. A reason is that there may be other group differences on the response variable that are not related to the covariate(s), which therefore cannot be handled in this way. Consequently, it is important to realize that ANCOVA (and for that matter MANCOVA as well) cannot guarantee the removal of prior existing differences between the groups. Indeed, there may be group differences other than those represented by the selected covariate(s), or not linearly related to the latter, and so the groups may still end up being different even after taking into account this covariate(s). This phenomenon is occasionally referred to as "Lord's Paradox," which states that no matter how many prior covariates a researcher considers in an ANCOVA, one can never be certain that they have totally adjusted for prior group differences. Furthermore, while ANCOVA offers a method for statistical adjustment of groups, no implication of causality can be made from its results because such an after-the-fact procedure cannot be a substitute for running an experiment (Maxwell & Delaney, 2004).

A third type of ANCOVA application is made in settings where one is concerned with assessing the contribution of various dependent variables to significant group differences found in a multivariate analysis of variance. In such circumstances, the effect of an ANCOVA is to remove the linear impact of a subset of dependent variables upon group differences on the remaining response measure(s). This is tantamount to carrying out ANCOVA with covariates being some of the outcome variables from the earlier MANOVA, and is discussed in more detail in Section 6.4.

From a conceptual viewpoint, and at the expense of some simplification, one may consider an ANCOVA application to proceed in the following steps. First, a regression of the response variable on the covariate(s) is conducted. Second, using the results from it, the effect of the covariate(s) on the dependent variable and its group means is removed, in order to furnish an adjusted mean per group (details follow below). The group mean is thereby adjusted for the linear effect of the covariate(s) upon the outcome measure. In the last step, one tests whether the resulting adjusted means differ across groups. (In actual fact, all these steps are simultaneously implemented in an ANCOVA utilization; see below for further details.) That is, the ANCOVA null hypothesis is

$$H_{0,\text{ANCOVA}}: \mu_{1,\text{a}} = \mu_{2,\text{a}} = \ldots = \mu_{g,\text{a}},$$

where
the subindex "a" stands for "adjusted"
g is the number of groups in a study under consideration ($g > 1$)

In other words, the null hypothesis states that there are no differences in the group means after the latter are adjusted for the effect of the covariate. We stress that this is not the null hypothesis of ANOVA, which is

$$H_{0,\text{ANOVA}}: \mu_1 = \mu_2 = \ldots = \mu_g,$$

where only the unadjusted group differences are of interest (i.e., the differences between the unadjusted means across groups or, in other words, the differences in group means before an ANCOVA adjustment is made). Conversely the group differences without the adjustment, which are of concern in ANOVA, are not really of interest in ANCOVA.

We observe that the preceding discussion is not dependent on particular design features in analysis of variance settings. Hence, ANCOVA could be used in any ANOVA design (as long as the assumptions of the former are fulfilled). For example, it can be used in between-subject, within-subject, or mixed designs, as long as there is a continuous covariate(s) that considerably relates to the response measure.

To elaborate further on the formal basis of ANCOVA, let us recall the general form of the regression analysis model with a single response variable (Hays, 1994; see also Chapter 2):

$$\underline{y} = \text{X}\,\underline{\text{b}} + \underline{\varepsilon}, \tag{6.1}$$

where the individual predictor values are contained in the matrix X (at times referred to also as "design" matrix), and as usual, the error vector $\underline{\varepsilon}$ is assumed unrelated with the predictors and distributed as $\text{N}_n(0, \sigma^2\text{I}_n)$, and $\underline{y}$ is the vector of dependent variable scores (as commonly used throughout the text, n denotes sample size).

In this model, no assumptions are made about the predictors. In particular, they could be group membership (i.e., group coding variables) and/or continuous variables. More explicitly, in a factorial design setting with (continuous) covariates, we could write Equation 6.1 as

$$\underline{y} = \text{X}_{(1)}\underline{\text{b}}_{(1)} + \text{X}_{(2)}\underline{\text{b}}_{(2)} + \underline{\varepsilon}, \tag{6.2}$$

where

the matrix $X_{(1)}$ encompasses the individual data on all categorical predictors

the matrix $X_{(2)}$ that data on all continuous predictors

the vectors $\underline{b}_{(1)}$ and $\underline{b}_{(2)}$ are the partial regression coefficients associated with the categorical and continuous predictors, respectively

That is, we could rewrite Equation 6.2 as

$$\underline{y} = [X_{(1)}X_{(2)}]\begin{bmatrix} \underline{b}_{(1)} \\ \underline{b}_{(2)} \end{bmatrix} + \underline{\varepsilon} = X\underline{b} + \underline{\varepsilon}, \tag{6.3}$$

where for compactness we use the notation $X = [X_{(1)}, X_{(2)}]$ for the set of individual scores on all predictors. Equation 6.3 may be viewed as representing the ANCOVA model. As readily seen from its right-hand side, the model of ANCOVA is just a special case of the general regression analysis model. Hence, all statistical methods available to us within the general linear model framework (with a single dependent variable) can be used in an application of ANCOVA. From this perspective, it is readily seen that what ANCOVA accomplishes is controlling for (continuous) covariate(s) while examining at the same time the effect of other predictors—specifically, categorical factors—upon a dependent variable.

To demonstrate this discussion, let us consider one of the simplest possible ANCOVA models. First, suppose that the following model were used to represent a response variable's scores in several groups:

$$y_{ij} = \mu + \alpha_j + \varepsilon_{ij},$$

where

$i = 1, \ldots, n_j$ is a subject index within each of $g > 1$ groups

$j = 1, \ldots, g$ is a group index

μ is the grand mean

α_j is the group (or treatment) effect

ε_{ij} denotes the error term

The last equation is of course none other than the well-known one-way ANOVA model (cf. Chapter 4). In this model, our main interest is to test the null hypothesis

$$H_0\colon \mu_1 = \mu_2 = \ldots = \mu_g,$$

where $\mu_i = \mu + \alpha_i$ $(i = 1, \ldots, g)$, or alternatively the hypothesis:

$$H_0\colon \alpha_1 = \alpha_2 = \ldots = \alpha_g = 0.$$

Let us consider a corresponding ANCOVA model in which a covariate, denoted *X*, is included in the model. Such a model looks as follows:

$$y_{ij} = \mu + \alpha_j + \beta X_{ij} + \varepsilon_{ij}, \tag{6.4}$$

where β reflects the effect of the covariate on the outcome variable *y*. (We note that the right-hand side of Equation 6.4 can be extended by as many additional terms as there are covariates under consideration [Maxwell & Delaney, 2004].)

With the ANCOVA model defined in Equation 6.4, the test of primary interest is that of group differences after accounting for individual differences on the covariate(s). That is, this test actually involves the comparison of two nested models: a full model that is next referred to as Model 1 and a restricted model denoted Model 2:

$$\text{Model 1: } y_{ij} = \mu + \alpha_j + \beta X_{ij} + \varepsilon_{ij}$$

$$\text{Model 2: } y_{ij} = \mu + \beta X_{ij} + \varepsilon_{ij}.$$

These two models can be compared with one another, since the restricted model results from the full model by imposing restrictions in the latter, namely, $\alpha_1 = \alpha_2 = \ldots = \alpha_g = 0$. Hence, the ANCOVA test of no group differences after accounting for individual differences in the covariate(s) that are related to such on the response variable, can be conducted using the corresponding *F* test (for dropping predictors) within the framework of a regression analysis (Maxwell & Delany, 2004; Pedhazur, 1997).

This test in fact accomplishes the same aims as a comparison across groups of the so-called adjusted means, denoted $\bar{Y}'_j$, which are defined as follows:

$$\bar{Y}'_j = \bar{Y}_j - b_W(\bar{X}_j - \bar{X}), \tag{6.5}$$

where

b_W is the pooled within-group slope of the relationship between covariate and dependent variable (under the assumption of regression homogeneity; see below)
$\bar{X}$ is the estimated grand mean on the covariate
$\bar{X}_j$ is its mean in the *j*th group
$\bar{Y}_j$ is the response variable mean in that group ($j = 1, \ldots, g$)

For example, consider a study in which two groups were measured, and the following group means were observed on the response variable and covariate, respectively: $\bar{Y}_1 = 60$, $\bar{X}_1 = 40$ and $\bar{Y}_2 = 40$, $\bar{X}_2 = 20$. Assuming for simplicity that $b_W = 1$ and $\bar{X} = 15$, then values for $\bar{Y}'_1 = 35$ and $\bar{Y}'_2 = 35$

would be obtained as adjusted means for each group. These means would obviously indicate that no group differences are present after the covariate is accounted for.

To illustrate the analysis of covariance approach further, consider the following empirical example. Two hundred and forty-eight seniors are involved in a two-group training program in which two assessments—pretest and posttest—of an induction reasoning test are obtained. The purpose of the study is to evaluate whether there is an effect of the training, after accounting for possible group differences at pretest on the inductive reasoning measure. (The data are available in the file ch6ex1.dat available from www.psypress.com/applied-multivariate-analysis; in that file, ir.1 and ir.2 denote the pretest and posttest with the inductive reasoning measure, respectively, and the values of 0 and 1 on the variable ''group'' correspond to the training vs. no training dichotomy.)

In this example, we are dealing with an ANCOVA setting in which a single dependent variable is to be examined (namely ir.2, the posttest score on inductive reasoning), the training (or group) condition is considered as an independent variable, and the pretest score on inductive reasoning (the variable ir.1) is viewed as a covariate. With this setup, we want to control for potential preexisting differences in inductive reasoning scores while testing for group differences in its posttest assessment, in an attempt to evaluate the possible effect of the training program.

To accomplish this goal with SPSS, we use the following menu options/sequence:

Analyze → General linear model → Univariate (choose posttest as dependent variable, group as between-subject factor, and pretest as covariate; Options: Homogeneity test).

To accomplish the same analysis with SAS, the following command statements within PROC GLM can be used:

```
DATA ANCOVA;
INFILE 'ch6ex1.dat';
INPUT ir1 ir2 group;
PROC GLM;
CLASS group;
MODEL ir2 = group ir1;
LSMEAN group / out = adjmeans;
MEANS group /hovtest = Levene;
RUN;
PROC PRINT data = adjmeans;
RUN;
```

In this SAS command file, a number of new statements are employed. In particular, the LSMEAN statement is used to compute the least-squares means for each group adjusted for the covariate, whereas the MEANS statement computes the unadjusted means (as can be recalled, the MEANS statement was previously used in Chapter 4). The PROC PRINT statement invokes separate printing of the values of the adjusted means, which were also saved using the earlier "out = adjmeans" option of LSMEANS. We note that although this statement is not essential, it would be needed if one wishes to obtain the variable means and standard deviations.

The outputs produced by these SPSS and SAS commands follow next. For ease of presentation, the outputs are organized into sections and clarifying comments are accordingly inserted after each of them.

SPSS output

Univariate Analysis of Variance

Between-Subjects Factors

		N
group	.00	87
	1.00	161

Levene's Test of Equality of Error Variances[a]

Dependent Variable: ir.2

F	df1	df2	Sig.
1.293	1	246	.257

Tests the null hypothesis that the error variance of the dependent variable is equal across groups.
a. Design: Intercept+ir. 1+group

As can be seen by examining this output, the assumption of variance homogeneity based on Levene's Test appears plausible. Because the test of variance homogeneity in SAS would obviously lead to identical results, we do not display this section of the SAS output, but move on to the next generated output parts.

SAS output

```
Dependent Variable: ir2
Source                DF   Sum of Squares   Mean Square   F Value    Pr > F

Model                  2    57964.74484     28982.37242   287.46    <.0001
Error                245    24701.08481       100.82075
Corrected Total      247    82665.82966

              R-Square    Coeff Var    Root MSE    ir2 Mean
              0.701194     21.62823    10.04095    46.42521

Source               DF      Type I SS     Mean Square   F Value     Pr > F

group                 1     3241.99714      3241.99714     32.16    <.0001
ir1                   1    54722.74771     54722.74771    542.77    <.0001

Source               DF    Type III SS     Mean Square   F Value     Pr > F

group                 1     2636.69601      2636.69601     26.15    <.0001
ir1                   1    54722.74771     54722.74771    542.77    <.0001

                             The GLM Procedure

                            Least Squares Means

                            group     ir2 LSMEAN

                            0         41.9883018
                            1         48.8227997

                             The GLM Procedure
Level of            ----------ir2-----------  -----------ir1----------
group        N          Mean        Std Dev          Mean        Std Dev
0           87    41.5067011     17.6944273    30.0911264     14.7645835
1          161    49.0830435     18.1138510    30.8255714     14.7667586
```

In this SAS output, "Type I SS" for group corresponds to the between-group sum of squares that are obtained for the ANOVA model disregarding the covariate. This "Type I SS" quantity is the one used when computing the corresponding F ratio along with its statistical significance, ignoring the covariate (in this case, $F = 32.16$, $p < .0001$). In contrast, the "Type III SS" for group reflects the between-groups sum of squares adjusted for the covariate. That is, the "Type III SS" quantities are the ones that should be used when computing the appropriate F ratio and its statistical significance (i.e., $F = 26.15$, $p < .0001$), if one wishes to account for the covariate. The above output also displays both the adjusted group means on the dependent variable "ir2" (i.e., the LS means in group $0 = 41.98$, and in group $1 = 48.82$), and the unadjusted group means (in group $0 = 41.50$, and in group $1 = 49.08$) on the same variable and on the variable "ir1." We note that there can be occasions when the computed F ratio based on the "Type I SS" values is significant, whereas the F ratio based on the "Type III SS" is not significant (or conversely). Such a result would indicate that once the group means are adjusted for the covariate, no significant differences between the groups are evident (or conversely).

SPSS output

Tests of Between-Subjects Effects

Dependent Variable: ir.2

Source	Type III Sum of Squares	df	Mean Square	F	Sig.
Corrected Model	57964.745[a]	2	28982.372	287.464	.000
Intercept	9718.490	1	9718.490	96.394	.000
ir.1	54722.748	1	54722.748	542.773	.000
group	2636.696	1	2636.696	26.152	.000
Error	24701.085	245	100.821		
Total	617180.346	248			
Corrected Total	82665.830	247			

a. R Squared = .701 (Adjusted R Squared = .699)

On the basis of the last table as well as the preceding SAS output part, one can conclude that the two groups differ on posttest induction scores even after accounting for group differences in pretest inductive reasoning, that is, after pretest induction is included as a covariate in the model. In other words, the adjusted means of posttest inductive reasoning scores (i.e., the means had both groups been on average the same at pretest) differ across groups. We, therefore, conclude that the analyzed data contain evidence warranting rejection of the null hypothesis of no training program effect after accounting for possible group differences in initial inductive reasoning.

In conclusion of this section, we mention that the approach used here to analyze data from pretest/posttest designs resembles an alternative procedure that was quite popular for a number of years in the past century among behavioral, educational, and social scientists. That procedure consisted of carrying out an ANOVA on difference scores, that is, the difference between the pretest and posttest scores. The ANCOVA approach utilized here is generally preferable to that difference score approach, however, because the ANOVA procedure underlying the latter is in fact based on the rather strong assumption $\beta = 1$ in the earlier considered Equation 6.4. When this population value assumption is fulfilled, which will most likely be only infrequently the case in empirical research, using ANOVA on difference scores will be a more sensible approach to follow; this is because one will then not need to use up sample information to evaluate this slope parameter, as one would have to do if the above ANCOVA procedure were followed. However, since

$\beta \neq 1$ is the case in general, the older ANOVA method for change scores is going to be generally less powerful than the one utilized in this section, due to the fact that the former method's application is tantamount to making the assumption $\beta = 1$ that misspecifies the underlying model (Maxwell & Delany, 2004).

6.2 Multivariate Analysis of Covariance

Multivariate analysis of covariance is analysis of covariance whereby at least two dependent variables are considered simultaneously. Hence, MANCOVA is applicable when a research question is concerned with whether there are group differences in the multivariate mean vectors on a given set of multiple response measures after accounting for differences on them due to considered covariate(s). That is, formally the MANCOVA query asks if the group differences remaining after controlling for such on the covariate(s) can be explained by chance fluctuations only. In other words, this is a question of whether all observed group differences in the multivariate mean vectors would be explainable as resulting from sampling error only, if all subjects had the same value on the covariate(s) under consideration.

Bearing in mind this definition of MANCOVA, Equation 6.3 can now be extended to accommodate in its left-hand side all dependent variables simultaneously (with corresponding extensions in the right-hand side to appropriately include the available data on their predictors [Timm, 2002]). In this way, it is observed that MANCOVA is also a special case of the general linear model with multiple response variables. Thereby, the distinctive feature of MANCOVA, just like with ANCOVA, is that there is at least one categorical predictor (group membership, factor) plus a continuous predictor(s). From this perspective, MANCOVA can be seen as regression analysis with multiple dependent variables and a mixture of categorical and continuous predictors. This leads us, somewhat loosely speaking, to another conceptual similarity of MANCOVA to ANCOVA—the former also could be viewed as proceeding in two steps (that are in fact carried out simultaneously). The two steps being (a) remove all group differences in the dependent variables that are due to group differences on the covariate(s), that is, obtain adjusted multivariate means on all response variables; and (b) test whether the resulting vectors of adjusted means are the same across groups.

We illustrate MANCOVA by returning to the earlier considered study in this chapter, in which $n = 248$ seniors were examined on inductive reasoning tests. However, we are now interested in how a set of three intelligence measures—inductive reasoning, figural relations, and culture-fair tests—differ across groups at posttest after accounting for the covariate, which is as before the pretest measurement on inductive reasoning. Hence, we are dealing here with a MANCOVA setup since we have three dependent variables, group membership as an independent variable (training vs. no training), and a single covariate—pretest on inductive reasoning. (The data for this example study are available in the file ch6ex2.dat available from www.psypress.com/applied-multivariate-analysis; in that file, ir.#, fr.#, and cf.# are the consecutive assessments with the inductive reasoning, figural relations, and culture-fair tests, respectively, while # stands for occasion, and "group" is the training vs. no training variable.)

To address the MANCOVA question about overall training effect with respect to the three intelligence tests, after accounting for the covariate, in SPSS we use the following menu options/sequence:

Analyze → General linear model → Multivariate (choose ir.2, fr.2, and cf.2 as dependent variables, group as independent variable, and ir.1 as a covariate; Options: Homogeneity test).

To accomplish the same MANCOVA analysis in SAS, the following command statements within PROC GLM can be used:

```
DATA ANCOVA2;
INFILE 'ch6ex2.dat';
INPUT ir1 ir2 fr2 cf2 group ir1group;
PROC GLM;
CLASS group;
MODEL ir2 fr2 cf2 = group ir1;
lsmeans group / out = adjmeans;
means group /hovtest = levene;
MANOVA h = group ir1;
RUN;
PROC PRINT data = adjmeans;
RUN;
```

The resulting outputs produced by SPSS and SAS follow next, again organized into sections and with clarifying comments accordingly provided after each section.

SPSS output

General Linear Model

Between-Subjects Factors

		N
group	.00	87
	1.00	161

Box's Test of Equality of Covariance Matrices(a)

Box's M	6.669
F	1.095
df1	6
df2	205176.504
Sig.	.363

Tests the null hypothesis that the observed covariance matrices of the dependent variables are equal across groups.
a. Design: Intercept + ir.1 + group

Levene's Test of Equality of Error Variances(a)

	F	df1	df2	Sig.
ir.2	.157	1	246	.693
fr.2	1.285	1	246	.258
cf.2	.010	1	246	.919

Tests the null hypothesis that the error variance of the dependent variable is equal across groups.
a. Design: Intercept + ir.1 + group

As can be seen by examining these output sections, according to Box's M Test the covariance matrix homogeneity assumption is also fulfilled (we note that this assumption underlies applications of MANOVA and MANCOVA; see discussion in Chapter 4 and in the next section). Since the pertinent null hypothesis is not rejected, we move on to the next generated output and note that we do not display these results again when presenting SAS output.

Multivariate Tests(b)

Effect		Value	F	Hypothesis df	Error df	Sig.
Intercept	Pillai's Trace	.795	313.354(a)	3.000	243.000	.000
	Wilks' Lambda	.205	313.354(a)	3.000	243.000	.000
	Hotelling's Trace	3.869	313.354(a)	3.000	243.000	.000
	Roy's Largest Root	3.869	313.354(a)	3.000	243.000	.000
ir.1	Pillai's Trace	.677	170.074(a)	3.000	243.000	.000
	Wilks' Lambda	.323	170.074(a)	3.000	243.000	.000
	Hotelling's Trace	2.100	170.074(a)	3.000	243.000	.000
	Roy's Largest Root	2.100	170.074(a)	3.000	243.000	.000
group	Pillai's Trace	.097	8.697(a)	3.000	243.000	.000
	Wilks' Lambda	.903	8.697(a)	3.000	243.000	.000
	Hotelling's Trace	.107	8.697(a)	3.000	243.000	.000
	Roy's Largest Root	.107	8.697(a)	3.000	243.000	.000

a. Exact statistic
b. Design: Intercept + ir.1 + group

The last table with multivariate tests suggests that even after accounting for possible group differences on the pretest of inductive reasoning (i.e., the covariate), there is evidence in the data for group mean differences on the posttest intelligence measures when considered simultaneously. For example, Pillai's Trace statistic is .097, its corresponding $F = 8.697$, and the associated $p < .0001$ (see last panel of that table); thus, it is suggested that there are group mean differences on the vector of posttest intelligence measures after controlling for possible pretest group differences on induction.

If a researcher were interested to begin with in evaluating individually each of the considered posttest intelligence tests, the following table could be examined:

Tests of Between-Subjects Effects

Source	Dependent Variable	Type III Sum of Squares	df	Mean Square	F	Sig.
Corrected Model	ir.2	57211.756(a)	2	28605.878	254.756	.000
	fr.2	23580.504(b)	2	11790.252	87.262	.000
	cf.2	63236.299(c)	2	31618.149	206.812	.000
Intercept	ir.2	14059.050	1	14059.050	125.206	.000
	fr.2	119103.059	1	119103.059	881.509	.000
	cf.2	16609.697	1	16609.697	108.643	.000
ir.1	ir.2	54459.849	1	54459.849	485.004	.000
	fr.2	20636.852	1	20636.852	152.738	.000
	cf.2	60518.313	1	60518.313	395.846	.000
group	ir.2	2198.065	1	2198.065	19.575	.000
	fr.2	2582.397	1	2582.397	19.113	.000
	cf.2	2139.819	1	2139.819	13.996	.000
Error	ir.2	27510.417	245	112.287		
	fr.2	33102.602	245	135.113		
	cf.2	37456.471	245	152.884		
Total	ir.2	685243.614	248			
	fr.2	1301038.435	248			
	cf.2	781205.247	248			
Corrected Total	ir.2	84722.173	247			
	fr.2	56683.106	247			
	cf.2	100692.770	247			

a. R Squared = .675 (Adjusted R Squared = .673)
b. R Squared = .416 (Adjusted R Squared = .411)
c. R Squared = .628 (Adjusted R Squared = .625)

As can be seen from this table, each intelligence test shows posttest group mean differences even after controlling for group differences on initial inductive reasoning test scores (see the three rows of the "group" panel in above table). On the basis of these results, we can conclude that the training group outperformed at posttest the control group, both overall as well as on each of the used intelligence tests, after accounting for possible group differences in initial induction score. We display next the corresponding results obtained with SAS for completeness of the discussion in this section.

SAS output

The GLM Procedure
Multivariate Analysis of Variance

MANOVA Test Criteria and Exact F Statistics for the Hypothesis of No Overall group Effect
H = Type III SSCP Matrix for group
E = Error SSCP Matrix
S = 1 M = 0.5 N = 120.5

Statistic	Value	F Value	Num DF	Den DF	Pr > F
Wilks' Lambda	0.90303888	8.70	3	243	<.0001
Pillai's Trace	0.09696112	8.70	3	243	<.0001
Hotelling-Lawley Trace	0.10737203	8.70	3	243	<.0001
Roy's Greatest Root	0.10737203	8.70	3	243	<.0001

The GLM Procedure
Multivariate Analysis of Variance

MANOVA Test Criteria and Exact F Statistics for the Hypothesis of No Overall ir1 Effect
H = Type III SSCP Matrix for ir1
E = Error SSCP Matrix
S = 1 M = 0.5 N = 120.5

Statistic	Value	F Value	Num DF	Den DF	Pr > F
Wilks' Lambda	0.32261383	170.07	3	243	<.0001
Pillai's Trace	0.67738617	170.07	3	243	<.0001
Hotelling-Lawley Trace	2.09968112	170.07	3	243	<.0001
Roy's Greatest Root	2.09968112	170.07	3	243	<.0001

The last output section with multivariate test results indicates (as seen above from the SPSS output) that even after controlling for group differences on the pretest of inductive reasoning, there is evidence in the data for group mean differences on the posttest intelligence measures when considered simultaneously. Again, if a researcher were interested to begin with in examining individually each of the posttest intelligence measures, the following displayed tables could be examined:

The GLM Procedure

Dependent Variable: ir2

Source	DF	Sum of Squares	Mean Square	F Value	Pr > F
Model	2	57211.75611	28605.87806	254.76	<.0001
Error	245	27510.41707	112.28742		
Corrected Total	247	84722.17318			

R-Square	Coeff Var	Root MSE	ir2 Mean
0.675287	21.53412	10.59658	49.20831

Source	DF	Type I SS	Mean Square	F Value	Pr > F
group	1	2751.90722	2751.90722	24.51	<.0001
ir1	1	54459.84889	54459.84889	485.00	<.0001

Source	DF	Type III SS	Mean Square	F Value	Pr > F
group	1	2198.06541	2198.06541	19.58	<.0001
ir1	1	54459.84889	54459.84889	485.00	<.0001

The GLM Procedure

Dependent Variable: fr2

Source	DF	Sum of Squares	Mean Square	F Value	Pr > F
Model	2	23580.50380	11790.25190	87.26	<.0001
Error	245	33102.60246	135.11266		
Corrected Total	247	56683.10626			

R-Square	Coeff Var	Root MSE	fr2 Mean
0.416006	16.40974	11.62380	70.83475

Source	DF	Type I SS	Mean Square	F Value	Pr > F
group	1	2943.65225	2943.65225	21.79	<.0001
ir1	1	20636.85155	20636.85155	152.74	<.0001

Source	DF	Type III SS	Mean Square	F Value	Pr > F
group	1	2582.39717	2582.39717	19.11	<.0001
ir1	1	20636.85155	20636.85155	152.74	<.0001

The GLM Procedure

Dependent Variable: cf2

Source	DF	Sum of Squares	Mean Square	F Value	Pr > F
Model	2	63236.2986	31618.1493	206.81	<.0001
Error	245	37456.4714	152.8836		
Corrected Total	247	100692.7700			

R-Square	Coeff Var	Root MSE	cf2 Mean
0.628012	23.60414	12.36461	52.38322

Source	DF	Type I SS	Mean Square	F Value	Pr > F
group	1	2717.98561	2717.98561	17.78	<.0001
ir1	1	60518.31301	60518.31301	395.85	<.0001

Source	DF	Type III SS	Mean Square	F Value	Pr > F
group	1	2139.81918	2139.81918	14.00	0.0002
ir1	1	60518.31301	60518.31301	395.85	<.0001

As can be seen from these tables, each of the three intelligence tests demonstrates posttest group mean differences even after controlling for group differences on initial inductive reasoning ability. Finally, the values of both the adjusted group means (i.e., LS means) on each intelligence test, the unadjusted group means, and the means on the pretest measure are displayed in the following tables:

The GLM Procedure

Least Squares Means

group	ir2 LSMEAN	fr2 LSMEAN	cf2 LSMEAN
0	45.1572240	66.4437615	48.3861742
1	51.3974007	73.2075202	54.5431170

The GLM Procedure

Level of		---------ir2---------		---------fr2---------		---------cf2---------	
group	N	Mean	Std Dev	Mean	Std Dev	Mean	Std Dev
0	87	44.6767816	18.0077469	66.1480115	15.0999695	47.8797126	20.2254060
1	161	51.6570186	18.3851632	73.3673354	14.6053657	54.8167950	19.8108270

Level of		--------------ir1------------	
group	N	Mean	Std Dev
0	87	30.0911264	14.7645835
1	161	30.8255714	14.7667586

6.3 Step-Down Analysis (Roy–Bargmann Analysis)

In empirical behavioral, social, and educational research, used dependent or response measures are typically correlated. As a consequence, when researchers are interested in following up a significant MANOVA finding using univariate F tests (i.e., conducting ANOVA's on the individual outcome variables), they must bear in mind that the response measures' relationships imply that these follow-up tests are correlated as well and hence there is some redundancy in their results (see Chapter 4). Therefore, in a sense, considered as a set, these univariate follow-up tests are not particularly insightful because they contain redundant information. To handle this issue, one can use the so-called step-down analysis (SDA). This approach represents a MANCOVA (or an ANCOVA, as a special case with a single dependent variable) where some of the response variables in the original MANOVA are being treated as covariates. In particular, "lower" priority original response variables are used as covariates when one is interested in testing for group mean differences on the remaining "higher" priority outcome measures, after accounting for group differences on the lower priority dependent variables.

For SDA purposes, one first assigns "priorities" to initially used dependent variables in a MANOVA, according to either theoretical or practical considerations, or a specific research question being asked (Tabachnick &

Fidell, 2007). For example, if some of the original outcomes causally affect other response variables, the former may be assigned lower priority. Alternatively, a research question may ask about group differences on a particular dependent variable(s), after accounting for such on other response measures. In this case, one can assign low priority to the latter variables. SDA is then a MANCOVA whereby high-priority, initial response variables are considered outcomes, while low-priority, original response variables are treated as covariates. From a formal viewpoint, the model that underlies SDA is a general linear model with (a) group membership variables (factors) used as predictors, and (b) low-priority initial response measures participating as additional explanatory variables, or covariates (cf. Timm, 2002).

Only for the purpose of giving an example here, let us consider again the study that was focused on in the preceding section, and suppose for this discussion that one were interested in examining whether there were group differences with respect to the induction, figural relations, and culture-fair test scores. (The data are available in the file ch6ex3.dat available from www.psypress.com/applied-multivariate-analysis, where the measures induction, figural relations, and culture-fair are denoted "ir," "fr," and "cf," respectively, and "group" is the experimental vs. control dichotomy). This research query is a conventional MANOVA question, whereby these three tests are the dependent variables and group is the independent variable. To accomplish this analytic goal, with SPSS we use the following menu options/sequence:

Analyze → General linear model → Multivariate (all three tests as dependent variables, group as independent variable, Options: homogeneity tests).

The following SAS command statements within the PROC GLM procedure achieve the same aim:

```
DATA INTELLIGENCE_TESTS;
INFILE 'ch6ex3.dat';
INPUT IR FR CF GROUP;
PROC GLM;
  class GROUP;
  model IR FR CF = GROUP;
  means GROUP/HOVTEST = LEVENE;
  manova h = GROUP;
RUN;
proc discrim pool = test;
class GROUP;
run;
```

Since our main interest in this section is to demonstrate an application of step-down analysis, we only provide the resulting output produced by SPSS (for relevant details on MANOVA, see Chapter 4) and consider those aspects that are necessary for the completion of an SDA.

SPSS output

General Linear Model

Between-Subjects Factors

		N
group	.00	87
	1.00	161

Box's Test of Equality of Covariance Matrices(a)

Box's M	6.561
F	1.077
df1	6
df2	205176.504
Sig.	.373

Tests the null hypothesis that the observed covariance matrices of the dependent variables are equal across groups.
a. Design: Intercept+group

Levene's Test of Equality of Error Variances(a)

	F	df1	df2	Sig.
ir	.013	1	246	.910
fr	.092	1	246	.762
cf	.228	1	246	.633

Tests the null hypothesis that the error variance of the dependent variable is equal across groups.
a. Design: Intercept+group

On the basis of these results, it appears that the covariance matrix homogeneity assumption is plausible.

Multivariate Tests(b)

Effect		Value	F	Hypothesis df	Error df	Sig.
Intercept	Pillai's Trace	.958	1856.382(a)	3.000	244.000	.000
	Wilks' Lambda	.042	1856.382(a)	3.000	244.000	.000
	Hotelling's Trace	22.824	1856.382(a)	3.000	244.000	.000
	Roy's Largest Root	22.824	1856.382(a)	3.000	244.000	.000
group	Pillai's Trace	.055	4.758(a)	3.000	244.000	.003
	Wilks' Lambda	.945	4.758(a)	3.000	244.000	.003
	Hotelling's Trace	.058	4.758(a)	3.000	244.000	.003
	Roy's Largest Root	.058	4.758(a)	3.000	244.000	.003

a. Exact statistic
b. Design: Intercept + group

These multivariate tests indicate that there is evidence for group differences on the dependent variable mean vectors.

Tests of Between-Subjects Effects

Source	Dependent Variable	Type III Sum of Squares	df	Mean Square	F	Sig.
Corrected Model	Ir	2751.907(a)	1	2751.907	8.259	.004
	Fr	2943.652(b)	1	2943.652	13.475	.000
	Cf	2717.986(c)	1	2717.986	6.824	.010
Intercept	Ir	524144.260	1	524144.260	1573.003	.000
	Fr	1099353.629	1	1099353.629	5032.448	.000
	Cf	595668.724	1	595668.724	1495.635	.000
group	Ir	2751.907	1	2751.907	8.259	.004
	Fr	2943.652	1	2943.652	13.475	.000
	Cf	2717.986	1	2717.986	6.824	.010
Error	Ir	81970.266	246	333.212		
	Fr	53739.454	246	218.453		
	Cf	97974.784	246	398.271		
Total	Ir	685243.614	248			
	Fr	1301038.435	248			
	Cf	781205.247	248			
Corrected Total	Ir	84722.173	247			
	Fr	56683.106	247			
	Cf	100692.770	247			

a. R Squared = .032 (Adjusted R Squared = .029)
b. R Squared = .052 (Adjusted R Squared = .048)
c. R Squared = .027 (Adjusted R Squared = .023)

An examination of the last table indicates that each intelligence test shows group differences as well. We emphasize, however, that as mentioned earlier, the pertinent three univariate tests are not independent of one another, and in this sense provide partly redundant information.

Given these findings and the fact that the more comprehensive, culture-fair test presumably encompasses components that also relate to the induction and figural relation subabilities of the cluster of fluid intelligence (Horn, 1982), it is of interest to examine whether once we partial out the group effect on the latter two tests, there still remains evidence in the data for training effect on the culture-fair test. That is, we are asking whether after controlling (adjusting) for group differences on the more specialized fluid abilities of induction and figural relations, there would be such remaining group differences on the more general culture-fair measure, which cannot be explained by chance effects only.

To this end, we follow up the last carried out MANOVA with a step-down analysis. For the latter, following the preceding discussion, we formally assign "higher priority" to the culture-fair test and "lower priority" to the induction and figural relations tests. That is, in the particular exemplary setting here, we have to deal with an ANCOVA having one dependent variable (viz. culture-fair test), group membership as independent variable, and the remaining two response variables from the previous MANOVA analysis (viz. induction and figural relations) as covariates.

This analysis is accomplished with SPSS using the following menu options/sequence:

Analyze → General linear model → Univariate (cf as dependent variable, group as independent variable, ir and fr as covariates; Options: homogeneity test)

To accomplish this ANCOVA analysis with SAS, the following command statements within the PROC GLM procedure would be used (see also previous section on ANCOVA):

```
DATA INTELLIGENCE_TESTS;
INFILE 'chap6ex3.dat';
INPUT ir fr cf group;
PROC GLM;
CLASS group;
MODEL cf = group ir fr;
LSMEAN group /out = adjmeans;
MEANS group /hovtest = Levene;
RUN;
```

The resulting outputs produced by SPSS and SAS follow next, with clarifying comments inserted accordingly.

SPSS output

Univariate Analysis of Variance

Tests of Between-Subjects Effects

Dependent Variable: cf

Source	Type III Sum of Squares	df	Mean Square	F	Sig.
Corrected Model	85385.896(a)	3	28461.965	453.699	.000
Intercept	850.173	1	850.173	13.552	.000
ir	25703.087	1	25703.087	409.721	.000
fr	2487.296	1	2487.296	39.649	.000
group	47.580	1	47.580	.758	.385
Error	15306.874	244	62.733		
Total	781205.247	248			
Corrected Total	100692.770	247			

a. R Squared = .848 (Adjusted R Squared = .846)

SAS output

The GLM Procedure

Dependent Variable: cf

Source	DF	Sum of Squares	Mean Square	F Value	Pr > F
Model	3	85385.8959	28461.9653	453.70	<.0001
Error	244	15306.8741	62.7331		
Corrected Total	247	100692.7700			

R-Square	Coeff Var	Root MSE	cf Mean
0.847984	15.12015	7.920422	52.38322

Source	DF	Type I SS	Mean Square	F Value	Pr > F
group	1	2717.98561	2717.98561	43.33	<.0001
ir	1	80180.61401	80180.61401	1278.12	<.0001
fr	1	2487.29631	2487.29631	39.65	<.0001

Source	DF	Type III SS	Mean Square	F Value	Pr > F
group	1	47.57954	47.57954	0.76	0.3847
ir	1	25703.08715	25703.08715	409.72	<.0001
fr	1	2487.29631	2487.29631	39.65	<.0001

The GLM Procedure
Least Squares Means

group	cf LSMEAN
0	52.9953018
1	52.0524704

An examination of the results of this SDA reveals that there are no remaining group differences on the presumably more general fluid measure (culture-fair test) once we adjust for group differences on the more specialized fluid tests (inductive reasoning and figural relations). This is because after controlling for the two fluid covariates, the corresponding test statistic is $F = 0.76$, with associated $p = .3847$, and hence nonsignificant. We conclude that there are no group differences in culture-fair test scores once accounting for group differences on the induction and figural relations tests. This finding may not be unexpected, given the fact that the training targeted these two fluid abilities (Baltes et al., 1986).

6.4 Assumptions of Analysis of Covariance

Analysis of covariance is a powerful methodology that has, however, certain limitations deriving from the assumptions it is based on. More specifically, since formally ANCOVA is carried out within the framework of ANOVA, all the assumptions of ANOVA (in particular, normality, variance homogeneity on dependent variable, and independent samples;

see Chapter 4) are also assumptions of ANCOVA. Furthermore, the credibility of ANCOVA results depends on a number of additional assumptions. These include the following ones (discussed in detail below): (a) the covariate(s) being perfectly measured; (b) the regression of the dependent variable on the independent one(s) is linear and with the same slope(s) across all involved groups (often referred to as regression homogeneity or parallelism—we note in passing that techniques generally known as hierarchical linear modeling and structural equation modeling allow different group-specific or random slopes); and (c) that there is no treatment by covariate interaction—in other words, that the covariate and treatment are independent of each other. The last assumption will be typically fulfilled if the covariate is measured before the treatment is begun in the groups receiving it, which is the requirement for ANCOVA that we have mentioned at the outset of the chapter.

The assumption of a perfect covariate(s) cannot generally be met in many behavioral, social, and educational studies since they typically involve variables that are measured with considerable error. In practical terms, one may "replace" this assumption with the requirement that whenever variables are to be used as covariates, they should exhibit high reliability (e.g., at least above .80 but preferably closer to or above .90). Otherwise, use of an alternative methodology, such as structural equation modeling, can be recommended that is developed to deal specifically with such fallible predictors/covariates (Raykov & Marcoulides, 2006).

The assumption of regression homogeneity is needed to ensure that the within-group regression coefficients for the relationship of dependent variable to covariates are the same, so that the groups can be pooled to estimate those common values (e.g., b_W in Equation 6.5, for the case of a single covariate). Recent research has found that the effect of violation of this assumption is oftentimes less pronounced than earlier thought (Maxwell & Delaney, 2004; cf. Huitema, 1980). Nonetheless, it is recommendable that the homogeneity assumption be tested before one decides to proceed with utilizing ANCOVA. This test can be accomplished using the well-known *F*-test for dropping predictors within the framework of regression analysis when comparing a full model to a restricted model, which we have referred to earlier in this chapter (Pedhazur, 1997). To this end, consider the following two models that differ only in the interaction term, denoted next IV*C, of group (or independent variable, designated IV) and covariate denoted C:

$$\text{Model 1 (full model): DV} = b_0 + b_1.\text{IV} + b_2.\text{C} + b_3.\text{IV*C} + e$$
$$\text{Model 2 (restricted model): DV} = b_0 + b_1.\text{IV} + b_2.\text{C} + e^*,$$

where e and e^* stand for their error terms (with the usual regression analysis assumptions of errors with zero means and unrelated to predictors, as well as normality and variance homogeneity). As indicated earlier in this chapter, if these two models explain significantly different proportions of dependent variable variance—that is, if they are associated with a significant difference in their R^2 indices—then the regression homogeneity assumption is violated; otherwise it is not.

To demonstrate this approach, consider again the earlier example of analysis of covariance aimed at evaluating group differences on posttest inductive reasoning after accounting for group differences on its pretest as a covariate (see file ch6ex2.dat available from www.psypress.com/applied-multivariate-analysis, where these variables are denoted ir.2 and ir.1, respectively). To test for homogeneity of regression, which stipulates the same slope of ir2 on ir1 in both groups, with SPSS or SAS we proceed as follows. First we compute the interaction of covariate and group membership, using the SPSS (syntax-based) command

COMP IR1GROUP = IR.1 * GROUP.,

or the SAS command

IR1GROUP = IR1*GROUP;.

Then in SPSS we employ the following sequence of menu options:

Analyze → Regression → Linear (group and ir.1 moved as predictors into block 1; add IR1.GROUP as predictor into block 2; in Statistics: check "R2-change").

To accomplish the same analysis with SAS, the following command statements within the regression analysis procedure PROC REG would be used (we note that the file ch6ex2.dat contains the earlier created interaction variable in its last column and so the above compute statement is not needed in the program listed below):

```
DATA slope homogeneity assumption test;
INFILE 'ch6ex2.dat';
INPUT ir1 ir2 fr2  cf2 group ir1group;
PROC REG;
m1: MODEL ir2 = group ir1;
m2: MODEL ir2 = group ir1 ir1group;
TEST ir1group = 0;
RUN;
```

In these SAS command statements, the two considered models (i.e., the restricted model labeled "m1", and the full model labeled "m2") are generated within the same session and the "TEST" statement provides a statistical test of the significance of this additional model regression coefficient, that is, provides a test of the regression homogeneity assumption.

The resulting outputs by SPSS and SAS follow next with clarifying comments inserted accordingly after each section.

SPSS output

Regression

Variables Entered/Removed[b]

Model	Variables Entered	Variables Removed	Method
1	group, ir.1[a]	-	Enter
2	ir1.group[a]	-	Enter

a. All requested variables entered.
b. Dependent Variable: ir.2

Model Summary

Model	R	R Square	Adjusted R Square	Std. Error of the Estimate	Change Statistics				
					R Square Change	F Change	df1	df2	Sig. F Change
1	.822(a)	.675	.673	10.59658	.675	254.756	2	245	.000
2	.822(b)	.675	.671	10.61734	.000	.043	1	244	.836

a. Predictors: (Constant), ir.1, group
b. Predictors: (Constant), ir.1, group, ir1.group

Since the full model (Model 2) and the restricted model (Model 1) explain proportions of dependent variance that are not significantly different from one another (see bottom entry in the right-most column of the last table), we conclude that the regression homogeneity assumption is not violated. We, therefore, can consider as plausible the assumption that the regression coefficient b_3 in the full model is 0 in the studied population. Thus, we select to use the more parsimonious model without the group by covariate interaction, that is, the restricted model. As mentioned, at the software level this model is referred to as Model 1 in the next table.

ANOVA(c)

Model		Sum of Squares	df	Mean Square	F	Sig.
1	Regression	57211.756	2	28605.878	254.756	.000(a)
	Residual	27510.417	245	112.287		
	Total	84722.173	247			
2	Regression	57216.580	3	19072.193	169.188	.000(b)
	Residual	27505.593	244	112.728		
	Total	84722.173	247			

a. Predictors: (Constant), ir.1, group
b. Predictors: (Constant), ir.1, group, ir1.group
c. Dependent Variable: ir.2

Coefficients(a)

Model		Unstandardized Coefficients		Standardized Coefficients	t	Sig.
		B	Std. Error	Beta		
1	(Constant)	14.356	1.785		8.042	.000
	group	6.240	1.410	.161	4.424	.000
	ir.1	1.008	.046	.802	22.023	.000
2	(Constant)	13.966	2.596		5.379	.000
	group	6.844	3.242	.177	2.111	.036
	ir.1	1.021	.078	.812	13.161	.000
	irl.group	−.020	.096	−.020	−.207	.836

a. Dependent Variable: ir.2

Given that due to model parsimony considerations we selected Model 1 (restricted model), in the last table we are concerned only with the first panel. Its results suggest that even after accounting for group differences on the covariate, pretest inductive reasoning score, there are significant group differences in the means of posttest induction.

Excluded Variables(b)

Model		Beta In	t	Sig.	Partial Correlation	Collinearity Statistics
						Tolerance
1	Irl.group	−.020(a)	−.207	.836	−.013	.138

a. Predictors in the Model: (Constant), ir.1, group
b. Dependent Variable: ir.2

An equivalent approach to testing regression homogeneity is to enter all predictors and the created interaction term as a single block into the model, and then check for significance (only) the interaction term (Pedhazur, 1997). If it is not significant, the regression homogeneity assumption can be considered plausible; if it is significant, however, this assumption is violated. This strategy is essentially the same as the approach that we followed with the last presented SAS command file, whose output is provided below.

SAS output

```
                    The REG Procedure
                       Model: m1
                 Dependent Variable: ir2

                  Analysis of Variance

Source           DF  Sum of Squares  Mean Square  F Value   Pr > F
Model             2           57212        28606   254.76   <.0001
Error           245           27510    112.28742
Corrected Total 247           84722
```

Root MSE	10.59658	R-Square	0.6753
Dependent Mean	49.20831	Adj R-Sq	0.6726
Coeff Var	21.53412		

Parameter Estimates

Variable	DF	Parameter Estimate	Standard Error	t Value	Pr > |t|
Intercept	1	14.35559	1.78501	8.04	<.0001
group	1	6.24018	1.41040	4.42	<.0001
ir1	1	1.00765	0.04575	22.02	<.0001

The last table furnishes the regression analysis results for the restricted model (Model 1), followed in the table below by the results for the full model (Model 2). One can readily see that there is no (population) change in the value of the critical R^2 index between the two models—the statistical test of the regression coefficient in which they differ (estimated at −.01989, with a standard error .09615), which is part of the full model (Model 2), is not significant ($F = 0.04$, $p = .836$).

The REG Procedure
Model: m2
Dependent Variable: ir2

Analysis of Variance

Source	DF	Sum of Squares	Mean Square	F Value	Pr > F
Model	3	57217	19072	169.19	<.0001
Error	244	27506	112.72784		
Corrected Total	247	84722			

Root MSE	10.61734	R-Square	0.6753
Dependent Mean	49.20831	Adj R-Sq	0.6714
Coeff Var	21.57631		

Parameter Estimates

Variable	DF	Parameter Estimate	Standard Error	t Value	Pr > |t|
Intercept	1	13.96629	2.59622	5.38	<.0001
group	1	6.84376	3.24202	2.11	0.0358
ir1	1	1.02058	0.07754	13.16	<.0001
ir1group	1	−0.01989	0.09615	−0.21	0.8363

The REG Procedure
Model: m2

Test 1 Results for Dependent Variable ir2

Source	DF	Mean Square	F Value	Pr > F
Numerator	1	4.82382	0.04	0.8363
Denominator	244	112.72784		

We note that all of the above discussed assumptions of ANCOVA along with all previously mentioned MANOVA assumptions (see Chapter 4) are also assumptions of multivariate analysis of covariance (MANCOVA). In particular, MANCOVA assumptions include the following: (a) all covariates are measured without error; (b) the regression of the each dependent variable on each covariate is the same across all considered groups; and (c) the covariates and treatment are independent of each other. We note that in a MANCOVA setting some of these assumptions can also be assessed using a testing approach developed within the framework of structural equation modeling (Raykov, 2001).

We also emphasize that like MANOVA is preferable to ANOVA when a research query is multivariate in nature, so is MANCOVA preferable then to ANCOVA if including covariates into one's modeling efforts. In particular, the goal of MANCOVA cannot be achieved then via a series of ANCOVAs on each dependent variable. The reason is that this multiple testing procedure leads to an inflated type I error rate (just like such an approach would lead to inflated type I error in the MANOVA vs. ANOVA setting). Rather, MANCOVA controls this error rate by incorporating a multivariate protection against its inflation. Further, it is possible that ANCOVAs on each dependent variable just fail to reject the null hypothesis of no group differences in adjusted means on these measures, while MANCOVA's test of group differences in adjusted mean vectors turns out to be significant. The reason for such a finding would be the fact that in MANCOVA one takes into account the interrelationship between the dependent variables. The converse finding is also possible—a nonsignificant MANCOVA that "washes out" (dilutes) significant findings in single response ANCOVAs. (This may be the case in some settings with opposite directions of group differences on adjusted means for some dependent variables.)

We conclude this chapter by stressing that even if particular assumptions of a MANCOVA are violated, results from a pertinent MANOVA (i.e., testing group differences after disregarding covariates) may still be of substantive interest (Tabachnick & Fidell, 2007). Hence, if in an empirical study a researcher has serious doubts whether ANCOVA or MANCOVA assumptions are plausible, they can always carry out a MANOVA and compare the unadjusted means across studied groups (as long as, of course, the MANOVA assumptions are plausible and the comparison of unadjusted means pertains to research questions asked).

7

Principal Component Analysis

7.1 Introduction

A widely recognized fact about social, behavioral, and educational phenomena is that they are exceedingly complex. To understand them well, therefore, it is typically necessary to examine many of their aspects that are usually reflected in studied variables. For this reason, researchers commonly collect data from a large number of interrelated measures pertaining to a phenomenon under investigation. To comprehend this multitude of variables, it becomes desirable to reduce them to more fundamental measures with the property that each of them represents a subset of the initial interrelated variables. For example, an opinion questionnaire may contain numerous questions pertaining to several personality characteristics. Reducing these questions so as to be able to collectively summarize each of the characteristics would clearly help to more readily understand the latter.

Principal component analysis (PCA) is a statistical technique that has been specifically developed to address this data reduction goal. PCA not only allows such data reduction, but also accomplishes it in a manner that permits its results to be used in applications of other multivariate statistical methods (e.g., analysis of variance or regression analysis). In general terms, the major aim of PCA is to reduce the complexity of the interrelationships among a potentially large number of observed variables to a relatively small number of linear combinations of them, which are referred to as principal components. This goal typically overrides in empirical research its secondary aim, that is, interpretation of the principal components. To a large extent, the interpretation of principal components is generally guided by the degree to which each variable is associated with a particular component. Those variables that are found to be most closely related to a component in question are used as a guide for its interpretation. While interpretability of principal components is always desirable, a PCA is worthwhile undertaking even when they do not have a clear-cut substantive meaning.

Consider the following empirical example. In a study of characteristics of high-school seniors who continue on to college, an investigator collects data on a large number of personality variables, motivation and aspiration measures, general mental ability, scholastic history, family history, health, and physical measures (Tabachnick & Fidell, 2007). The researcher is interested in finding, if possible, more fundamental dimensions that characterize behavior of students entering college. For instance, some personality characteristics may combine with several motivation and aspiration variables into a derived measure that assesses the degree to which a student is able to work independently. In addition, several general mental ability measures may combine with some of scholastic history to form a derived variable assessing intelligence, etc. As indicated above, these derived measures, called principal components, represent weighted sums—that is, linear combinations—of the original observed variables.

This example demonstrates that a PCA of an initial, potentially large set of observed variables aims at "discovering" such linear combinations. The goal of PCA will be achieved if a relatively small number of principal components are "disclosed," which possess the property that they account for most of the variability in the original data set. If they are also substantively interpretable, it will be an added benefit that will contribute to better understanding of a studied phenomenon.

In the behavioral, social and educational sciences, PCA has a relatively long history of application in the development of objective tests for measuring specific abilities, motivation, personality, intelligence, and other related constructs or latent (unobservable, hidden) dimensions. In these applications, one typically starts with a large set of measures aimed at assessing those constructs. Then PCA is employed on the obtained data with the aim of reducing their multitude to a few meaningful components that represent "pure" measures of the underlying latent dimensions or variables.

From a conceptual viewpoint, PCA begins by focusing on the correlations or variances and covariances among a given set of observed variables. That is, one may view the empirical correlation or covariance matrix, R or S (see Chapter 1), as a starting point of PCA. The essence of PCA is the summarization of the pattern of interrelationships in the matrix R (or the matrix S), as a result of which one obtains particular linear combinations of subsets of the initial variables. It is desirable that these weighted combinations, which represent the sought components, reduce substantially the complexity of the original data set. Each of them combines a subset of initial variables that are markedly interrelated. In addition, the principal components have the property that they are uncorrelated with one another. Furthermore, each component is obtained in such a way as to account for as much as possible variance in the initial data set. It is thereby hoped that a relatively small number of components account for most of the variance in the original data, and thus accomplish marked data reduction.

We stress that what counts in PCA is accounting for variance—the principal components to be further considered must explain as much as possible variance of the initial variables. This is distinct from the central goal of a related statistical technique, that of factor analysis (FA), where one is concerned predominantly with explaining the structure of variable interrelationships. In fact, because PCA and FA share a number of conceptual similarities, in the past there has been a considerable degree of confusion among empirical researchers in their applications for purposes of data reduction. We emphasize, however, that PCA and FA are distinct from one another in more ways than those in which they are similar. We will discuss in greater detail their distinctions in Chapter 8 that deals with factor analysis.

7.2 Beginnings of Principal Component Analysis

Principal component analysis is fundamentally based on the notions of *eigenvalue* and *eigenvector*. To understand them, we need to revisit first the concept of a vector and linear transformation (see also the discussion in Chapters 1, 2, and 4).

Suppose we are given the data resulting from a study on p interrelated variables (where $p > 1$). For any individual in this investigation, let us think of his/her data on the p variables as a point in the p-dimensional space generated by the latter. For example, if we collected scores from each individual on a set of five mathematics and computation-related tests, his/her data could be represented as a point in the five-dimensional space spanned by these variables. To make matters more transparent, let us view his/her data on all five variables as represented by an arrow starting from the origin and ending in that point, which arrow we will call a vector. In fact, to simplify things even further, we will consider all vectors in the p-dimensional space that are parallel to a particular one and are of the same length, regardless of their location, as being the same vector. We will look at the one of them, which starts at the origin, as the representative of all these vectors.

As another example, assume we are dealing with a study of the relationship between motivation and aspiration in middle school students, that is, with $p = 2$ observed measures. The two coordinate axes in Figure 7.1 below stand for these variables (the horizontal axis corresponding to Aspiration and the vertical axis to Motivation). A given subject's data are represented thereby with the vector starting at the origin and ending at the point in the plane, which has as abscissa his/her motivation score and as ordinata his/her aspiration score.

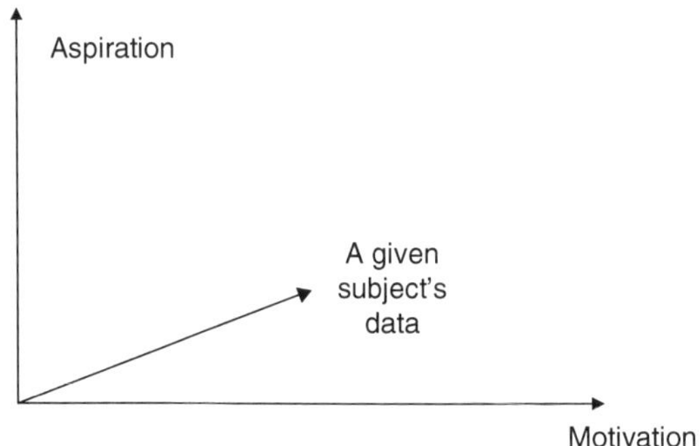

FIGURE 7.1
A subject's data representation.

More generally, any given subject's data in the p-dimensional space is a $p \times 1$ vector, denoted by $\underline{x}$. As we discussed in Chapter 2, for a particular $p \times p$ matrix, $A = [a_{ij}]$, the product $\underline{y} = A\ \underline{x}$ is:

$$\underline{y} = \begin{bmatrix} y_1 \\ y_2 \\ \dots \\ y_p \end{bmatrix} = A \quad \underline{x} = \begin{bmatrix} a_{11} & a_{12} & \dots & a_{1p} \\ a_{21} & a_{22} & \dots & a_{2p} \\ \cdot & \cdot & \cdot & \cdot \\ a_{p1} & a_{p2} & \dots & a_{pp} \end{bmatrix} \begin{bmatrix} x_1 \\ x_2 \\ \dots \\ x_p \end{bmatrix}$$

$$= \begin{bmatrix} a_{11}x_1 + a_{12}x_2 + \dots + a_{1p}x_p \\ a_{21}x_1 + a_{22}x_2 + \dots + a_{2p}x_p \\ \dots \\ a_{p1}x_1 + a_{p2}x_2 + \dots + a_{pp}x_p \end{bmatrix}.$$

That is, $\underline{y} = A\ \underline{x}$ belongs to the same p-dimensional space as $\underline{x}$. Hence, the vector $\underline{y}$ can be looked at as the result of a particular transformation of the vector $\underline{x}$, which is carried out by the matrix A. That is, the matrix A operates on the vector $\underline{x}$ and produces as a result the vector $\underline{y}$ that lies in a multidimensional space of interest. Note that the result of this transformation is a vector $\underline{y}$ with location, length, and orientation (direction) that is determined by A as well as $\underline{x}$, but other than that has unrestricted location, length, and orientation.

In PCA, FA, and a number of other multivariate statistical techniques (e.g., MANOVA, MANCOVA, and discriminant function analysis), of particular relevance are the cases where for a given square matrix A the transformed vector $\underline{y}$ is collinear with the original vector $\underline{x}$; that is, $\underline{y}$ as a vector lies on the same line in the multivariate space as $\underline{x}$ and starts at the same point but may be of different length than that of $\underline{x}$. These are the cases where $\underline{y}$ is a constant multiple of $\underline{x}$: $\underline{y} = A\ \underline{x} = \lambda\ \underline{x}$, where λ is a scalar, that is, a number that can be positive, 0, or negative; we exclude thereby from consideration the case $\underline{x} = \underline{0}$, for triviality reasons (as then $\underline{y} = \underline{0}$ as well). In other words, when $\underline{y} = A\ \underline{x} = \lambda\ \underline{x}$, the matrix A operates on $\underline{x}$ and produces a vector $\underline{y}$ that is collinear with $\underline{x}$, i.e., is a constant multiple of $\underline{x}$. With this in mind, we are now ready for the following more formal definition.

Definition: If $\underline{y} = \mathrm{A}\,\underline{x} = \lambda\,\underline{x}$, where A is a $p \times p$ matrix (with $p > 1$) and $\underline{x} \neq \underline{0}$, then the scalar λ is called *eigenvalue* of A and the vector $\underline{x}$ is called the *eigenvector* of A pertaining to this eigenvalue.

In general, for the contexts of concern to us in the rest of the book, there are as many eigenvalues of a given square matrix as is its size (i.e., p in the present case). The eigenvalues are real-valued (see below) and commonly denoted by $\lambda_1, \lambda_2, \ldots, \lambda_p$ from largest to smallest, whereby the case of some of them being equal to one another is not excluded (i.e., $\lambda_1 \geq \lambda_2 \geq \ldots \geq \lambda_p$; see also discussion in Chapter 4). To each one of these eigenvalues, a pertinent eigenvector can be found. As indicated in Chapter 4, eigenvalues are also referred to as "latent roots" or "characteristic roots" in the literature, and the eigenvectors as "latent vectors" or "characteristic vectors." In this text, for ease of presentation we will only use the references "eigenvalue" and "eigenvector." Occasionally, we will refer to the analytic activities for obtaining them as "eigenanalysis." In the rest of this chapter, the following proposition will be of particular importance (see also Chapters 2 and 4).

Proposition 1: The eigenvalues of a square matrix A are the solutions of the equation $|\mathrm{A} - \lambda\, I_p| = 0$, that is, nullify the determinant of the matrix $\mathrm{A} - \lambda\, I_p$.

We demonstrate this proposition with the following simple example. In it, we are interested in finding the eigenvalues and eigenvectors of the 2×2 matrix $\mathrm{A} = \begin{bmatrix} 1 & 0 \\ 1 & 3 \end{bmatrix}$. First, using Proposition 1, the eigenvalues of A are the solutions of the following equation (see Chapter 2 for pertinent operations regarding determinant evaluation):

$$0 = |\mathrm{A} - \lambda I_2| = \left| \begin{bmatrix} 1 & 0 \\ 1 & 3 \end{bmatrix} - \lambda \begin{bmatrix} 1 & 0 \\ 0 & 1 \end{bmatrix} \right|$$
$$= \begin{vmatrix} 1-\lambda & 0 \\ 1 & 3-\lambda \end{vmatrix} = (1-\lambda)(3-\lambda).$$

Therefore, the eigenvalues are $\lambda_1 = 1$ and $\lambda_2 = 3$, since as can be readily found out these values are the roots of the quadratic equation of concern, $\lambda^2 - 4\lambda + 3 = 0$.

Second, to find the eigenvectors, for each of the eigenvalues λ_1 and λ_2 we need to solve the equations $\mathrm{A}\,\underline{x} = \lambda_i\,\underline{x}$ (with $i = 1, 2$). The first equation, $\mathrm{A}\,\underline{x} = 1\,\underline{x}$, is actually a system of two equations:

$$\mathrm{A}\,\underline{x} = 1\underline{x} = \begin{bmatrix} 1 & 0 \\ 1 & 3 \end{bmatrix} \begin{bmatrix} x_1 \\ x_2 \end{bmatrix} = \begin{bmatrix} x_1 \\ x_2 \end{bmatrix},$$

that is (see Chapter 2),

$$(1)\ x_1 + (0)\ x_2 = (1)\ x_1,$$
$$(1)\ x_1 + (3)\ x_2 = (1)\ x_2,$$

where multiplying numbers—the respective elements of the matrix A—are placed in parentheses (to emphasize the multiplication operation), or simply

$$x_1 = x_1,$$
$$x_1 + 3x_2 = x_2.$$

A solution of the last system of equations, among infinitely many pairs of values for x_1 and x_2 that satisfy them, is the couple $x_1 = -2$ and $x_2 = 1$ say. That is, the following vector

$$\underline{e}_1 = \begin{bmatrix} -2 \\ 1 \end{bmatrix}$$

is an eigenvector of A pertaining to the eigenvalue $\lambda_1 = 1$. Similarly, it can be determined that

$$\underline{e}_2 = \begin{bmatrix} 0 \\ 1 \end{bmatrix}$$

is an eigenvector of A pertaining to the eigenvalue $\lambda_2 = 3$.

We observe that the found eigenvectors are not unique, since the system of equations we had to solve in order to obtain them does not have a unique solution. This fact is not unexpected, as can be seen from the definition of eigenvector. Indeed, if $\underline{x}$ is an eigenvector for a given matrix, so is also the vector $(c\ \underline{x})$ for any constant c (with $c \neq 0$). While the components of the eigenvector are not unique due to this feature, their ratios—that is, their relative values—are, however, uniquely determined (see, e.g., the last considered system of a pair of equations from which the eigenvector e_1 was obtained). This means that the direction of the eigenvector is unique in the multidimensional space of concern, but not its length. Given this fact, a generally convenient feature is to choose eigenvectors whose length is 1, which are referred to as normalized eigenvectors (see Chapter 2 for a discussion of the concept of vector length). That is, after an eigenvector has been found, we can divide its components by its length in order to furnish the normalized eigenvector. This normalization implies that the product of the so-determined eigenvectors satisfies the equation $\underline{e}_1'\underline{e}_1 = 1$, sometimes also called unit-norm condition. We also note in passing here that for the matrices of interest in the rest of this chapter, two normalized eigenvectors, say $\underline{e}_1$ and $\underline{e}_2$, associated with two distinct eigenvalues will be also orthogonal (unrelated) to one another, that is, in addition to the normalization condition they will also fulfill the following one: $\underset{\sim}{e}_1'\underset{\sim}{e}_2 = 0$. (We will return to this issue in a later section of the chapter.)

Hence, for the earlier considered matrix A, the normalized eigenvector corresponding to its eigenvalue $\lambda_1 = 1$ is $\underset{\sim}{e}_1^* = [-2/\sqrt{5}, 1/\sqrt{5}]' = [-0.89, 0.44]'$, since the squared length of its pertinent eigenvector $\underset{\sim}{e}_1$ is $(-2)^2 + 1^2 = 5$. Similarly, the normalized eigenvector corresponding to the second eigenvalue of A, $\lambda_2 = 3$, is readily found to be $\underset{\sim}{e}_2^* = [0, 1]'$, because the squared length of its eigenvector $\underset{\sim}{e}_2$ is $0^2 + 1^2 = 1$.

This example was relatively easy to handle by hand due to the manageable size of the involved matrix A. When a matrix under consideration is of larger size, however, statistical software is better used to conduct the eigenanalysis. We demonstrate this activity next. To this end, consider the following covariance matrix S obtained in a study of $n = 144$ sophomores on $p = 5$ personality measures (only its main diagonal elements and those below it are given next, due to symmetry):

$$S = \begin{bmatrix} 44.23 & & & & \\ 22.42 & 55.13 & & & \\ 31.23 & 22.55 & 61.21 & & \\ 32.54 & 25.56 & 29.42 & 57.42 & \\ 19.22 & 23.15 & 27.35 & 33.34 & 72.36 \end{bmatrix}.$$

To find out its eigenvalues and eigenvectors, that is, to carry out eigenanalysis of S, with SPSS we can use the following command file (obtained in a newly opened syntax window, after clicking on "New" in the toolbar and choosing the "Syntax" option therein).

```
MATRIX.
COMPUTE S = {44.23, 22.42, 31.23, 32.54, 19.22;
             22.42, 55.13, 22.55, 25.56, 23.15;
             31.23, 22.55, 61.21, 29.42, 27.35;
             32.54, 25.56, 29.42, 57.42, 33.34;
             19.22, 23.15, 27.35, 33.34, 72.36}.
CALL EIGEN(S, A, B).
PRINT A.
PRINT B.
END MATRIX.
```

In this command sequence, we first start the matrix operations module of the software, using the command MATRIX, and then define in the next five lines the matrix to be eigenanalyzed, S. (Note that we provide to SPSS the full matrix, rather than only its nonduplicated elements as given in its first numerical definition above.) To invoke then eigenanalysis of S, we use the command CALL EIGEN, giving in parentheses the storage matrices for the output of these activities—in the columns of A, the eigenvectors of S will be provided, and in those of B its eigenvalues. This input file produces the following output, with descending eigenvalues presented in B, and their pertinent eigenvectors correspondingly positioned from left to right in A (the headings of the columns of A are added by the authors).

```
MATRIX
Run MATRIX procedure:
A
       e1            e2            e3            e4            e5
 -.3977961896  -.3798563266   .1107024739  -.2688749789  -.7828911614
 -.3888835095  -.1821674251  -.8489872117   .3015193573   .0623813810
 -.4657607067  -.3499436354   .5100831291   .5652534735   .2844470570
 -.4863576069  -.0449173210   .0197923851  -.7049942258   .5138386084
 -.4868788862   .8354915240   .0799320931   .1423515897  -.1955754302

B
 166.0219255
  45.3261932
  35.8827780
  28.5780586
  14.5410447

—— END MATRIX ——
```

That is, the eigenvalues of S, in decreasing order, are: $\lambda_1 = 166.02$, $\lambda_2 = 45.33, \ldots, \lambda_5 = 14.54$; and the eigenvectors pertaining to them are the successive columns of A (from left to right).

In order to accomplish the same goal with SAS, the following program file utilizing PROC IML can be employed (for additional details, see Chapter 2).

```
PROC IML;
S =  {44.23 22.42 31.23 32.54 19.22,
      22.42 55.13 22.55 25.56 23.15,
      31.23 22.55 61.21 29.42 27.35,
      32.54 25.56 29.42 57.42 33.34,
      19.22 23.15 27.35 33.34 72.36};
CALL EIGEN(lambda, e, S);
PRINT lambda;
PRINT e;
QUIT;
```

We note that with SAS we use the command line CALL EIGEN to generate the eigenvalues and eigenvectors of the matrix S—giving in parenthesis the storage matrices for the output of these activities—whereby in "lambda" its eigenvalues will be presented and in "e" its corresponding eigenvectors. The two PRINT statements provide the following output that is essentially identical to the previously considered SPSS output (with descending eigenvalues given in the vector called lambda and their respective eigenvectors in the following matrix E):

```
                        The SAS System

                           LAMBDA
                          166.02193
                          45.326193
                          35.882778
                          28.578059
                          14.541045

                              E
0.3977962  0.3798563  -0.110702  -0.268875  0.7828912
0.3888835  0.1821674  0.8489872  0.3015194  -0.062381
0.4657607  0.3499436  -0.510083  0.5652535  -0.284447
0.4863576  0.0449173  -0.019792  -0.704994  -0.513839
0.4868789  -0.835492  -0.079932  0.1423516  0.1955754
```

In the remainder of this book, when it comes to eigenanalysis, we will be mainly interested in eigenvalues and eigenvectors of symmetric matrices—in particular, of correlation or covariance matrices—which consist of real numbers. They have the following important property that we will use repeatedly in the sequel.

Proposition 2: All eigenvalues of a symmetric matrix A are real numbers, and their pertinent eigenvectors can be chosen so as to contain only real elements. In addition, the eigenvectors corresponding to different eigenvalues are orthogonal (unrelated) to one another, and eigenvectors pertaining to equal eigenvalues can be chosen so as to be orthogonal to one another.

This result allows us to introduce now the formal basis of PCA as a data reduction technique. This is the following decomposition that is valid for any symmetric matrix A (of size $p \times p$; $p > 1$). Due to its special relevance, the representation in the next Equation 7.1 is called the spectral decomposition of A.

Proposition 3 (Spectral Decomposition): If $\lambda_1, \ldots, \lambda_p$ and $\underline{e}_1, \ldots, \underline{e}_p$ are correspondingly the eigenvalues ($\lambda_1 \geq \lambda_2 \geq \ldots \geq \lambda_p$) and associated eigenvectors of the symmetric matrix A, then this matrix can be represented as the following sum:

$$A = \lambda_1 \underline{e}_1 \underline{e}'_1 + \lambda_2 \underline{e}_2 \underline{e}'_2 + \ldots + \lambda_p \underline{e}_p \underline{e}'_p. \tag{7.1}$$

The spectral decomposition of the matrix A is a very important representation, in particular in the context of PCA, because oftentimes when measuring multiple interrelated variables several of the smallest eigenvalues are likely to be close to 0. Under such circumstances, one can choose to truncate the decomposition in Equation 7.1 without loosing much of the original data interrelationships. This truncation is accomplished by disregarding those terms in the right-hand side of Equation 7.1, which correspond to the smallest eigenvalues, and is the essence of PCA. We will also use this decomposition later in the book for other related purposes (e.g., Chapter 8).

The concept of eigenvalue has also an important analytic meaning that was hinted to in Chapter 4. In particular, it can be shown that the determinant of a given matrix is the product of its eigenvalues, i.e., $|A| = \lambda_1\ \lambda_2 \ldots \lambda_p$, in the above notation; in addition, its trace is the sum of these eigenvalues: $\text{tr}(A) = \lambda_1 + \lambda_2 + \ldots + \lambda_p$. As we know from Chapter 2, determinant and trace can be viewed as extensions of the concept of univariate variance to the multivariate case; thus, if we now take A to be the sample covariance matrix for a given set of observed variables, we see from the last two equations that the notion of eigenvalue is inextricably related to that of generalized variance.

In addition to this analytic meaning, the notions of eigenvalue and eigenvector have distinctive geometric meaning as well. In particular, if Σ (or S) is the covariance matrix for a given set of variables following a multinormal distribution in a studied population (or sample), then the eigenvectors of Σ (or S) give the directions of variability of the associated data points in the multivariate space, while the eigenvalues give the extent of variability along these axes. That is, for a given sample, the eigenvalues of the covariance matrix S describe the shape of the region in which the collected data fall, while its eigenvectors show the directions of the axes along which variability occurs (Johnson & Wichern, 2002). In particular, eigenvalues that are similar in magnitude indicate comparable variability in several dimensions (determined by the associated eigenvectors). Conversely, eigenvalues that are quite dissimilar in magnitude indicate very different variability along these dimensions. We will return to these issues later in the chapter.

7.3 How Does Principal Component Analysis Proceed?

At this point, it is helpful to recall that the goal of a PCA is to explain the variability in a set of observed measures through as few as possible linear combinations of them, which combinations are the principal components. For example, suppose the following $n \times p$ matrix

$$X = \begin{bmatrix} x_{11} & x_{12} & . & . & x_{1p} \\ x_{21} & x_{22} & . & . & x_{2p} \\ . & . & . & . & . \\ . & . & . & . & . \\ x_{n1} & x_{n2} & . & . & x_{np} \end{bmatrix}$$

represents the data obtained from a sample of n persons on a set of p variables ($p > 1$). Then the end-product of a PCA is that the original $n \times p$ data matrix X can be viewed as having been reduced to a data set consisting of n measurements on k principal components (with $k < p$), which exhibits nearly the same variability as the original data set (details follow below). Specifically, these new k derived variables demonstrate together variability that is nearly equal to the total variability of the initial p observed variables (the columns of X).

How is this goal achieved? Assume we observed the above $n \times p$ data matrix X from a population with covariance (or correlation) matrix Σ for the p variables involved. Let us think, for ease of reference, of the variables in this data set as elements of the random vector $\underline{\mathrm{x}} = [x_1, x_2, \ldots, x_p]$. A PCA will attempt to construct the following linear combinations:

$$\begin{aligned} y_1 &= \underline{\mathrm{a}}_1'\underline{\mathrm{x}} = \mathrm{a}_{11}\mathrm{x}_1 + \mathrm{a}_{12}\mathrm{x}_2 + \ldots + \mathrm{a}_{1p}\mathrm{x}_p \\ y_2 &= \underline{\mathrm{a}}_2'\underline{\mathrm{x}} = \mathrm{a}_{21}\mathrm{x}_1 + \mathrm{a}_{22}\mathrm{x}_2 + \ldots + \mathrm{a}_{2p}\mathrm{x}_p \\ &\ldots \\ y_p &= \underline{\mathrm{a}}_p'\underline{\mathrm{x}} = \mathrm{a}_{p1}\mathrm{x}_1 + \mathrm{a}_{p2}\mathrm{x}_2 + \ldots + \mathrm{a}_{pp}\mathrm{x}_p, \end{aligned} \tag{7.2}$$

where $\underline{\mathrm{a}}_1, \ldots, \underline{\mathrm{a}}_p$ are at this time unknown, $p \times 1$ vectors containing the corresponding linear combination weights. We want the first component y_1 to explain as much as possible variability, that is, to possess the highest possible variance that could be obtained by linearly combining the initial variables in $\underline{\mathrm{x}}$. However, we immediately see a problem here. If for a given vector of weights $\underline{\mathrm{a}}_1$ we multiply all its components with one and the same number c say (with $c \neq 0$), then the resulting linear combination $(c\underline{\mathrm{a}}_1)'\ \underline{\mathrm{x}}$ will have c^2 times larger variance than this linear combination before multiplication, $\underline{\mathrm{a}}_1'\underline{\mathrm{x}}$. Hence, we need to impose a constraint on the magnitude of the elements of $\underline{\mathrm{a}}_1$ in order for this maximization task to be meaningful; this is because otherwise we will not be able to find a vector $\underline{\mathrm{a}}_1$ such that $\underline{\mathrm{a}}_1'\underline{\mathrm{x}}$ possesses the highest possible variance across all choices of $\underline{\mathrm{a}}_1$, as that number c could be chosen larger than any number we could think of. Although many constraints are meaningful, the choice typically made for convenience reasons is to assume that the length of $\underline{\mathrm{a}}_1$ is 1, that is, $\|\underline{\mathrm{a}}_1\| = 1$. Now the above maximization task has a solution, and once this optimal vector of weights $\underline{\mathrm{a}}_1$ is found, the linear combination $\underline{\mathrm{a}}_1'\underline{\mathrm{x}}$ is called the first principal component (of the analyzed data set).

As a next step, we want to find a second vector $\underline{a}_2$, such that the second component $y_2 = \underline{a}_2'\underline{x}$ has the following two properties: (a) it explains as much as possible of the remaining variability in the data (assuming again $\|\underline{a}_2\| = 1$, to make this maximization task meaningful); and (b) it is positioned in a direction unrelated to the first principal component already furnished, i.e., y_2 is uncorrelated with y_1. We wish y_2 to be uncorrelated to y_1 since we do not want the variability in y_2 to be related to that in y_1, as we desire to end up with the smallest number of linear combinations that explain as large as possible portion of the initial variables' variances. Once we have found this set of weights $\underline{a}_2$, the linear combination $\underline{a}_2'\underline{x}$ is called the second principal component. We can continue in this manner until we find all p vectors $\underline{a}_1, \ldots, \underline{a}_p$ that appear in the right-hand sides of Equation 7.2. We emphasize that successive principal components are unrelated to one another by construction, and that formally there are as many principal components as there are initial studied variables.

After we find the vectors $\underline{a}_1, \ldots, \underline{a}_p$, we will be able to perfectly explain all variance in the data. In that case, we will not really accomplish any reduction of data complexity, but only redistribute the variance in the original variables in the vector $\underline{x}$ into variance of the derived measures in the vector $\underline{y}$. However, oftentimes in empirical research we do not need, for most practical purposes, to go beyond the first few linear combinations in Equation 7.2 if we want to approximate with their total variance that of the variables in the initial sample. In those cases, it is possible to explain a sufficiently large amount of variance in the original variables with only a few principal components. At that point, the (practical) goal of PCA has been accomplished.

Now, how does all this relate to the notions of eigenvalue and eigenvector on which we have spent a considerable amount of discussion earlier in this chapter? It turns out that the linear combination of the original variables that furnishes the first principal component has weights that are exactly the elements of the (normalized) eigenvector pertaining to the largest eigenvalue, λ_1, of Σ, the analyzed population covariance (or correlation) matrix of the observed variables. Similarly, the linear combination of the original variables that yields the second principal component has weights that are the elements of the eigenvector corresponding to the second largest eigenvalue, λ_2, of Σ, and so on; the kth principal component results as a linear combination of the initial variables with weights being the elements of the eigenvector associated with the kth eigenvalue, λ_k, of Σ $(1 \leq k \leq p)$. Thus, all one needs to do in order to obtain the principal components is find out the eigenvalues and (normalized) eigenvectors of the analyzed matrix of variable interrelationship indices. We should keep in mind also that in the context of interest to this chapter, these eigenvectors are all orthogonal to one another (see Proposition 2).

In an application, we typically do not know the population covariance (or correlation) matrix Σ (or P). In order to proceed with PCA, therefore, we need to estimate it from a given sample. As is typical, we estimate this population matrix with the empirical covariance (or correlation) matrix S (or R). Once having done so, we carry out the PCA as just outlined on S (or R).

This procedure, if followed all the way, yields as mentioned all p principal components for a given set of observed variables. In their totality, these p components account for the variance in the entire data set, and hence do not furnish any data reduction. In order to accomplish the latter, we need a rule that would allow us to make a well-informed choice as to when to stop extracting principal components, after we do this for the first few of them. Such a stopping rule can be obtained after noting the following equality:

$$\text{Var}(y_i) = \lambda_i,$$

where Var(.) denotes variance and λ_i is the ith eigenvalue of the analyzed matrix S (or R; $i=1,\ldots, p$). That is, the ith eigenvalue represents the variance of the ith principal component ($i=1,\ldots, p$). Further, one can also show that:

$$\begin{aligned}\text{Var}(y_1) + \text{Var}(y_2) + \ldots + \text{Var}(y_p) &= \text{Var}(x_1) + \text{Var}(x_2) + \ldots + \text{Var}(x_p)\\ &= \lambda_1 + \lambda_2 + \ldots + \lambda_p.\end{aligned}$$

The last equations state that the sum of the variances of all principal components is equal to the sum of the variances of all observed variables, which in turn is equal to the sum of all eigenvalues. That is, the original variance in the data set, being the sum of all eigenvalues, is "redistributed" among the principal components. Hence, the ratio

$$r_1 = \frac{\lambda_1}{\lambda_1 + \lambda_2 + \ldots + \lambda_p} \tag{7.3}$$

can be used to represent the proportion of variance explained by the first principal component. More generally, one can consider the ratio

$$r_k = \frac{\lambda_1 + \lambda_2 + \ldots + \lambda_k}{\lambda_1 + \lambda_2 + \ldots + \lambda_p} \tag{7.4}$$

as the proportion of variance explained by the first $k\,(k<p)$ principal components. Equation 7.4 helps one to try answering the question, Until when should the search for principal components continue? We address this issue further in the next section.

7.4 Illustrations of Principal Component Analysis

There are two formal answers to the last question about stopping rule in PCA, which depend on what ''data'' are being used for PCA. We discuss them in turn.

7.4.1 Analysis of the Covariance Matrix Σ (S) of the Original Variables

When the covariance matrix of a given set of observed variables is being analyzed, and a desirable proportion of explained original variance is settled on before the data are looked at, then one keeps extracting principal components until the value of the ratio in Equation 7.4 exceeds this specified proportion for the first time. This recommendation is referred to as Rule 1 in the rest of the chapter. For example, one may elect to specify that the ratio be equal or exceed a value of 0.80 (i.e., at least 80% of the variance should be explained before stopping to extract components).

In case this desirable proportion cannot be specified before looking at the data, one can keep extracting principal components until one is found whose eigenvalue is smaller than the average eigenvalue of the analyzed covariance matrix. (This average is readily obtained by consulting the software output containing the eigenvalues.) Then all components extracted before that point can be seen as ones to keep subsequently. We refer to this recommendation as Rule 2.

7.4.2 Analysis of the Correlation Matrix P (R) of the Original Variables

When the correlation matrix of the initial variables is analyzed, Rule 1 remains unchanged. Rule 2 is in this case modified to the following one, however: Keep extracting principal components as long as their eigenvalues are larger than 1. We note that the latter rule is in fact identical to the earlier mentioned Rule 2 in the present case, since the variances of all variables are 1 here (because analyzed is the correlation matrix) and hence their average is 1 as well. We note that this version of Rule 2 was proposed by Henry Kaiser and is often referred to as Kaiser's eigenvalue criterion (Kaiser, 1960).

In addition to these two rules, whether one analyzes the covariance or correlation matrix, one can also inspect the so-called scree plot. This method was popularized by Cattell (1966) and is usually referred to as Cattell's scree plot. The latter represents pictorially the magnitude of the eigenvalues in successive order. The term scree is used in the geographical sense to refer to the rubble at the foot of a steep mountain slope. Thereupon, when examining a scree plot, one looks for when an ''elbow'' shape (i.e., a slightly tilted ''flat'' pattern) begins to emerge—this is considered the point from which ''all else is mere rubble or scree at the foot of the slope." According to the scree plot rule, one can extract then one fewer principal components.

To demonstrate these procedures, let us consider the following empirical example. In this study, a sample of one hundred and sixty one sophomores were given five intelligence tests. These were measures that evaluate inductive reasoning ability (for letters and for numbers) and figural relations ability, as well as Raven's advanced matrices test, and a culture-free test. The question that will be addressed by this analysis is, "How many principal components can one extract from the data, that is, what data reduction could be accomplished in this set?" (The data are available in the file named "ch7ex1.dat" available from www.psypress.com/applied-multivariate-analysis, with corresponding variable names; see below.)

To conduct a PCA and accomplish this goal using SPSS, the following series of menu options can be used. (Note that formally we use the SPSS procedure Factor; this fact is not really important at the moment, and we will return to it in a later section as well as in Chapter 8.)

Analyze → Data reduction → Factor (Descriptives: Bartlett's test of sphericity; Extraction: Scree plot; Options: Sorted by size).

To accomplish the same activity in SAS, one of two options can be employed—either PROC PRINCOMP or PROC FACTOR. Whereas PROC PRINCOMP was specifically developed to conduct principal component analysis, PROC FACTOR can be used to conduct either PCA or factor analysis. A somewhat distracting feature that appears when using PROC FACTOR to conduct PCA is that the principal components extracted are labeled in the output as "factors"; that is, in the output sections "Component 1" will be referred to a "Factor 1," "Component 2" as "Factor 2," and so forth. (As an aside, this might explain some of the confusion often encountered in the literature with respect to the similarities and differences between the two procedures; see discussion in the next chapter.) When utilized for conducting PCA, both SAS procedures of course will provide identical results. We use PROC FACTOR in this chapter to maintain formal comparability with the SPSS analysis and in particular the next chapter, and due to the fact that it can be readily used to obtain a scree plot. (We note in passing that the scree plot, as we will see in the next chapter, can be straightforwardly utilized in a factor analysis session to help determine how many factors may be extracted.) The following program file can be submitted to SAS for performing a PCA with the FACTOR procedure.

```
DATA PCA;
INFILE 'ch7ex1.dat';
INPUT IRLETT FIGREL IRSYMB CULTFR RAVEN;
PROC FACTOR SCREE;
RUN;
```

The outputs produced by SPSS and SAS when carrying out PCA on the correlation matrix of the analyzed variables are as follows (provided in segments to simplify the discussion).

SPSS output

Factor Analysis

KMO and Bartlett's Test

Kaiser-Meyer-Olkin Measure of Sampling Adequacy.		.859
Bartlett's Test of Sphericity	Approx. Chi-Square	614.146
	df	10
	Sig.	.000

The first piece of information that is recommended to examine when conducting a PCA is Bartlett's test of sphericity. This test evaluates (based on the assumption of normality) the evidence in the analyzed data set with regard to the hypothesis that the population correlation matrix is diagonal. That is, the null hypothesis tested here is that the population correlation matrix is an identity matrix. According to this hypothesis, the analyzed variables are unrelated to one another (which with normality implies that they are independent). In that case, there is no point in carrying out a PCA since the observed variables cannot be really transformed into linear combinations in a lower-dimensional space. This further implies that due to their independence, the same number of principal components as analyzed variables would be needed to explain the variability in the data; hence, no data reduction could be achieved, and, therefore, an application of PCA is pointless. When a PCA is to be undertaken, therefore, Battlett's test needs to be significant. If its null hypothesis is not rejected, there is no point in carrying out PCA since there cannot be any redundancy in the analyzed data, yet redundancy is what is essential for PCA as an analytic procedure. Hence, Bartlett's test can be seen as a means of ascertaining whether it is meaningful to proceed with PCA at all. We stress that, as can be deduced from the preceding discussion in this chapter, PCA can only then be successful when there are marked interrelationships among the original variables—in which case data reduction makes sense. We note in passing that the Kaiser-Meyer-Olkin measure of sampling adequacy also looks at the degree of overlap among the variables, but because it does not come with a statistical test and is usually evaluated somewhat subjectively (in particular, it has been proposed that it should ideally be larger than .6), it is not commonly used or reported.

SAS output

In order to generate Bartlett's test of sphericity with SAS, the following statement must be added to the PROC FACTOR command line: METHOD = ML. This statement invokes the maximum likelihood estimation method (under the assumption of normality), in order to provide Bartlett's test statistic. We note, however, that this statement is normally used in the context of conducting a factor analysis, but is presented here in order to keep comparability with the output obtained with the above SPSS session for PCA. We also note that SAS does not provide the Kaiser-Meyer-Olkin measure of sampling adequacy as part of its PROC FACTOR output.

```
                    The FACTOR Procedure

         Significance Tests Based on 161 Observations

              Test                   DF   Chi-Square   Pr > ChiSq

HO: No common factors                10    614.1460      <.0001
HA: At least one common factor
```

SPSS output

Total Variance Explained

Component	Initial Eigenvalues			Extraction Sums of Squared Loadings		
	Total	% of Variance	Cumulative %	Total	% of Variance	Cumulative %
1	3.805	76.096	76.096	3.805	76.096	76.096
2	.429	8.576	84.673			
3	.398	7.962	92.634			
4	.236	4.726	97.361			
5	.132	2.639	100.000			

Extraction Method: Principal Component Analysis.

Communalities

	Initial	Extraction
v1	1.000	.765
v2	1.000	.746
v3	1.000	.854
v4	1.000	.781
v5	1.000	.659

Extraction Method: Principal Component Analysis.

SAS output

```
                 The FACTOR Procedure
      Initial Factor Method: Principal Components

 Eigenvalues of the Correlation Matrix: Total = 5 Average = 1

         Eigenvalue    Difference    Proportion    Cumulative

  1      3.80481831    3.37599426      0.7610        0.7610
  2      0.42882404    0.03074308      0.0858        0.8467
  3      0.39808097    0.16177602      0.0796        0.9263
  4      0.23630495    0.10433322      0.0473        0.9736
  5      0.13197173                    0.0264        1.0000

    1 factor will be retained by the MINEIGEN criterion.

            Variance Explained by Each Factor

                      Factor1
                      3.8048183

     Final Communality Estimates: Total = 3.804818

  IRLETT        FIGREL        IRSYMB        CULTFR        RAVEN
0.76459241    0.74575524    0.85401043    0.78127123    0.65918901
```

In the above tables output by SPSS and SAS, we will not be concerned with the so-called communalities at this moment but will return to this topic in greater detail when we discuss factor analysis, since this is a notion that is characteristic of that analytic approach (see Chapter 8). We include the table of communalities for completeness only (so that no part of the output is omitted) and move on to the next output sections.

From the above tables, we also see that a single principal component (with an eigenvalue equal to 3.805) accounts for more than 76% of the variance in the data. In fact, there is only one eigenvalue found that is larger than 1, and the next largest eigenvalue is much smaller than 1. According to Kaiser's criterion, it may be suggested that it is worthwhile considering only a single linear combination of the five measures under consideration, the one being expressed by their first principal component.

This empirical example conforms with a wide-spread practice in empirical social and behavioral research to analyze the correlation matrix of a given data when conducting PCA. The reason is that frequently in these disciplines the units of measurement are meaningless and likely irrelevant, in addition to being quite different from one variable to another. In those cases, differences in scales across variables are essentially arbitrary, and analysis of the correlation matrix is an appropriate decision. However, in some settings it may be the case that units of measurement are indeed meaningful. Under such circumstances, preserving their differences may be desirable while carrying out PCA. Then it would be appropriate to analyze the covariance matrix of the initially recorded variables. We stress, however, that the results of these two analyses—PCA of the correlation matrix and PCA of the covariance matrix—need not be similar. In fact, in general there is no functional way to obtain any of them from the other: results from PCA on the covariance matrix differ in general from those of PCA on the correlation matrix, and there is no functional relationship between the two sets of results. Although it might seem that principal components of a correlation matrix could be furnished by a transformation of these components obtained when analyzing the covariance matrix, this is actually not the case (Jolliffe, 1972, 2002; Marcoulides & Hershberger, 1997). Indeed, because of the sensitivity of composites to units of measurement or scales of the variables, weighted composites resulting from correlation and covariance matrix analyses do not give equivalent information, nor can they be mathematically derived directly from each other.

Returning to the empirical example still under consideration, we turn next to the scree plot as one additional piece of information (of primarily visual nature) that could influence our decision with regard to the number of principal components to extract.

SPSS output

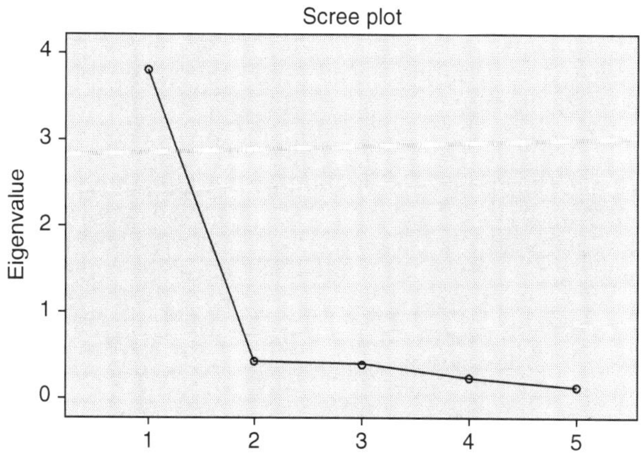

SAS output

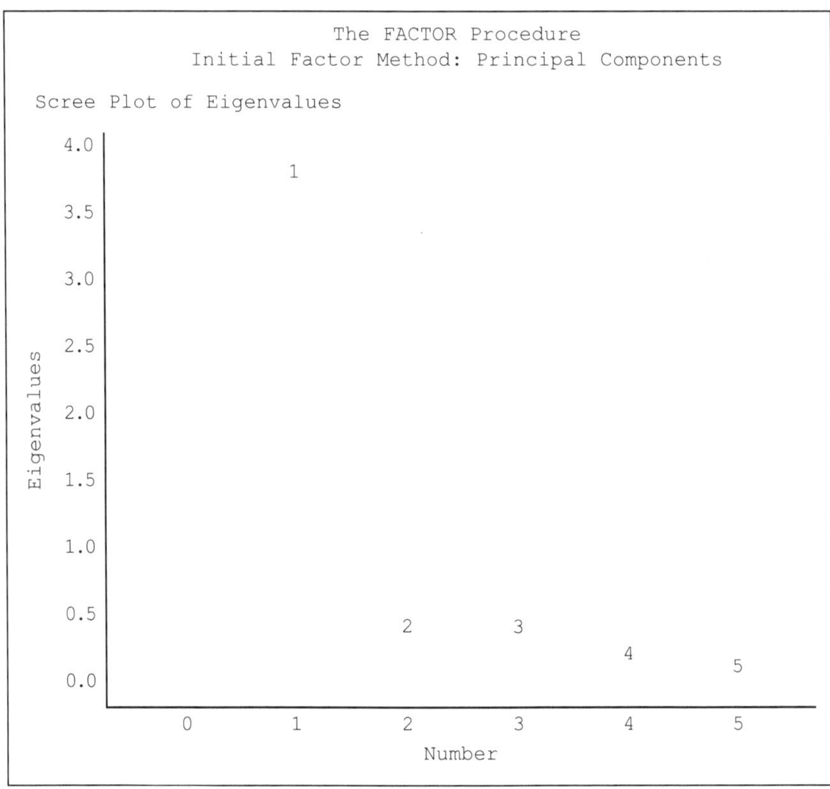

As expected, the above scree plots conform to the earlier suggestion to extract just 1 principal component, which explains satisfactorily the variance in the data. Indeed, as can be easily discerned, the "elbow" in the plot starts at 2, and subtracting 1 from 2 to follow the above scree plot rule, we do get 1.

SPSS output

Component Matrix[a]

	Component
	1
v3	.924
v4	.884
v1	.874
v2	.864
v5	.812

Extraction Method: Principal Component Analysis.
[a] 1 components extracted.

SAS output

```
                 The FACTOR Procedure
 Initial Factor Method: Principal Components

                    Factor Pattern

                                          Factor1
 IRLETT                                   0.87441
 FIGREL                                   0.86357
 IRSYMB                                   0.92413
 CULTFR                                   0.88390
 RAVEN                                    0.81190
```

Finally, the last tables give the weights of the first principal component (denoted by y_1 and labeled in SAS as Factor1), which is accordingly constructed as follows (after rounding off the corresponding coefficients):

$$y_1 = .87\,x_1 + .86\,x_2 + .92\,x_3 + .88\,x_4 + .81\,x_5.$$

As can be seen from its definition, this principal component evidently loads quite highly on each of the five tests administered (they were all assessed in the metric of percentage correct; this relationship can also be observed by taking its correlations with the observed variables). Therefore, this component could be loosely interpreted as reflecting general mental ability, which is the feature that these five measures have in common. This is not an unusual finding with many mental ability tests, where a single dominant principal component explains a high percentage of variance. In other subject-matter fields, such a finding may or may not be commonly the case, however.

We now turn to the situation in which more than a single principal component is meaningful to extract. In this example, four hundred and eighty seniors were tested on three mental ability measures and three educational motivation subscales. It is of interest here to find out whether the resulting variability in the 480 × 6 observed data matrix could be reduced to fewer, more fundamental components. (The data are available in the file "ch7ex2.dat" available from www.psypress.com/applied-multivariate-analysis, with corresponding variables names.) To this end, we proceed with conducting a PCA in the same manner as illustrated with the previous example, and obtain the following output. To save space, only the SPSS output is presented next, since the output created by submitting a similarly specified SAS program would lead to essentially the same results. To simplify matters further, we also ask for the suppression of any component weights that are found to be lower than .20—of course, other values for this suppression cutoff could also be selected. (We note again that we do not focus on variable communalities, a notion of relevance in factor analysis that we will be concerned with in the next chapter, but simply present their values next for completeness.)

SPSS output

Communalities

	Initial	Extraction
ability.1	1.000	.651
ability.2	1.000	.699
ability.3	1.000	.664
ed.motiv.1	1.000	.595
ed.motiv.2	1.000	.642
ed.motiv.3	1.000	.590

Extraction Method: Principal Component Analysis.

Total Variance Explained

	Initial Eigenvalues			Extraction Sums of Squared Loadings		
Component	Total	% of Variance	Cumulative %	Total	% of Variance	Cumulative %
1	2.036	33.929	33.929	2.036	33.929	33.929
2	1.805	30.083	64.013	1.805	30.083	64.013
3	.627	10.451	74.463			
4	.562	9.362	83.825			
5	.522	8.704	92.529			
6	.448	7.471	100.000			

Extraction Method: Principal Component Analysis.

As can be seen by examining the "Total Variance Explained" table, two components are suggested for extraction, as this is the number of eigenvalues that are larger than 1. Together these two components explain nearly two thirds of the total variance of the analyzed variables. The scree plot given next corroborates visually this suggestion. (Note that the "elbow" in the scree plot begins at 3, i.e., suggested is extraction of $3 - 1 = 2$ components.)

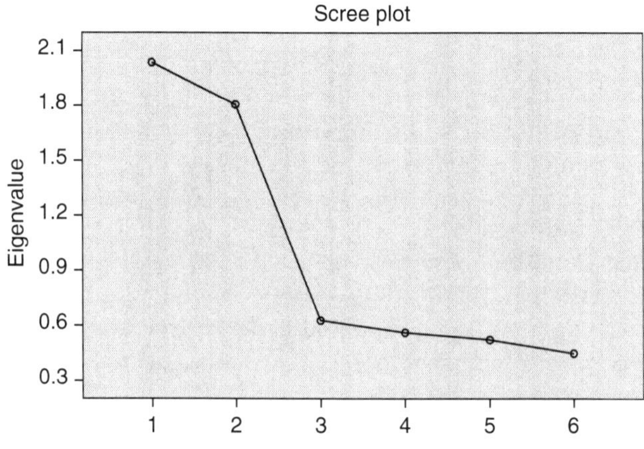

Component Matrix[a]

	Component	
	1	2
ability.2	.813	
ability.3	.787	.212
ability.1	.776	.221
ed.motiv.2	−.277	.752
ed.motiv.3		.750
ed.motiv.1	−.223	.738

Extraction Method: Principal Component Analysis.
a. 2 components extracted.

According to the last table, the first component loads strongly on the ability measures and could be interpreted as a composite representing Intelligence. Alternatively, the second component loads strongly on the educational motivation measures and could be viewed as a composite of Educational motivation. We note that because the weights for the variables "ed.motiv.3" and "ability.2" on the first and second component, respectively, are below our prespecified cutoff of .20, they are not provided in the output. From this table, fairly rough approximations of the two extracted components, in terms of original measures saliently contributing to them, are obtained as follows ("$\times$" denotes multiplication):

$$C1 = .81 \times \text{ability.2} + .79 \times \text{ability.3} + .78 \times \text{ability.1}$$
$$C2 = .75 \times \text{ed.motiv.2} + .75 \times \text{ed.motiv.3} + .74 \times \text{ed.motiv.1}$$

A more precise definition of the components is furnished by adding the correspondingly weighted sums of the remaining initial variables. (To this end, one would need to conduct the same analysis again but this time without suppressing the weights that are smaller than .2; from the results obtained then, one can compute for each subject his/her scores on the two components using a simple "Compute" statement in SPSS or SAS; see preceding example and Chapter 2.)

This analysis suggests that the overall variability in the original data set could be reduced, without much loss, to variability along two principal components. These components happen to have an easy and meaningful interpretation in this example, but this need not be the case in general. (To enhance interpretation, one may also examine the correlations between each component and the original observed variables; e.g., Johnson & Wichern, 2002.) Nonetheless, we stress that the more important goal of PCA is data reduction. We also note that in the last considered example, the two components accounted for most of the variability in the original data matrix on the initial six measures. The reason was that those six measures in fact consisted of two subsets of three highly interrelated measures each. The findings we obtained, therefore, were not unexpected, but this need not be often the case in the behavioral and social sciences. In fact, the example considered next is more representative of many studies in empirical research where the underlying data structure is likely to be less clear-cut. Still, PCA is a very useful way to look then at the data, in particular as an early step of a data analysis endeavor.

We emphasized previously in the chapter that principal components are uncorrelated with one another by construction. Behavioral and social phenomena, however, are well-known for being multifaceted, with multiple interrelated aspects. For this reason, in some multivariate contexts it is possible that due to complex patterns of interrelationships between the observed variables, it is simply not feasible to come up with a small number of principal components to explain satisfactorily the variability in the data. This difficulty is in part due to the fact that the components are defined as being uncorrelated to one another. In such empirical cases, a finding of overall percentage explained variance that is not very high should not be unexpected (although discomforting). In those settings, the technique of factor analysis considered in the next chapter—in particular with its feature of allowing for interrelated latent dimensions underlying observed behavior—may be more appropriate. Therefore, if one is primarily interested in interpretation of the components or underlying dimensions, PCA may not necessarily be the best approach to use. Then it would typically be worthwhile considering use of factor analysis instead, or perhaps in addition to PCA. We will return to a more detailed discussion on this matter in the next chapter.

7.5 Using Principal Component Analysis in Empirical Research

In this section, we discuss a number of ways in which PCA can be helpful when conducting behavioral and social research. We also consider some potentially problematic utilizations of PCA that sometimes appear in the literature. We emphasize that the following should not be considered an exhaustive list of practically relevant issues, but certainly represents a fair number of common applications and possible misuses.

7.5.1 Multicollinearity Detection

PCA can provide important information bearing upon the question of whether multicollinearity (MC) occurs among a given set of predictor or explanatory variables in a regression analysis or some other multivariate statistical method. To this end, one needs to look at the smallest eigenvalues of the analyzed matrix of interrelationships. If at least one of them is very close to 0, MC is likely to occur among the predictors. This is because then $\text{Var}(Y_p) = \lambda_p \approx 0$, for the smallest eigenvalue, and hence a nearly perfect linear combination exists among the explanatory variables ($p > 1$; the symbol "$\approx$" is used to designate "approximately equal"). (We also observe that if one or more eigenvalues are indeed equal to zero, then the analyzed matrix is singular—see Chapter 3.) In such a case, the elements of the pertinent eigenvector, denoted by $e_{p,1}, e_{p,2}, \ldots, e_{p,p}$, give the weights to be used in furnishing the nearly perfect linear combination in the data:

$$e_{p,1}x_1 + e_{p,2}x_2 + \ldots + e_{p,p}x_p \approx c,$$

where c denotes an appropriate constant. We note in passing that in order to obtain all eigenvectors and corresponding principal components, one may eigenanalyze the correlation or covariance matrix as carried out earlier in the chapter. (Alternatively, one can ask for factor analysis with the principal component method for extraction and request as many factors as there are analyzed variables; see next chapter.)

In the face of MC, one may consider dropping from subsequent analyses some of the variables appearing in the left-hand side of the last equation (which have a nonzero weight in it). As another option, one may use as predictors the principal components obtained in this PCA session, but it is stressed that subject-matter interpretation of these derived measures/predictors may be difficult.

We note in this context that if the number of subjects in some cells of a study design is smaller than the number of dependent variables, PCA may permit one to still carry out a contemplated MANOVA (or use another statistical technique that assumes more subjects than dependent variables, e.g., discriminant function analysis; see Chapter 9). In those cases, if dependent variables under consideration are markedly interrelated, one may use the first several principal components rather than the original variables, as formal outcome or response measures in a subsequent MANOVA or other analyses. (It should then be kept in mind that interpretation of any found mean differences may be complicated due to possible difficulties in interpretation of the principal components.)

7.5.2 PCA With Nearly Uncorrelated Variables Is Meaningless

As indicated in an earlier part of the chapter, PCA aims at "transforming" an original set of correlated variables, denoted by $x_1, x_2, \ldots, x_p$, into a new set of uncorrelated variables or components, denoted by $y_1, y_2, \ldots, y_k$ (with $k < p$), which account for nearly as much variability as there is in the original data. Hence, if the initial variables are already uncorrelated or nearly so, there is nothing to be really gained by carrying out PCA. This is because no data reduction is possible then, since there is no redundancy in the data to begin with, yet as mentioned earlier it is redundancy that is essential to be present in a set of analyzed variables in order for PCA to be meaningful. For this reason, it is strongly recommended that one checks regularly Bartlett's test of sphericity in the software output (on the assumption of normality), in order to find out whether PCA is worthwhile at all to proceed with.

7.5.3 Can PCA Be Used as a Method for Observed Variable Elimination?

Principal component analysis cannot be used to eliminate variables from subsequent analyses. Even though in an empirical setting one may find that the first or only few principal components account for much of the variance in the initially observed p variables, all p original variables are needed to score each individual on any of these components. Once this is done, a researcher can use the k say principal components as new variables in other analyses (e.g., as predictors in a regression analysis; $k < p$), which will then utilize these k rather than the original p variables. This does not mean, however, that any of the original variables have been eliminated from the initial data set, or that any of them is irrelevant for obtaining the principal components. In fact, as seen from the instrumental set of Equations 7.2 defining the principal components, in general all observed variables are needed in order to compute any of the components.

7.5.4 Which Matrix Should Be Analyzed?

As indicated in an earlier section, two matrices of variable interrelationships can be used as starting point of a PCA, the covariance or the correlation matrix. However, the results obtained by an analysis of each type of matrix are in general not similar and not even functionally related. Hence, the choice of which matrix to analyze has potentially very serious consequences with regard to end results and their interpretation, and thus must be made after careful considerations. If the original variables do not occur on an "equal footing" (i.e., have very different and largely arbitrary units of measurement), it may be more meaningful to apply PCA on the correlation matrix that is not dependent on these essentially arbitrary differences. When the units of measurement are interpretable, and if on substantive grounds it is decided that their differences need to be preserved rather than given up, then analysis of the covariance matrix may be recommended. In either case, careful consideration should also be given to the particular research question being examined.

7.5.5 PCA as a Helpful Aid in Assessing Multinormality

If the original data are normally distributed, then as mentioned in Chapter 3 any linear combination of them must be (univariate) normally distributed. Thus, each principal component—being by construction a linear combination of the initial variables—should then be normal as well. As illustrated in Chapter 3, there are a number of approaches that can be used to assess the univariate normality assumption, which are thus all available to evaluate this assumption for any of the principal components. Consequently, by examining the distribution(s) across the available sample of the individual scores on each component, one can readily perform a simple check of a necessary condition of multinormality. Similarly, scatterplots of pairs of principal components should exhibit an elliptically shaped region, which gives a convenient way of assessing another implication of multivariate normality, viz. bivariate normality. (This discussion is not to be interpreted as implying that one could give up use of the normality assessment approaches discussed in Chapter 3, in favor of normality evaluation based merely on principal components, but aims only to indicate yet another way of examining consequences of the multinormality assumption.)

7.5.6 PCA as "Orthogonal" Regression

It is well known that if one carries out a PCA on $p=2$ initial observed variables, the first principal component is an "orthogonal regression" line. The latter results when one chooses the associated intercept and slope in such a way that the resulting line minimizes the sum of squared orthogonal projections of each point onto it. This type of regression has actually a rather long history and differs from the conventional regression approach, where the sum of squared distances in the vertical direction is minimized (rather than in the orthogonal direction to the line). It is interesting to note that the orthogonal regression line also results as the outcome of a more general regression analysis with a single explanatory variable, where a main assumption of conventional regression is given up, viz. that of a perfectly measured predictor.

7.5.7 PCA Is Conducted via Factor Analysis Routines in Some Software

Some widely circulated statistical packages, including SPSS, do not have a separate program for conducting PCA but instead use that for the method called factor analysis that is the subject of the next chapter. (Some also blend the two into a single program, or share some of the analytical tools of each—e.g., SAS.) The reason is that there are a number of different procedures that can be used for the so-called initial factor extraction, one of which is PCA—a topic we return to in the next chapter. Furthermore, we note that default options in many statistical analysis programs carry out a PCA on the correlation matrix. If for some of the above reasons one decides to carry out a PCA on the covariance matrix, however, these defaults need to be overridden.

7.5.8 PCA as a Rotation of Original Coordinate Axes

The result of an application of PCA, as mentioned previously, is a set of principal components that are uncorrelated to one another. As such, these components span on their own a space whose dimensionality equals their number—viz. the number of components that a researcher decides to extract, or retain out of all p possible. For this reason, one can view the effect of carrying out a PCA as a rigid rotation of the original coordinate axes (possibly after placing the origin in the data centroid, i.e., at the mean, of all analyzed variables). In this way, the first principal component will represent the axis in the new coordinate system that spans the "longest" dimension of the data scatter. Further, the second principal component will represent the axis that is located orthogonally to the first component and in addition spans the longest dimension of data scatter in the space orthogonal to the first component, and so on with the remaining principal components if extracted. Consequently, as discussed previously, the principal component decomposition achieved by an eigenanalysis is unique, in that each successive component has maximum variance. In so doing, a principal component analysis in effect precludes the need for rotation of resulting principal components. Possible confusion among researchers about rotation in PCA is mainly due to some of the conceptual similarities between principal component analysis and factor analysis where rotation is often routinely conducted to ease interpretation (see Chapter 8).

7.5.9 PCA as a Data Exploratory Technique

In many empirical settings, it is highly recommendable that PCA be performed as a first exploratory step in a data analysis session. This activity can be conducted even before other, possibly earlier planned multivariate analyses are carried out. In this role, PCA will provide the researcher with a first "view" of the data and help them understand their correlation (covariance) structure. In particular, when used as a first step, PCA will help one to discover the "true" dimensionality of the data, and in this way is likely to contribute to the generation of substantively meaningful hypotheses that can be evaluated in subsequent analyses. This benefit from PCA will be maximized if it is carried out on a random subsample out of an initially available large sample that has been drawn from a studied population.

We wish to emphasize in conclusion that PCA is useful regardless of whether or not the principal components can be interpreted. The reason is that PCA helps one "discover" the actual dimensionality of the data, i.e., the dimensionality of the space in which the data practically lie—this is the number of eigenvalues (considerably) larger than 0. In addition, PCA can produce new measures that are fewer in number than the original ones analyzed, which exhibit in total nearly the same variability as the initial set of measures but are not redundant with regard to one another. Principal component analysis also gives a "new" meaning to the notions of eigenvalue and eigenvector for an analyzed matrix of interrelationships, viz. as the amounts of true variability practically underlying the data and the directions along which this variability occurs. With this in mind, if some of the eigenvalues of the analyzed matrix are close to 0, then the data are not really p-dimensional for most practical purposes, but in fact of lower dimension. (Note that this finding is indicative of redundancy in the data.) In this sense, the eigenvalues disclose the true dimensionality of the data (dimensionality of the variability), and the eigenvectors tell us the directions along which this true variability actually occurs.

8

Exploratory Factor Analysis

8.1 Introduction

The previous chapter introduced principal component analysis (PCA) as a method for data reduction. As we emphasized repeatedly throughout the chapter, interpretation of the obtained components, while often desirable, is not necessarily of primary relevance in most applications of PCA. In fact, when in this regard substantive meaningfulness of the results is of particular importance, a related approach may be more appropriate to use. This approach is called factor analysis (FA), and is the topic of the present chapter. Although as we will see, FA represents a statistical technique that shares a number of similarities with PCA, it is distinct from PCA in several consequential points.

FA was originally developed as a procedure for "disclosing" (discovering) unobserved traits or latent traits, typically referred to as factors, which presumably underlie subjects' performance on a given set of observed variables and explain their interrelationships. These factors are not directly measurable, but are instead latent or hidden random variables or constructs, with the observed measures being their indicators or manifestations in overt behavior. Statistically, a goal of FA is to explain the manifest (observed) variable interrelationships—that is, the pattern of manifest variable interrelations—with as few as possible factors. Thereby, the factors are expected to be substantively interpretable and to explain why certain sets (or subsets) of observed variables are highly correlated among themselves. Last but not least, if needed, one can also provide estimated subject scores for each of these factors, which are akin to the principal component scores and may similarly be used in subsequent statistical analyses.

More specifically, the aims of FA can be summarized as follows: (a) to determine if a smaller set of factors can explain the interrelationships among a number of original variables (a similar data reduction concern as in PCA); (b) to find out the number of these factors; (c) to interpret the factors in subject-matter terms; as well as possibly (d) to evaluate the studied persons on the factors, by providing estimates of their individual factor scores that could be used in subsequent analyses.

FA has a long history. In terms of systematic treatments, it is the English psychologist Charles Spearman (1904) who is generally credited with having developed it into a form that resembles closely the method used today in the behavioral and social sciences. The strongest impetus for the early development of FA came from psychology, in particular in connection with the study of human intelligence. Spearman (1904) originally introduced the FA idea in his seminal article "General intelligence: Objectively determined and measured," published in the *American Journal of Psychology*. He formulated notions concerning individual abilities as presumed manifestations of an underlying general ability factor that he referred to as general intelligence or simply g, and needed a methodology to tap into it—the FA method. Later applications used FA for modeling purposes in a variety of substantive areas across the behavioral, social, educational, and biomedical disciplines, as well as outside them.

8.2 Model of Factor Analysis

A characteristic feature of FA, especially when compared to PCA, is that FA is based on a specific statistical model. Broadly speaking, a statistical model typically consists of (a) a set of equations defining formally the assumed variable relationships, and (b) associated distributional assumptions concerning involved variables. To define the FA model, let us denote a vector of p observed variables by $\underline{x} = (x_1, x_2, \ldots, x_p)'$. For example, these variables may be the subtest scores in an achievement test battery, the items in a scale of self-esteem, or these in a depression inventory. For convenience, we assume that they all have zero mean (e.g., have been centered or standardized beforehand; we relax this assumption in Chapter 9). The FA model is then defined as follows:

$$\begin{aligned}
x_1 &= \lambda_{11}f_1 + \lambda_{12}f_2 + \cdots + \lambda_{1m}f_m + \varepsilon_1 \\
x_2 &= \lambda_{21}f_1 + \lambda_{22}f_2 + \cdots + \lambda_{2m}f_m + \varepsilon_2 \\
&\vdots \\
&\vdots \\
x_p &= \lambda_{p1}f_1 + \lambda_{p2}f_2 + \cdots + \lambda_{pm}f_m + \varepsilon_p,
\end{aligned} \tag{8.1}$$

where

$f_1, \ldots, f_m$ (typically with $m < p$) are the factors

λ_{ij} are called factor loadings (viz. of the ith observed measure on the jth factor)

$\varepsilon_1, \ldots, \varepsilon_p$ are the error terms (also occasionally called residual terms or alternatively uniqueness terms or uniqueness factors; $i = 1, \ldots, p$; $j = 1, \ldots, m$)

The error terms contain both random error of measurement and what is called variable specificity. Variable specificity comprises all sources of variance (other than pure measurement error) in the pertinent observed variable, which do not originate from the factors. Variable specificity is often also referred to as specificity factor. Its variance is the amount of variability in a given manifest variable, which is not shared with the other observed variables in the model and is unrelated to that of random measurement error for this variable.

To complete the definition of the FA model, in addition to its equations in Equation 8.1 we assume (a) $M(\varepsilon_i) = 0$ and $M(f_j) = 0$ (where $M(.)$ denotes mean or expectation); (b) the two vectors $\underline{\varepsilon}$ and $\underline{f}$ of error terms and factors are uncorrelated; and (c) the covariance matrix Ψ of the residuals is diagonal (i.e., no two error terms are correlated), with diagonal elements denoted $\psi_i = \text{Var}(\varepsilon_i)$ (with i, $j = 1, \ldots, p$ or m, respectively, and Var(.) standing for variance). When it comes to testing hypotheses within the framework of FA, we also assume normality for the random variables in the model. For the development of the estimation approach underlying FA, however, it is not necessary to assume any particular manifest variable distribution.

An important feature of the FA framework needs to be emphasized at this point. Looking at Equation 8.1, we see that there is no way that we could obtain directly the factors $\underline{f}$ from the observed variables $\underline{x}$, since we do not know the error terms $\underline{\varepsilon}$. Note also that, strictly speaking, the errors are latent variables as well, since they are not directly observed or measured. We, however, keep referring to them as errors (or error terms, residuals, or uniqueness factors), to separate them from the variables of actual interest—the "proper" latent variables collected in the vector $\underline{f}$. In this regard, the equations in Equation 8.1 and some reflection upon them reveal that there is no linear combination of the observed variables in $\underline{x}$ that can furnish any factor in $\underline{f}$. That is, the factors are not linear combinations of the observed variables, unlike principal components discussed in Chapter 7. This is a major distinction between FA and PCA, which stems from the fact that the concepts in which they are developed differ in their nature—unobserved factors (FA) versus observed principal components (PCA). This will become particularly relevant in the context of evaluating factor scores, a topic to be dealt with later in the chapter.

This main point of difference between PCA and FA is closely related to a second discrepancy between the two techniques. As seen from Equation 8.1, FA is a model-based method whereas PCA is not a model-based procedure. A third point of difference between the two emerges nearly immediately from this observation. Specifically, in PCA there is no explicit error term, while in FA there is—the last terms in the right-hand side of Equation 8.1. We hasten to add that the framework of PCA is not devoid of error in principle but it stems from the degree to which the approximation of the variability of $\underline{x}$ by the set of extracted (retained) components is imperfect. This error of approximation does not share the same nature as the model error in FA; however, with the latter being an essential element of the FA model, there is no such element in PCA, just as there is no model on which PCA is based.

We use Equations 8.1 in the rest of this book in their following compact form:

$$\underline{x} = \Lambda \underline{f} + \underline{\varepsilon}, \tag{8.2}$$

where $\underline{x}$ is the $p \times 1$ vector of observed or manifest variables, $\Lambda = [\lambda_{ij}]$ is the $p \times m$ matrix of factor loadings, $\underline{f}$ is the $m \times 1$ vector of factors, and $\underline{\varepsilon}$ is the $p \times 1$ vector of error terms assumed unrelated among themselves and with the factors in $\underline{f}$, as well as with zero mean. For example, in a single-factor setting with $p = 4$ variables (i.e., $m = 1$), the FA model is

$$\begin{bmatrix} x_1 \\ x_2 \\ x_3 \\ x_4 \end{bmatrix} = \begin{bmatrix} \lambda_{11} \\ \lambda_{21} \\ \lambda_{31} \\ \lambda_{41} \end{bmatrix} [f_1] + \begin{bmatrix} \varepsilon_1 \\ \varepsilon_2 \\ \varepsilon_3 \\ \varepsilon_4 \end{bmatrix},$$

with the above mentioned assumptions about the error term.

We note that in cases with more than a single factor, depending on what is assumed about the factors in $\underline{f}$, there are two types of FA models. In the first case, the factors are assumed to be uncorrelated and with variances equal to 1 (for reasons to be discussed shortly); this is the so-called orthogonal factor model. If on the other hand the factors are not assumed uncorrelated, but can instead be interrelated, one is dealing with the so-called oblique factor model. (These two cases are considered in greater detail later in this chapter.) We emphasize that this is yet another feature in which PCA and FA differ. Specifically, in PCA the components are unrelated to one another by construction (since they are defined as orthogonal to each other), while in FA the factors may be related.

At this point, it is useful to recall some well-known rules for working out variable variances including the case of linear combinations of random variables. For ease of presentation, in case of four random variables—designated X, Y, U, V—and constants denoted a, b, c, d (Raykov & Marcoulides, 2006), the following four simple rules will turn out to be quite helpful in the rest of this chapter (next "Cov" denotes covariance).

Rule 1: $\text{Cov}(X,X) = \text{Var}(X)$,

that is, the covariance of a variable with itself is that variable's variance.

Rule 2: $\text{Cov}(aX + bY, cU + dV) = ac\ \text{Cov}(X,U) + ad\ \text{Cov}(X,V) + bc\ \text{Cov}(Y,U) + bd\ \text{Cov}(Y,V)$.

Rule 2 resembles the algebraic procedure for disclosing brackets, and together with Rule 1 lets us obtain the next statement that we formulate also as a rule.

Rule 3: $\text{Var}(aX + bY) = a^2\,\text{Var}(X) + b^2\,\text{Var}(Y) + 2ab\,\text{Cov}(X,Y)$.

A compact representation of these three rules, which we use below, is the following one that due to its relevance in the sequel we would like to refer to as a rule in its own right.

Rule 4: For random vectors $\underline{z}$ and $\underline{w}$, and appropriately sized matrices A and B, $\text{Cov}(A\underline{z}, B\underline{w}) = A\,\text{Cov}(\underline{z}, \underline{w})\,B'$,

where $\text{Cov}(\underline{z}, \underline{w})$ denotes the matrix consisting of the covariances of all possible variable pairs comprising one element of $\underline{z}$ and one of $\underline{w}$.

As a special case of this rule when $A = B$, we obtain the following important relationship: The covariance matrix of a set of linear combinations of the elements (random variables) in a random vector $\underline{z}$, is as follows

$$\text{Cov}(A\underline{z}) = A\,\text{Cov}(\underline{z})A', \tag{8.3}$$

where $\text{Cov}(A\underline{z})$ and $\text{Cov}(\underline{z})$ denote the covariance matrices of that set and vector, respectively.

Using the above rules, it can be shown from Equations 8.1 that the orthogonal factor model implies the following set of equalities:

$$\begin{aligned} \text{Var}(x_i) &= \lambda_{i1}^2 + \lambda_{i1}^2 + \ldots + \lambda_{im}^2 + \psi_i, \\ \text{Cov}(x_i,x_k) &= \lambda_{i1}\lambda_{k1} + \lambda_{i2}\lambda_{k2} + \ldots + \lambda_{im}\lambda_{km}, \end{aligned} \tag{8.4}$$

and

$$\text{Cov}(x_i,f_j) = \lambda_{ij}(i,k = 1,\ldots,p; j = 1,\ldots,m).$$

For instance, in the earlier considered example of a single-factor model with $p = 4$ observed variables, under the above assumption of unitary factor variance,

$$\begin{aligned} \text{Cov}(x_1,x_2) &= \text{Cov}(\lambda_{11}f_1 + \varepsilon_1, \lambda_{21}f_1 + \varepsilon_2) \\ &= \lambda_{11}\lambda_{21}\text{Cov}(f_1,f_1) + \lambda_{11}\text{Cov}(f_1,\varepsilon_1) + \lambda_{21}\text{Cov}(f_1,\varepsilon_2) + \text{Cov}(\varepsilon_1,\varepsilon_2) \\ &= \lambda_{11}\lambda_{21} \end{aligned}$$

due to the previously indicated assumptions that error terms are uncorrelated among themselves and with the factor.

Equations 8.4 relate observed variable interrelationship indices on the left, with factor loadings on the right. In fact, when the population covariance matrix Σ of a given set of variables is considered, the first two of Equations 8.4 state that if the FA model is correct then each element of Σ is a nonlinear function of factor loadings and possibly a uniqueness variance. Hence, Equations 8.4 mean then that the covariance matrix of the original set of observed variables has been expressed in terms of more fundamental unknowns that desirably should be fewer in number than that of the original variables. In other words, and more generally speaking, the implication of the FA model in Equation 8.1 is that the matrix of interrelationship indices of a considered set of observed variables is parameterized in terms of fewer, more fundamental parameters. These are the unknowns of the (orthogonal) FA model, namely the factor loadings and error term variances. In addition, Equations 8.4 highlight the particular role and relevance of the factor loading parameters when a set of observed variables is submitted to a FA.

We also note from Equations 8.1 that the FA model is a linear model, because it is based on linear relationships between the observed variables and underlying factors. In fact, the FA model is very similar to the regression analysis model, in particular the general linear model (Timm, 2002). The important distinction is, however, that in FA the role of predictors is played by the factors that are not observed rather than by predictors as in the regression model that are observed. Thus, we may view the FA model as a general linear model with latent explanatory or predictor variables.

Returning now to Equations 8.4, one readily notices that they can also be rewritten in the following instructive way (for the ith observed variable):

$$\begin{aligned} \text{Var}(x_i) &= \lambda_{i1}^2 + \lambda_{i1}^2 + \ldots + \lambda_{im}^2 + \psi_i \\ &= (\lambda_{i1}^2 + \lambda_{i1}^2 + \ldots + \lambda_{im}^2) + \psi_i \\ &= h_i^2 + \psi_i \\ &= \text{communality} + \text{uniqueness variance} \end{aligned}$$

where $i = 1, \ldots, p$. Thereby the term h_i^2, referred to as communality, denotes the amount of variance in a given manifest measure x_i, which is explained in terms of the m factors $f_1, f_2, \ldots, f_m$; the remaining variance in x_i, that is, ψ_i, is called uniqueness variance and represents the variability in this measure that is unexplained by the factors ($i = 1, \ldots, p$). We stress that if $\text{Var}(x_i) = 1$, then its communality h_i^2 is in fact the R^2 index of the regression of the observed variable x_i upon the m factors ($i = 1, \ldots, p$). It is instructive to think conceptually of this regression, but it should be emphasized that it cannot be conducted in practice due to unavailable individual scores on the factors, as these are not observed.

This discussion highlights the unknown elements (parameters) of the orthogonal FA model, which we indicated earlier, namely the factor loadings (λ_{ij}) and the variances of the error terms (ψ_i) (i, $j = 1, \ldots, p$).

In total, there are therefore $pm + p$ parameters in the FA model, and we could place them for convenience in a vector of model parameters $\underline{\theta} = (\lambda_{11}, \ldots, \lambda_{pm}, \psi_1, \ldots, \psi_p)'$. The collection of the first two equations of Equation 8.4 across all observed variable pairs x_i and x_k, can now be compactly written as follows:

$$\Sigma = \Lambda\,\Lambda' + \Psi, \tag{8.5}$$

where $i, k = 1, \ldots, p$. Equation 8.5 states that when the orthogonal factor model is correct, then the $p(p+1)/2$ nonredundant elements of the covariance matrix Σ can be reproduced exactly by (a) the pm factor loadings in the matrix Λ, and (b) the p uniqueness variances in the matrix Ψ (its diagonal elements). Thus, if this model holds, in general a substantial reduction is accomplished since $p(p+1)/2$ is much larger than $pm + p$ in many empirical settings, especially when a large number p of observed variables are used in a study while m is comparatively small.

To illustrate this discussion, consider an example in which data from $p = 4$ observed variables were collected and yielded the following covariance matrix:

$$\Sigma = \begin{bmatrix} 19 & & & \\ 30 & 57 & & \\ 2 & 5 & 38 & \\ 12 & 23 & 47 & 68 \end{bmatrix}.$$

It can be readily shown, by direct multiplication and addition, that the following Λ and Ψ matrices fulfill Equation 8.5:

$$\Lambda = \begin{bmatrix} 4 & 1 \\ 7 & 2 \\ -1 & 6 \\ 1 & 8 \end{bmatrix} \text{ and } \Psi = \begin{bmatrix} 2 & & & \\ 0 & 4 & & \\ 0 & 0 & 1 & \\ 0 & 0 & 0 & 3 \end{bmatrix}.$$

In other words, the decomposition of the given variance–covariance matrix Σ can be written as follows:

$$\Sigma = \Lambda\,\Lambda' + \Psi$$

$$= \begin{bmatrix} 19 & 30 & 2 & 12 \\ 30 & 57 & 5 & 23 \\ 2 & 5 & 38 & 47 \\ 12 & 23 & 47 & 68 \end{bmatrix} = \begin{bmatrix} 4 & 1 \\ 7 & 2 \\ -1 & 6 \\ 1 & 8 \end{bmatrix} \begin{bmatrix} 4 & 7 & -1 & 1 \\ 1 & 2 & 6 & 8 \end{bmatrix} + \begin{bmatrix} 2 & 0 & 0 & 0 \\ 0 & 4 & 0 & 0 \\ 0 & 0 & 1 & 0 \\ 0 & 0 & 0 & 3 \end{bmatrix}.$$

The communality of say x_1 is thereby easily found to be $4^2 + 1^2 = 17$, and its uniqueness variance is determined as $19 - 17 = 2$, which is the first diagonal entry in the error covariance matrix Ψ. Hence, the orthogonal FA model in Equations 8.1 holds for the population with this covariance matrix of four variables under consideration.

When $p = m$, that is, one is dealing with as many factors as there are observed variables, any covariance matrix Σ can be exactly reproduced as $\Lambda\ \Lambda'$, which is a general property of positive definite matrices (like Σ is—see discussion in Chapters 2 and 7; this reproduction can be seen as following from the spectral decomposition of Σ then). However, in such cases the FA model is typically not very useful because it does not entail any data reduction. In contrast, the FA model is most useful when m is much smaller than p, in which case a substantially simpler explanation of the interrelationships among the observed variables $\underline{x}$ is accomplished in terms of the m factors. For example, if $p = 12$ and a FA model with $m = 2$ is appropriate, then the $p(p + 1)/2 = 12 \times 13/2 = 78$ elements of Σ are described in terms of only $pm + p = 36$ model parameters. Thereby, as in any other case when $m < p$, it is hoped that matrices Λ and Ψ can be found with the property that the discrepancy between Σ and $\Lambda\ \Lambda' + \Psi$ is "small" (we return to this issue in Section 8.3 and offer a more detailed discussion on it).

8.3 How Does Factor Analysis Proceed?

Conducting a FA for a given set of observed variables consists of two main steps. In the first step, the factors are extracted or "disclosed." This results in the so-called initial factor solution that, however, is often not easily interpretable. In the second step, in the search for a better and simpler means of interpretation, a factor rotation is carried out. We consider each of these steps in turn below.

8.3.1 Factor Extraction

In the factor extraction step, an initial attempt is made to "disclose" one or more latent variables that are able to explain the interrelationships among a given set of observed variables (or measures). Although to date there are several different algorithms that can be used to carry out factor extraction, the two that are most instructive for our purposes in this chapter are (a) the principal component method, and (b) maximum likelihood (ML).

8.3.1.1 Principal Component Method

Recall from Chapter 7 that the spectral decomposition of a covariance or correlation matrix, denoted Σ, is as follows:

$$\Sigma = v_1 \underline{e}_1 \underline{e}_1' + v_2 \underline{e}_2 \underline{e}_2' + \ldots + v_p \underline{e}_p \underline{e}_p', \tag{8.6}$$

where v_i and e_i are used to denote the eigenvalue and eigenvectors, respectively ($i = 1, \ldots, p$). Equation 8.6 can also be rewritten as follows (noting that $v_1\underline{e}_1\underline{e}_1' = [\sqrt{v_1}\underline{e}_1][\sqrt{v_1}\underline{e}']$, and similarly for the remaining eigenvalues):

$$\Sigma = \left[\sqrt{v_1}\underline{e}_1, \sqrt{v_2}\underline{e}_2, \ldots, \sqrt{v_p}\underline{e}_p\right] \begin{bmatrix} \sqrt{v_1}\underline{e}_1' \\ \sqrt{v_2}\underline{e}_2' \\ \vdots \\ \sqrt{v_p}\underline{e}_p' \end{bmatrix} = \Lambda\Lambda' + O, \tag{8.7}$$

where O is a matrix consisting of zeros only and Λ is the matrix within brackets, which has as its columns the successive eigenvectors after rescaling by their corresponding eigenvalues. Although Equation 8.6 and the decomposition of Σ following it in Equation 8.7 are exact, they are not really useful as they are consistent with the existence of as many factors as there are observed variables. However, when the last $p - m$ eigenvalues of Σ are small ($m < p$), one may disregard their contribution to the spectral decomposition in Equation 8.6 and obtain the following approximation of Σ (as in Chapter 7 when using PCA):

$$\begin{aligned} \Sigma &\approx v_1\underline{e}_1\underline{e}_1' + v_2\underline{e}_2\underline{e}_2' + \cdots + v_m\underline{e}_m\underline{e}_m' \\ &= \left[\sqrt{v_1}\underline{e}_1, \sqrt{v_2}\underline{e}_2, \ldots, \sqrt{v_m}\underline{e}_m\right] \begin{bmatrix} \sqrt{v_1}\underline{e}_1' \\ \sqrt{v_2}\underline{e}_2' \\ \vdots \\ \sqrt{v_m}\underline{e}_m' \end{bmatrix} \\ &= \Lambda^*\Lambda^{*\prime}, \end{aligned} \tag{8.8}$$

where now $\Lambda^* = [\sqrt{v_1}\underline{e}_1, \sqrt{v_2}\underline{e}_2, \ldots, \sqrt{v_m}\underline{e}_m]$ has similar structure to, but is different from, the matrix Λ in Equation 8.5 (as before, "$\approx$" denotes approximate equality). We next define Ψ as a diagonal matrix that has as its main diagonal elements the differences between the observed variances, $\text{Var}(x_i)$, and the communality of x_i according to Equation 8.8, that is,

$$\psi_i = \sigma_{ii} - \lambda_{i1}^2 - \lambda_{i2}^2 - \ldots - \lambda_{im}^2,$$

with σ_{ii} being the ith diagonal element of $\Sigma (i = 1, \ldots, p)$. Then the approximation under consideration is

$$\Sigma \approx \Lambda\Lambda' + \Psi.$$

This approximation is flawless with respect to the diagonal elements of Σ, namely $\sigma_{11}, \sigma_{22}, \ldots, \sigma_{pp}$, in that it reproduces them perfectly, but may not be nearly as good with regard to the off-diagonal elements of Σ. Typically, when one increases the number m of factors, the approximation of these off-diagonal elements becomes better. In fact, it can be shown that the sum of the squared entries of the residualized matrix $\Sigma - \Lambda\Lambda' - \Psi$ does not exceed the sum of the squares of the last $p - m$ eigenvalues, $v_{m+1}^2 + \cdots + v_p^2$ (Johnson & Wichern, 2002). That is, the overall lack of perfect approximation of Σ by the right-hand side of Equation 8.8 is no higher than the sum of the squared left-out eigenvalues.

The discussion so far in this section assumed of course that we knew m, the number of factors. To obtain a reasonable idea about it, one could use the same rules that were applied in PCA (see Chapter 7). Specifically, when the correlation matrix is analyzed one could look at the number of eigenvalues that are larger than 1, percentage of explained variance, and the scree plot. As discussed in greater detail below, in FA there is also another rule that is based on a chi-square test of "significance" of the number of factors and is applicable when a particular method of analysis is used—maximum likelihood (ML) (see Section 8.3.1.2).

To illustrate this discussion, let us consider a study in which $n = 500$ college sophomores were given measures of neuroticism, extraversion and agreeableness, and three abstract reasoning ability tests of algebra, geometry, and trigonometry. A researcher is interested in answering the following questions: (a) How many factors could explain well the interrelationships among these six test scores?; and (b) How could these factors be interpreted? (The data are provided in the file ch8ex1.dat available from www.psypress.com/applied-multivariate-analysis.) We begin by taking a look at the correlation matrix of these data, which is presented next (with the variable names abbreviated accordingly) and shows some interesting features worth examining at this point.

This correlation matrix exhibits a pattern suggesting (informally) that it would be meaningful to carry out FA. Indeed, there are two groups of three variables each—figuratively speaking, one in the upper-left corner and one in the lower-right corner—such that within each group the variables are notably correlated while across groups they are not so. In general, when FA is suitable, variables across groups need not be uncorrelated, but should at least be markedly correlated within groups. We note in passing that in practice one should not expect analyzed sets of observed variables to come in a continuous sequence like in Table 8.1; this simple example is used here for didactic purposes only and in particular to highlight the feature of groups of notably correlated observed variables when FA is appropriate (which groups need not be easily discernible in general, however, when looking at a given correlation matrix).

To carry out FA with SPSS, we use the following sequence of menu options:

TABLE 8.1

Correlation Matrix of Eight Observed Variables

Variable	Extravsn	Neurotic	Agreeabl	Algebra	Geomet	Trigono
Extravsn	1.000	.521	.472	−.001	−.026	.018
Neurotic	.521	1.000	.534	−.035	−.036	−.016
Agreeabl	.472	.534	1.000	−.010	−.040	.039
Algebra	−.001	−.035	−.010	1.000	.430	.369
Geomet	−.026	−.036	−.040	.430	1.000	.427
Trigono	.018	−.016	.039	.369	.427	1.000

Analyze → Data reduction → Factor (Descriptives: initial solution, KMO and Bartlett's test of sphericity; extraction: principal components, unrotated solution, and scree plot; options: sorted by size, suppress values less than .30)

To accomplish the same goal with SAS, the following command statements within the PROC FACTOR procedure can be used:

```
DATA FA;
INFILE 'ch8ex1.dat';
INPUT extravsn neurotic agreeabl algebra geomet trigono;
PROC FACTOR SCREE;
RUN;
```

The resulting outputs by SPSS and SAS follow next. For ease of presentation, the outputs are organized into sections and clarifying comments are accordingly inserted after each section.

SPSS output

KMO and Bartlett's Test

Kaiser-Meyer-Olkin Measure of Sampling Adequacy.		.676
Bartlett's Test of Sphericity	Approx. Chi-Square	592.344
	df	15
	Sig.	.000

SAS output

The FACTOR Procedure

Significance Tests Based on 500 Observations

Test	DF	Chi-Square	Pr > ChiSq
H0: No common factors HA: At least one common factor	15	592.3443	<.0001

As was the case when carrying out a PCA, it is important to ensure that the Bartlett's test is significant (under the assumption of normality) because it examines whether the correlation matrix is an identity matrix. (Note that to generate Bartlett's test in SAS, the statement METHOD = ML must be added to the program; we return to this particular method later and discuss it in more detail, but use it here only to accomplish comparability of the outputs obtained with the two software.) In the case that Bartlett's test it is not significant, a meaningful version of the factor model in Equation 8.1 is not likely to hold and FA is not appropriate to carry out, since there would be no variable interrelationships in the studied population on which FA could capitalize. (As we mentioned in Chapter 7, in such a case also PCA would not be appropriate.)

SPSS output

Communalities

	Initial	Extraction
extravsn	1.000	.651
neurotic	1.000	.704
agreeabl	1.000	.664
algebra	1.000	.588
geomet	1.000	.644
trigono	1.000	.588

Extraction Method: Principal Component Analysis.

SAS output

The FACTOR Procedure

Initial Factor Method: Principal Components

Final Communality Estimates: Total = 3.839164

extravsn	neurotic	agreeabl	algebra	geomet	trigono
0.65080582	0.70394582	0.66397752	0.58814070	0.64393714	0.58835655

The initial solution communality values provided in the above tables (with SPSS, see the right-most column of the output section titled so) give the sum of squared factor loadings for each manifest variable, h_i^2. These values let one readily obtain the error term variance for a given variable, that is, ψ_i, in this solution—to this end, the entry in the pertinent row is simply subtracted from one (as $\psi_i = 1 - h_i^2$, see earlier discussion in this chapter; $i = 1, \ldots, p$). As seen from the last presented output sections, in the initial solution the algebra and trigonometry measures exhibit the highest amount of unexplained variance by considered factors (see below).

SPSS output

Total Variance Explained

Component	Initial Eigenvalues			Extraction Sums of Squared Loadings		
	Total	% of Variance	Cumulative %	Total	% of Variance	Cumulative %
1	2.027	33.790	33.790	2.027	33.790	33.790
2	1.812	30.196	63.986	1.812	30.196	63.986
3	.633	10.551	74.537			
4	.558	9.297	83.834			
5	.525	8.743	92.577			
6	.445	7.423	100.000			

Extraction Method: Principal Component Analysis.

SAS output

```
                         The FACTOR Procedure

            Initial Factor Method: Principal Components

                 Prior Communality Estimates: ONE
     Eigenvalues of the Correlation Matrix: Total = 6 Average = 1

        Eigenvalue    Difference    Proportion    Cumulative

1       2.02740470    0.21564585      0.3379        0.3379
2       1.81175885    1.17869565      0.3020        0.6399
3       0.63306320    0.07524984      0.1055        0.7454
4       0.55781336    0.03324762      0.0930        0.8383
5       0.52456575    0.07917161      0.0874        0.9258
6       0.44539413                    0.0742        1.0000

         2 factors will be retained by the MINEIGEN criterion.
```

These output sections suggest extracting two factors since (a) Kaiser's eigenvalue criterion indicates that there are two eigenvalues larger than one; and (b) the total variance explained by them is just under two thirds, which may appear satisfactory for many practical purposes. The scree plot presented below also corroborates the suggestion of extracting two factors in this study.

SPSS output

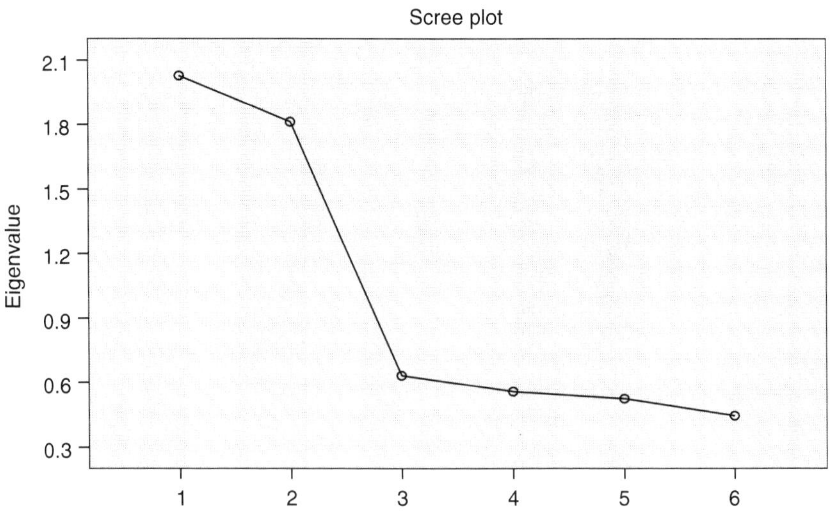

SAS output

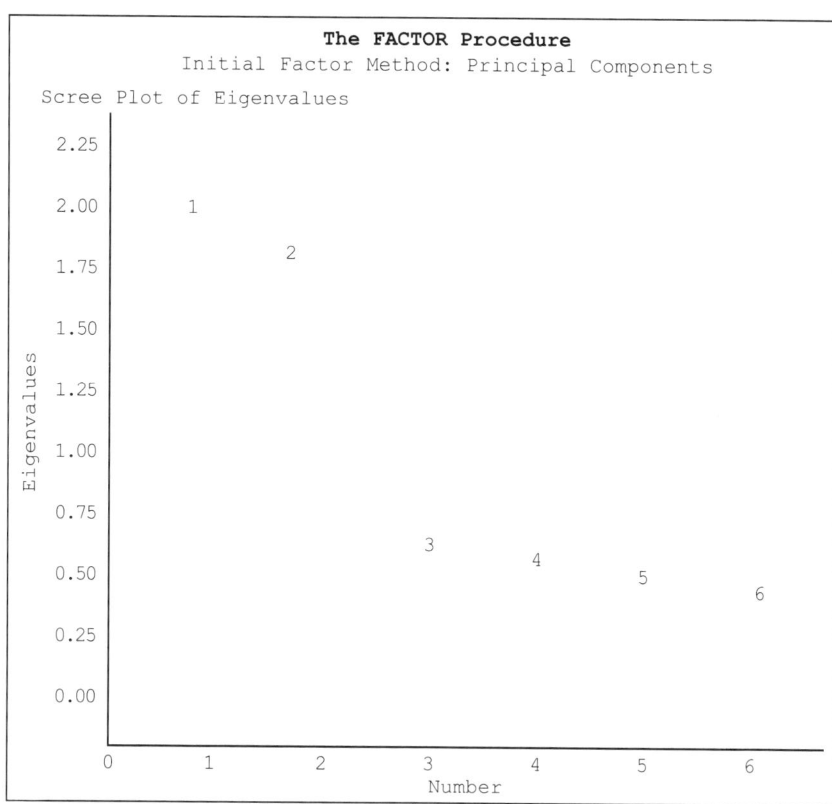

SPSS output

Component Matrix[a]

	Component	
	1	2
V2	.832	
V3	.801	
V1	.793	
V5		.781
V6		.761
V4		.752

Extraction Method: Principal Component Analysis.
a. 2 components extracted.

SAS output

The FACTOR Procedure
Initial Factor Method: Principal Components

Factor Pattern

	Factor1	Factor2
extravsn	0.79285	0.14897
neurotic	0.83159	0.11140
agreeabl	0.80118	0.14860
algebra	−0.14843	0.75240
geomet	−0.18346	0.78120
trigono	−0.09834	0.76071

As indicated earlier in this section, apart from a multiplicative constant—the square root of the respective eigenvalue—the component matrix elements can be viewed as factor loadings of the initial solution (as long as the principal component method is used for factor extraction). Usually, this solution is not going to be easily interpreted, however, but in the current case it happens to be quite meaningful. (We return to interpretation matters in greater detail in the next section.) In general, in order to interpret a factor in any solution—whether initial or after rotation (see below)—the same principle is used as when interpreting principal components in PCA. To this end, one needs to find out which observed variables load highly on a factor under consideration, and try to interpret the latter in terms of the common features that these saliently loading measures share with each other. Bearing this in mind, from the last presented output sections follows that in the initial solution Factor 1 could be interpreted as a ''personality'' factor, whereas Factor 2 could be viewed as a ''mathematics ability'' factor.

We hasten to add, however, that in empirical research one usually need not spend much time on the initial solution since it will generally be rotated to enhance interpretability (unless of course there is only one factor extracted when it cannot be rotated). We revisit these matters later in the chapter, after we consider another method for factor extraction that has some special desirable properties.

8.3.1.2 Maximum Likelihood Factor Analysis

This approach is based on the ML method of parameter estimation, which we have already discussed earlier in the book. Accordingly, factor loadings and error term variances, being model parameters, are estimated in such a way that the observed data has the highest likelihood ("probability") to be collected if one were to sample from the studied population again. In order to achieve this goal in the context of FA, a particular restriction is imposed on the loading and error covariance matrices, namely that $\Lambda' \Psi^{-1}\Lambda$ is a diagonal matrix (Johnson & Wichern, 2002).

An added advantage of the maximum likelihood factor analysis (MLFA) method is that it provides a goodness of fit test of the factor model for a given number of factors. The corresponding null hypothesis H_0 is that the extracted number of factors is sufficient to explain the analyzed population matrix of variable interrelationship indices, that is, H_0: "Number of factors $= m$." This hypothesis is tested against the alternative that more factors are needed. In order to carry out this test, the assumption of multivariate normality (MVN) needs to be made. Typically, one applies the MLFA by starting with a relatively small number of factors and uses this test as a guide with respect to needed number of factors. Specifically, if at a given number of factors the test is significant, one increases their number by 1 and applies the test again; one continues to do so until lack of significance is obtained with it. Statistical software will typically intertwine these steps, by first extracting as many factors as there are eigenvalues larger than 1 and then providing the result of that statistical test with as many factors. (In some software, the user can also prespecify the selection of a particular number of factors to consider for the analysis.)

To illustrate MLFA, let us use the same data as in the last example, but now employ the ML method of FA. To this aim, with SPSS we proceed as above, with the only difference that now one chooses ML as the method of extraction rather than principal component method. To accomplish the same activity in SAS, the command METHOD = ML is simply added to the PROC FACTOR statement line. (To prespecify the number of factors, one can also add the command NFACTOR = #, where # is this number of factors; this subcommand can also be shortened to N = #.) The resulting outputs produced by SPSS and SAS are as follows (presented again in segments to simplify the discussion). We note that since the ML method does not affect the test statistics concerned with the analyzed correlation matrix, the Bartlett's test results obtained earlier when using the principal component method for factor extraction will be identical, and consequently we dispense with repeating that section of the resulting output.

SPSS output

Communalities

	Initial	Extraction
V1	.324	.461
V2	.380	.587
V3	.340	.486
V4	.227	.372
V5	.271	.495
V6	.228	.370

Extraction Method: Principal Component Analysis.

SAS output

```
                      The FACTOR Procedure

           Initial Factor Method: Maximum Likelihood

               Prior Communality Estimates: SMC

  extravsn    neurotic    agreeabl    algebra     geomet      trigono
 0.32449413  0.38008655  0.33979248  0.22728099  0.27057031  0.22829183

        Final Communality Estimates and Variable Weights
   Total Communality: Weighted=5.387560  Unweighted=2.772249

 Variable                  Communality                    Weight

 extravsn                   0.46133150                1.85642930
 neurotic                   0.58758773                2.42475803
 agreeabl                   0.48594284                1.94530910
 algebra                    0.37175937                1.59174671
 geomet                     0.49516855                1.98085905
 trigono                    0.37045861                1.58845802
```

The above communalities obtained with MLFA are different to the communalities from the previous output we saw for the same data, because now the method of factor extraction is different. It is important to note that here, instead of the value 1.0 being used as initial communality for each variable, estimated values are provided, which is a distinct feature of MLFA.

SPSS output

Total Variance Explained

Factor	Initial Eigenvalues			Extraction Sums of Squared Loadings		
	Total	% of Variance	Cumulative %	Total	% of Variance	Cumulative %
1	2.027	33.790	33.790	1.541	25.679	25.679
2	1.812	30.196	63.986	1.231	20.524	46.202
3	.633	10.551	74.537			
4	.558	9.297	83.834			
5	.525	8.743	92.577			
6	.445	7.423	100.000			

Extraction Method: Maximum Likelihood.

SAS output

The FACTOR Procedure

Initial Factor Method: Maximum Likelihood

Prior Communality Estimates: SMC

Preliminary Eigenvalues: Total = 2.56906787 Average = 0.42817798

	Eigenvalue	Difference	Proportion	Cumulative
1	2.11316034	0.71318442	0.8225	0.8225
2	1.39997593	1.56935744	0.5449	1.3675
3	−.16938152	0.04770306	−0.0659	1.3015
4	−.21708457	0.02363775	−0.0845	1.2170
5	−.24072232	0.07615766	−0.0937	1.1233
6	−.31687998		−0.1233	1.0000

2 factors will be retained by the PROPORTION criterion.

The results provided by SPSS and SAS are somewhat different to each other. The output results provided by SPSS are identical to those seen earlier for the same data, whereas the results provided by SAS differ. This is because SPSS reports the eigenanalysis of the unadjusted correlation matrix (i.e., the initial correlation matrix, with 1s on the main diagonal). In contrast, SAS reports the eigenanalysis on the adjusted correlation matrix in which the squared multiple correlation-based communality estimates replace the constant 1 on the main diagonal. We note that despite these differences, the same subsequent suggestion is made for two factors to be extracted.

SPSS output

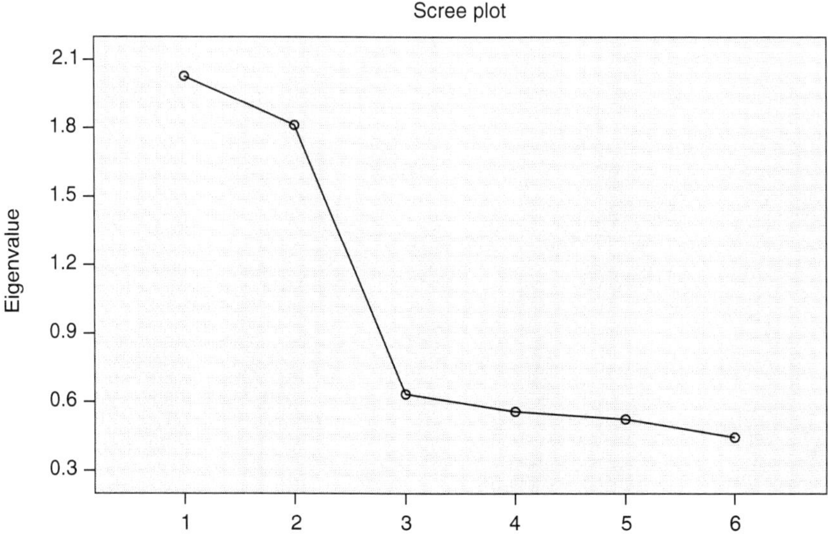

SAS output

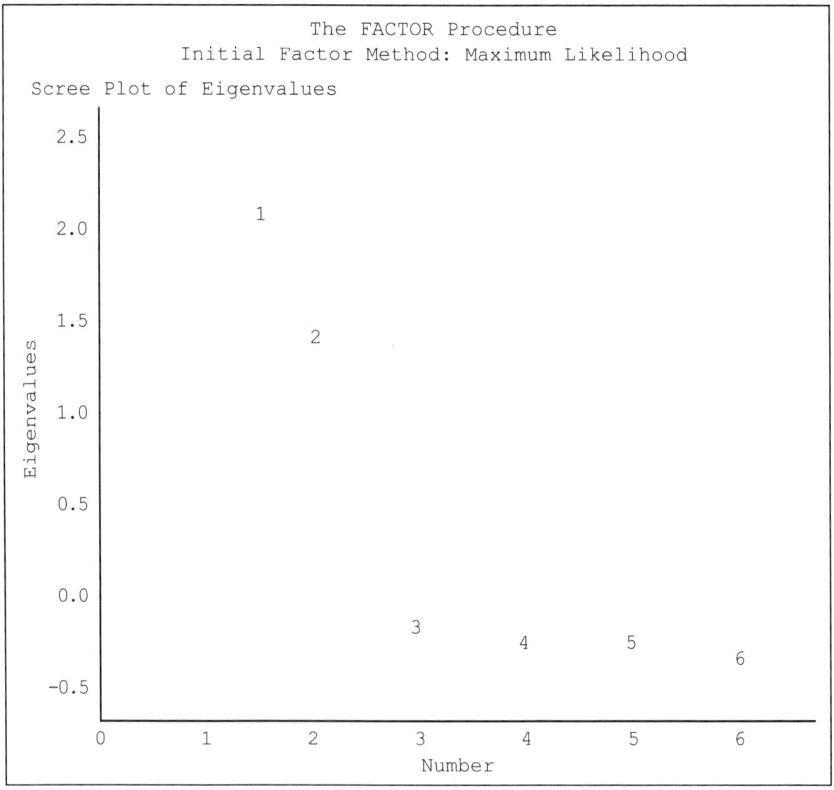

```
                 The FACTOR Procedure
        Initial Factor Method: Maximum Likelihood

 Scree Plot of Eigenvalues

     2.5

                     1
     2.0


     1.5
                           2

E    1.0
i
g
e
n
v    0.5
a
l
u
e
s    0.0

                                      3
                                                4           5
                                                                       6
    -0.5

         0          1          2          3          4          5          6
                                        Number
```

As could be expected, the above scree plots furnished by SPSS and SAS are somewhat different from each other. The scree plot by SPSS is based on the eigenvalues of the unadjusted correlation matrix, while the SAS scree plot is based on the adjusted correlation matrix. Both scree plots suggest, however, that two factors be extracted.

SPSS output

Factor Matrix[a]

	Factor	
	1	2
V2	.766	
V3	.694	
V1	.676	
V5		.697
V6		.608
V4		.606

Extraction Method: Maximum Likelihood.
a. 2 factors extracted. 3 iterations required.

SAS output

The FACTOR Procedure

Initial Factor Method: Maximum Likelihood

Factor Pattern

	Factor1	Factor2
extravsn	0.67627	0.06315
neurotic	0.76573	0.03539
agreeabl	0.69440	0.06125
algebra	−0.06755	0.60597
geomet	−0.09749	0.69690
trigono	−0.02899	0.60796

These tables contain the factor loading matrices obtained with the ML method of estimation. As expected, on the surface these are numerically different results from the ones obtained with the principal component method, but the same substantive conclusions are suggested with regard to salient variables loadings on each of the factors and their tentative interpretation. We return to the issue of interpretation in greater detail in the next section.

SPSS output

Goodness-of-fit Test

Chi-Square	df	Sig.
2.182	4	.702

SAS output

```
                    The FACTOR Procedure

         Initial Factor Method: Maximum Likelihood

       Significance Tests Based on 500 Observations

                                                              Pr >
        Test                        DF     Chi-Square        ChiSq

H0: No common factors               15       592.3443       <.0001
HA: At least one common factor
H0: 2 Factors are sufficient         4         2.1821       0.7023
HA: More factors are needed
```

As indicated previously, the ML method provides a goodness of fit test of the factor model for a given number of factors. The results of this test furnished by both SPSS and SAS suggest that the null hypothesis of two factors being sufficient to account for the analyzed variable interrelationships can be considered plausible, that is, H_0: $m = 2$ factors suffice can be retained.

8.3.2 Factor Rotation

One of the main critical points of FA, especially in its early years, was that the initial factor solution is generally not unique (Hayashi & Marcoulides, 2006, and references therein). In fact, there are infinitely many solutions directly related to the initial one (and to each other), which explain the variable interrelationships just as well, when there are $m > 1$ extracted factors. This is because the initial factor solution has only determined the m-dimensional space containing the factors, but not the exact position of those factors in it. To understand this issue in greater detail, we first need to become familiar with a particular kind of matrix.

We define a $k \times k$ matrix T as orthogonal, if $\text{T T}' = \text{T}' \text{T} = \text{I}_k$ (with $k > 1$). In other words, a square matrix is orthogonal if its rows (columns) are orthogonal to one another and of unit length. An important feature that we capitalize on soon is that for any such matrix, its postmultiplication with a vector represents an orthogonal transformation of that vector in the multivariate space. In fact, for many such matrices, this multiplication geometrically represents a rigid rotation of its underlying axes, that is, a rotation that keeps their angle the same while changing their positioning.

A related property that follows from the definition of an orthogonal matrix, is that for any such matrix T the transformation of the vector $\underline{x}$ into the vector $\underline{y} = T\,\underline{x}$ preserves lengths, angles, and distances. That is, using notation from Chapter 2, $\|y\| = \|x\|$, and the angle between any two vectors $\underline{x}_1$ and $\underline{x}_2$ is the same as that between the vectors $\underline{y}_1$ and $\underline{y}_2$ that are their respective images with this transformation, that is, $y_1 = Tx_1$ and $y_2 = Tx_2$. (Invariance of distances between points follows then from that of lengths and angles—recall the "cosine" law from high school geometry.)

To give an example, the matrix

$$T = \begin{bmatrix} \cos(\alpha) & \sin(\alpha) \\ -\sin(\alpha) & \cos(\alpha) \end{bmatrix}$$

is an orthogonal matrix, for a given angle α $(0 < \alpha < 2\pi)$, which represents a rigid rotation (transformation) in the two-dimensional space. Specifically, if a point $\underline{x} = (x_1, x_2)$ is given beforehand and the axes of the two-dimensional space are rotated by an angle α, the coordinates of $\underline{x}$ in the new, transformed system will be

$$\underline{y} = T\underline{x} = [\cos(\alpha)x_1 + \sin(\alpha)x_2, -\sin(\alpha)x_1 + \cos(\alpha)x_2]'.$$

For example, supposed that in the figure below the point $\underline{x} = (x_1, x_2) = (2, 2)$ is prespecified and the axes (denoted I and II) of the initially considered two-dimensional space are rotated by the angle $\alpha = 45°$ (counterclockwise) (see Figure 8.1). Then its coordinates in the new system with axes denoted I′ and II′, which are obtained through the transformation

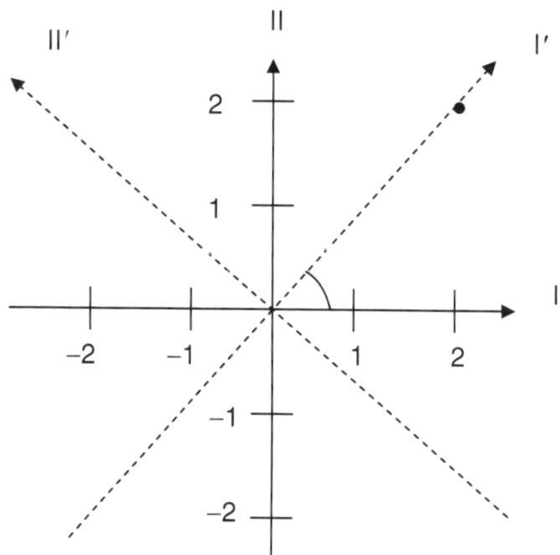

FIGURE 8.1
Graphical representation of an axis rotation (by 45°).

defined by the last presented matrix T, will be (2.83, 0). (The initial system axes are denoted with solid lines, the transformed ones with dashed lines.)

To discuss now the concept of rotation in greater detail, let us go back to the FA model in Equation 8.2, that is,

$$\underline{x} = \Lambda\underline{f} + \underline{\varepsilon}.$$

Obviously, we can rewrite the right-hand side of Equation 8.2 using also the matrix T as

$$\underline{x} = \Lambda\underline{f} + \underline{\varepsilon} = \Lambda I\underline{f} + \underline{\varepsilon} = \Lambda(TT')\underline{f} + \underline{\varepsilon} = (\Lambda T)(T'\underline{f}) + \underline{\varepsilon}, \tag{8.9}$$

where we employed the following property for transposition of matrices: $(AB)' = B'\,A'$ (for such matrices that AB exists; see discussion in Chapter 2, where it was mentioned first). That is, Equation 8.9 shows that the factor loading matrix is determined only up to multiplication with an orthogonal matrix (orthogonal transformation). Furthermore, Equation 8.9 demonstrates that the error terms (and hence their variances as well) are not changed by this transformation, and thus variable communalities are unaltered thereby. That is, the fit of the FA model is invariant to an orthogonal transformation of the factor loadings, which as mentioned before represents a rigid rotation of the underlying axes in the multivariate space (possibly followed by reflection around an axis).

Because, in Equation 8.9, one can choose the orthogonal matrix T in infinitely many ways, it follows that there are infinitely many solutions for the FA model, which are identical to one another with regard to their fit to a given data set (interrelationship indices matrix). These solutions differ from one another only by an orthogonal transformation of their factor loading matrix. That is, we can rigidly rotate the coordinate axes of the underlying multivariate space, and this will not change the fit of the FA model while producing a new solution for the FA model. Consequently, we can selectively rotate an initial FA solution to a (transformed) solution that is easier to interpret, by using a suitable orthogonal matrix.

Ease of interpretation is obtained in a solution that exhibits what is referred to as a "simple structure." The concept of simple structure in FA was introduced by Thurstone (1947), and is characteristic of attempts to create the following pattern of factor loadings: (a) each observed variable loads highly if possible on a single factor, and has small loadings on the remaining factors; and (b) for each factor there is if possible at least one variable saliently loading on it. (As readily available examples of what is meant empirically by simple structure, see Sections 8.3.1.1. and 8.3.1.2.) We emphasize that with empirical data it is usually quite difficult to achieve perfect or close to simple structure. For this reason, most factor rotations attempt to determine a solution that is as close as possible to simple structure.

This second step of rotation in FA is in practice motivated also by the observation that many times an initial solution is not easily interpretable. Typically, there are multiple variables with moderate loadings on a number of factors, and for some of the factors there are no (really) saliently loading variables. The aim of the rotation stage then is to obtain an equivalently well fitting solution that is closer to simple structure, and is thus meaningfully interpretable in terms of the substantive area of application.

To illustrate the last discussion with an example, consider the following factor loading matrix that is obtained as an initial solution in a FA of 12 observed variables (empty cells below imply that the respective loading is smaller than .2 in absolute value, for simplicity of presentation). This matrix resulted from a study of $n=300$ fourth-grade students that included observations on three measures of reading comprehension (words, sentences, paragraphs; denoted next V10 through V12), three measures of parental domination (V1 through V3), three measures of physical fitness (V7 through V9), and reaction times on three cognitive tests (V4 through V6).

Example Factor Loading Matrix

	Factor			
	1	2	3	4
V11	.596		.254	.393
V12	.547		.238	.301
V10	.511		.222	.406
V9		.559	.398	
V7		.547	.431	
V8		.533	.471	
V1	.249	.515	−.354	
V2	.397	.500	−.463	
V3	.361	.376	−.407	
V5	−.436		−.226	.587
V4	−.362		−.215	.454
V6	−.390		−.212	.452

As can be seen by examining this factor loading matrix, it is very hard to come up with a reasonable interpretation of the associated FA solution, especially with regard to the meaning of the underlying factors. In particular, each observed variable loads markedly on more than one factor, if one were to adopt a widely circulated guideline considering a loading prominent when larger than .3 (as then its communality is at least some 10% of observed variance; see Rule 2 for variance computation earlier in this chapter). This example clearly demonstrates the necessity of factor rotation, as a means of obtaining an easier to interpret solution with identical fit to an initial one.

There are two types of rotation that can be carried out: (a) orthogonal rotation, and (b) oblique rotation. The orthogonal rotation keeps the angle between the initially extracted factors the same; that is, they are orthogonal before the rotation, due to their construction, and remain orthogonal also after the rotation. The oblique rotation can change the angle between the factors; that is, if they are orthogonal before rotation, they need not be so after the rotation anymore. That is, in general, an oblique rotation alters an orthogonal coordinate (reference axes) system to a nonorthogonal system (affine system), thereby rotating axes differentially and thus changing both angles between and lengths of vectors under consideration. Although to date there are quite a number of rotational schemes available in the literature, below we discuss only the most popular rotation methods for orthogonal and oblique rotation.

8.3.2.1 Orthogonal Rotation

Orthogonal rotation in FA is mostly carried out using the so-called *Kaiser's varimax criterion*. This criterion is based on a measure, denoted V, of closeness to simple structure, which needs to be maximized across all possible rigid rotations and is defined as follows:

$$
\begin{aligned}
V &= \text{sum of variances of squared scaled loadings on factors} \\
&= \frac{1}{p}\sum_{j=1}^{m}\left\{\sum_{i=1}^{p}(\lambda_{ij}/h_i)^4 - \frac{1}{p}\left[\sum_{i=1}^{p}(\lambda_{ij}/h_i)^2\right]^2\right\}.
\end{aligned}
$$

This maximization has the tendency to polarize the factor loadings so that they are either high or low, thereupon making it somewhat easier to identify factors with particular observed variables. The result of this rotation is typically the creation of groups of large and of negligible loadings in any column of the rotated factor loading matrix Λ. This eases interpretation substantially for the factors, since one can focus then on the common characteristics of the variables loading highly on a given factor and disregard the contribution in this respect of the manifest variables with negligible loadings on it.

To illustrate, consider again the last example study that had a difficult to interpret initial solution. After using an orthogonal rotation on that solution (with SPSS, one needs to choose rotation: varimax; whereas with SAS one must add the command ROTATE = VARIMAX to the PROC FACTOR statement), the following factor loading matrix results. (Since both programs would provide identical output, to save space below we only present the rotation output generated using SPSS.)

Rotated Component Matrix[a]

	Component			
	1	2	3	4
V7	.829			
V8	.825			
V9	.817			
V11		.847		
V10		.812		
V12		.796		
V5			.830	
V6			.801	
V4			.789	
V2				.846
V1				.785
V3				.785

Extraction Method: Principal Component Analysis.
Rotation Method: Varimax with Kaiser Normalization.
a. Rotation converged in 4 iterations.

An examination of this rotation output indicates that the resulting solution is now readily interpretable. Indeed, Factor 1 is what is common to V7 through V9, which are physical characteristics variables; hence, Factor 1 would be considered a physical fitness factor. Further, Factor 2 would be what is common in the reading measures, so it could be viewed as a reading ability factor. In the same way, Factor 3 would be considered cognitive speed, and Factor 4 as parental style.

8.3.2.2 Oblique Rotation

Orthogonal rotations are appropriate for a factor model in which the common factors are assumed to be uncorrelated. When this is (nearly) the case, with this rotation a relatively parsimonious means of data description and explanation is found, which is generally a desirable outcome of FA. However, the assumption of uncorrelated factors is usually rather strong in many empirical social and behavioral research studies, where underlying latent variables typically tend to be related. To allow the extracted factors to be related in the second step of a factor analytic session, one may use an oblique rotation that as mentioned represents a nonrigid rotation of the coordinate system. For this reason, the rotated axes are no longer perpendicular to one another, but instead could be thought of as "passing through" the cluster of observed variables in the multivariate space (the n-space; see Chapter 2). The results of an oblique rotation typically allow an easier substantive interpretation than those of an orthogonal rotation. Furthermore, oblique rotations are especially recommendable for solutions resulting from the ML method, because the latter imposes a restriction for mathematical convenience, that is, the product $\Lambda' \Psi^{-1} \Lambda$ is a diagonal matrix, while extracting the latent factors (Johnson & Wichern, 2002); for this reason, the ML method tends to yield initial solutions that are difficult to interpret.

Oblique rotations are more general than orthogonal rotations. In addition to the rotated factor loading matrix, another part of their associated output is the correlation matrix of the factors, denoted Φ. If Φ is close to diagonal, an orthogonal rotation will likely provide nearly as good fit to the data and similar interpretability as the oblique rotation. In this sense, orthogonal rotations may be considered special case of oblique rotations, and in fact result from the latter when all angles between all pairs of factors are 90°.

As a way of illustration, consider a study in which data were obtained from $n = 300$ fifth-grade students on three reading comprehension tasks (denoted V1 through V3 next), three parental style measures (designated V4 through V6), and three college aspiration subscales (denoted V7 through V9 below). (The data are provided in the file ch8ex2.dat available from www.psypress.com/applied-multivariate-analysis.) For this study we wish to use the ML method, followed by an oblique rotation, since we do not anticipate that resulting factors could be uncorrelated, given the marked correlations between the observed variables. To this end, after employing the above indicated sequence of menu options with SPSS, choose for Extraction: Maximum Likelihood, and for Rotation: Direct oblimin. In order to carry out an oblique rotation in SAS, the following statement must be added to the PROC FACTOR command line: ROTATE = PROMAX. In addition, in order to suppress loadings that are say below .3 (see Section 8.3.2.1), the statement FUZZ = .3 is added to the same line. We note that SPSS and SAS do not offer the same oblique rotation options. For example, SPSS relies heavily on the so-called ''oblimin'' method, whereas SAS relies on the ''promax'' method (for complete details on the historical developments of these rotation procedures, see Harman, 1976). In studies with relatively clear latent structure, researchers can expect that these rotation methods would converge to similar solutions.

The resulting output produced by SPSS and SAS follows. To simplify the presentation and discussion we only provide the output related to the factor loading matrices based on a three-factor solution. (We note that this factor number was determined by examining the various criteria for it. Again, we suppress from numerical presentation next factor loadings below .3; see preceding subsection.)

SPSS output

Factor Matrix[a]

	Factor		
	1	2	3
V2	.789		
V3	.713		
V1	.640		
V7		.706	
V8		.678	
V9		.669	
V5			.701
V4			.588
V6			.571

Extraction Method: Maximum Likelihood.
a. 3 factors extracted. 4 iterations required.

SAS output

The FACTOR Procedure

Initial Factor Method: Maximum Likelihood

Factor Pattern

	Factor1	Factor2	Factor3
V1	0.63956	.	.
V2	0.78916	.	.
V3	0.71265	.	.
V4	.	.	0.58782
V5	.	.	0.70101
V6	.	.	0.57077
V7	.	0.70575	.
V8	.	0.67765	.
V9	.	0.66862	.

Values less than 0.3 are not printed.

An examination of these results indicates that even though the initial solution appears to be relatively straightforwardly interpretable, since it is not unlikely that the factors are correlated, oblique rotation may improve matters. For this reason we examine below the oblique rotated solution before making any final conclusions.

SPSS output

Pattern Matrix[a]

	Factor		
	1	2	3
V2	.797		
V3	.715		
V1	.650		
V7		.743	
V8		.727	
V9		.701	
V5			.771
V4			.642
V6			.636

Extraction Method: Maximum Likelihood.
Rotation Method: Oblimin with Kaiser Normalization
a. 3 factors extracted. 4 iterations required.

SAS output

```
                    The FACTOR Procedure

             Rotation Method: Promax (power=3)

  Rotated Factor Pattern (Standardized Regression Coefficients)

              Factor1          Factor2          Factor3
V1                  .          0.65024                .
V2                  .          0.79654                .
V3                  .          0.71444                .
V4                  .                .          0.63626
V5                  .                .          0.77083
V6                  .                .          0.64249
V7            0.72728                .                .
V8            0.74227                .                .
V9            0.70100                .                .

            Values less than 0.3 are not printed.
```

Each of the above factor pattern matrices presented by SPSS and SAS show the factor loadings after rotation, and the interpretation of their entries is done using the same principle as after orthogonal rotation—interpret a factor in terms of the common features of variables loading saliently on it (e.g., with loadings above .3). Oftentimes, however, the following so-called factor structure matrix may be easier to interpret, as it contains the correlations between the measured variables and the factors. Alternatively, the loadings in the factor pattern matrix convey the unique relationship between the variable and the factor, much in the same way that a partial regression coefficient does in multiple regression. The factor structure matrix coefficients are computed using the factor pattern matrix and the factor correlation matrix (see matrix displayed below). When interpreting the results from an oblique rotation, it is recommended that both the factor pattern and factor structure matrices be examined.

SPSS output

Structure Matrix

	Factor		
	1	2	3
V2	.796		
V3	.713		
V1	.652		
V8		.737	
V7		.731	
V9		.702	
V5			.770
V6			.646
V4			.634

Extraction Method: Maximum Likelihood.
Rotation Method: Oblimin with Kaiser Normalization.

SAS output

The FACTOR Procedure

Rotation Method: Promax (power = 3)

Factor Structure (Correlations)

	Factor1	Factor2	Factor3
V1	.	0.65108	.
V2	.	0.79625	.
V3	.	0.71400	.
V4	.	.	0.63376
V5	.	.	0.77034
V6	.	.	0.64548
V7	0.73096	.	.
V8	0.73765	.	.
V9	0.70210	.	.

Values less than 0.3 are not printed.

According to the structure matrix results, and consistent with the factor pattern matrix, the three factors can be interpreted here as the common features across the elements of each consecutive triples of observed variables, which correlate at least moderately with the respective factor. That is, Factor 1 can be interpreted as a reading ability, Factor 2 as aspiration, and Factor 3 as parental style.

SPSS output

Factor Correlation Matrix

Factor	1	2	3
1	1.000	.029	−.012
2	.029	1.000	.081
3	−.012	.081	1.000

Extraction Method: Maximum Likelihood.
Rotation Method: Oblimin with Kaiser Normalization.

SAS output

Inter-Factor Correlations

	Factor1	Factor2	Factor3
Factor1	1.00000	0.01323	0.07845
Factor2	0.01323	1.00000	−0.02311
Factor3	0.07845	−0.02311	1.00000

The final output section to consider is the correlation matrix among the extracted factors, which we denoted Φ above. We observe that they do not show particularly high interrelationships. Thus, it may be possible that an orthogonal rotation could have provided very similar if not essentially the same interpretation as above. (This is indeed the case with the present data, as could be verified following the same software analytic steps as above and choosing varimax rotation instead.)

In conclusion of this section, and as a follow up to an earlier remark we made in passing, we would like to comment on a descriptive rule that has been available and used for many years with respect to when a factor loading could be considered worthy of attention ("salient"). This rule proposes that if a factor loading is at least .3, then it may be taken into account when interpreting the corresponding factor. As mentioned before, following this rule of thumb when analyzing the correlation matrix implies that the factor contributes at least about 10% to the variance of that variable that is shared with other variables. In recent decades, methods for testing significance of loadings within the presently considered FA framework have been developed and made widely available in software. While in this chapter we used the above mentioned descriptive rule of thumb, we add here that a more rigorous approach would rely on use of such formal statistical tests. We return in greater detail to some aspects of this issue in Chapter 9.

8.4 Heywood Cases

As indicated in a previous section, the communality index reflects the R^2 coefficient for the regression of an observed variable on the factors in a FA model, and hence should lie between 0 and 1. However, sometimes communality estimates may exceed the value of 1, and hence the uniqueness variance will be negative then (see Equation 8.4). If this occurs when fitting to data a given FA model, it is commonly referred to as a "Heywood case." The presence of a Heywood case is a clear indication that the resulting solution is not admissible, and that possibly the underlying factor model is misspecified, that is, incorrect. More generally, a Heywood case may be the result of one or more of the following conditions: (a) not enough data have been sampled from the population to provide stable parameter estimates, (b) too many factors have been extracted in the factor solution, (c) too few factors have been extracted in the factor solution, (d) the initial communality estimates are not appropriate, (e) in the studied population, a corresponding variance or correlation parameter is very small or very close to 1 (or -1), respectively, and (f) an FA model is not appropriate for the observed data considered. In particular, a Heywood case may occur during the iteration process for obtaining an ML solution. If attempts to correct for the problems indicated in points (a) through (d) do not help, it is possible that no data reduction may be meaningfully accomplished using FA with fewer factors than observed variables, and in this case FA may be pointless to pursue.

8.5 Factor Score Estimation

Once an interpretable rotated solution is found and extracted factors are given substantive meaning, it may be of particular interest to evaluate each of the studied subjects/participants on these unobservable dimensions. This is the goal of the procedure called factor score estimation that produces estimates of the factors for each person. We stress that these are not the scores of the subjects on the factors, but only estimates of them. For the goals of this chapter, we consider two methods for obtaining factor score estimates, and mention that none of them can be viewed as uniformly better than the other.

8.5.1 Weighted Least Squares Method (Generalized Least Squares Method)

Let us return for a moment to the factor model Equation 8.2, that is,

$$\underline{x} = \Lambda \underline{f} + \underline{\varepsilon}.$$

Once we carry out a FA, we in fact furnish estimates of the factor loading matrix Λ and the error covariance matrix Ψ. Then, in order to obtain factor score estimates, we may consider these two matrices as known. With this in mind, up to notation, Equation 8.2 is the same as that of the general linear model with arbitrary (positive definite) error covariance matrix, which can be fitted to data using a procedure called generalized least squares (also sometimes referred to as weighted least squares; Timm, 2002). Hence, considering $\underline{f}$ as a set of unknown parameters, given any of the subjects studied (the ith say), we can correspondingly estimate his or her factor scores $\underline{f}_i$ as follows:

$$\hat{\underline{f}}_{i,B} = \left(\hat{\Lambda}'\hat{\Psi}^{-1}\hat{\Lambda}\right)^{-1}\hat{\Lambda}'\hat{\Psi}^{-1}\underline{x}_i, \tag{8.10}$$

where a caret ($^\wedge$) means estimate ($i = 1, \ldots, n$; n being sample size). The factor score estimates in Equation 8.10 are widely known as Bartlett estimates. We emphasize that in order to obtain them, we need to use the raw data, since it is the recorded set of variable scores for the ith subject that appears in the right-hand side of Equation 8.10 ($i = 1, \ldots, n$). That is, in order to furnish Bartlett's factor score estimates, it is not sufficient to have access to the correlation or covariance matrix only (or even the means in addition to any of these matrices). In an application, these estimates are obtained for each subject with SPSS say if one chooses the option Save: Bartlett method. In SAS, this can be accomplished by the following procedure statements:

```
PROC FACTOR nfactor = 3 out = scores;
PROC PRINT;
```

The resulting factor score estimates (as well as the ones discussed next) are then appended as new variables to the original data set.

8.5.2 Regression Method

Consider again, after estimation of the factor model, the matrices Λ and Ψ as known. Another method of factor score estimation begins by asking the question, "What is the best guess for $\underline{f}$ given knowledge of the observed variables $\underline{x}$?" Assuming $\underline{f}$ and $\underline{\varepsilon}$ are jointly normally distributed (multinormal), it can be shown that the best guess is the mean of $\underline{f}$ at that value of $\underline{x}$, denoted $M(\underline{f}|\underline{x})$ (the vertical bar denotes conditioning, that is, fixing of the value of $\underline{x}$; Johnson & Wichern, 2002). This value represents the factor score estimate for the respective subject with the regression method:

$$\hat{\underline{f}}_{i,R} = \hat{\Lambda}'\left(\hat{\Lambda}\hat{\Lambda}' + \hat{\Psi}\right)^{-1}\underline{x}_i \tag{8.11}$$

($i = 1, \ldots, n$). This method is analogous to predicting a response value in regression analysis by the mean of the dependent variable (DV) at a given set of scores for the predictor, $\underline{x}$, but here is used for purposes of predicting scores on unobserved variables, namely the factors.

We notice several features from Equations 8.10 and 8.11 that are worth stressing at this point. First, the factor score estimates are linear combinations of observed variables, whereby the weights of these combinations are functions of (estimated) factor loadings and error term variances. Hence, being linear combinations of observed variables, the factor score estimates cannot really represent individual scores on the unknown factors that are latent (i.e., unobserved) variables, and are only estimates of the latter. Second, the factor score estimates can be used to examine the assumption of multinormality, because they should be normally distributed if the original variables are multinormal. Third, like principal component scores, the factor score estimates can be used in subsequent analyses as values on (newly derived rather than initially observed) variables that can be treated, for example, as explanatory with respect to other measures.

In conclusion of this section, we emphasize that its developments—like those in the entire chapter—were based on the assumption that the analyzed variables were approximately continuous. If they are distinctly discrete, for example, with no more than say 3–4 values possible for subjects to take on them, then more specialized methods of FA need to be used. These methods are for instance available within the framework of latent variable modeling (Muthén & Muthén, 2006; see also Chapter 9). Last but not least, we also stress that the FA we dealt with in this chapter is of exploratory nature. In particular, when starting the analysis we did not have any idea concerning how many factors to extract or which factor was measured by which manifest variables. As a result, the FA approach of this chapter has the limitations that (a) hypotheses about factor correlations cannot be tested; (b) restrictions cannot be imposed on model parameters; and (c) the assumption of diagonal error covariance matrix cannot be relaxed. Also, adding further measures to an initial set of observed variables and then factor analyzing all of them may lead to considerably different results from those of FA carried out on the initial variables, even if the added measures are thought to be indicators of the same factors as those extracted from the original variable set. Limitations (a) through (c) are dealt with in an alternative approach to FA, which is of confirmatory nature and is the subject of Chapter 9.

8.6 Comparison of Factor Analysis and Principal Component Analysis

PCA and FA are methods that are quite frequently used in behavioral, social, educational, and biomedical research. Possibly due to a number of similarities that they exhibit, the two are at times the subject of considerable confusion in applied literature. It is therefore deemed appropriate to discuss explicitly their distinctions in a separate section of this chapter, in addition to some remarks we made on this issue in Section 8.1 and in Chapter 7.

To this end, we find it worthwhile to reiterate that the goal of PCA is explaining observed measure variances with as few as possible principal components that need not be substantively interpretable, while that of FA is explaining variable interrelationships (correlations) with as few as possible factors that should be meaningful in a subject-matter area of application. Further, PCA can actually be considered a mathematical technique, in part because it does not allow for error terms as random variables and in addition there is no model underlying PCA. Alternatively, FA is a statistical technique that utilizes specific error terms within a particular model (see Equations 8.1 and 8.2 and the discussion following them). From this perspective, PCA is essentially assumption-free, while FA has model-related assumptions. Moreover, in PCA the resulting principal components are linear combinations of the observed variables and thus share the same observed variable status. By way of contrast, in FA the factors are not linear combinations of the manifest measures (even though their estimates are), and are not observed and perhaps not even observable. In PCA, because principal components are uncorrelated by definition and are unique, a change in the number of extracted principal components does not change individual scores on the components (that have been extracted before further components are considered). This is in contrast to FA where the factors need not be uncorrelated and are unique only up to a rotation; thus, a change in the number of factors may well alter the estimated factor scores (see Equations 8.10 and 8.11).

We conclude this chapter by emphasizing that there is no method for proving the existence of factors. Even the "best" (i.e., most interpretable and with best statistical indices) FA results do not provide evidence for the existence of these latent variables. Latent variables are typically abstractions of common types of behavior (measurements) that help researchers advance the behavioral, social, and educational sciences by allowing them to develop and evaluate useful theories about studied phenomena.

9

Confirmatory Factor Analysis

9.1 Introduction

In Chapter 8, the exploratory factor analysis (EFA) approach was presented as a method for discovering how many factors could be used to explain the relationships among a given set of observed measures, and which variables load considerably on which factor. In Section 8.5, we emphasized that EFA has some potentially serious limitations. They are no longer applicable when one uses another mode of factor analysis, which is called confirmatory factor analysis (CFA) and is the topic of this chapter.

In CFA, one is not concerned with "discovering" or "disclosing" factors as in EFA, but instead with quantifying, testing, and confirming an a priori proposed (preconceived) or hypothetical structure of the relationships among a set of considered measures. In this respect, the main interest in CFA lies in examining the pattern of relations among the factors as well as those between them and the observed variables. For this reason, in order to conduct a CFA, one must have a clear initial idea about the composition of the set of analyzed variables. This analytic method is, therefore, a modeling approach that is designed to test hypotheses about the underlying factorial structure, when given in advance are the number of factors involved and their interpretation in terms of the manifest measures. Hence, one may view CFA as starting where a theory under consideration ends, quantifying all unknown aspects of a model derived from that theory, and testing whether the model is consistent with the available data.

9.2 A Start-Up Example

Let us begin with an illustration of a relatively simple CFA model. Consider a study concerned with examining the relationships between the latent dimensions of intelligence, aspiration, and academic achievement in high school students. A researcher hypothesizes that these three factors underlie the interrelationships between a set of measures obtained in this investigation. Accordingly, his or her proposed model includes three factors: intelligence, aspiration, and achievement. In addition, it posits that the intelligence factor is assessed with three specific IQ tests (referred to also as indicators), the aspiration factor is evaluated by another triple of variables (also sometimes called proxies), and finally that three other manifest measures indicate the aspiration factor. The main interest in this study lies in estimating the relationships among the intelligence, aspiration, and achievement factors. Assume also that data are available from a sample of $N = 350$ high school juniors. As in most chapters throughout the book, we presume here that the observed variables are multinormally distributed.

When employing CFA models in behavioral and social research, it is very helpful to make use of specially developed graphical devices called path diagrams. The path diagrams represent pictorially most details of a considered CFA model, and usually consist of circles (ellipses) and rectangles (squares) interconnected through one- and two-way arrows. In these diagrams, circles or ellipses are used to symbolize the factors, or latent variables of main interest. Rectangles or squares are used to designate the observed variables that assess their corresponding factors. One-way arrows connect the factors with the measures evaluating them, on the assumption that the factors play explanatory role for the latter, whereas two-way arrows stand for unspecified relationships, that is, correlations or covariances. In the rest of the book we also use short one-way arrows, with no attached symbols, to denote error or residual terms pertaining to the observed variables (see discussion in Chapter 8). The path diagram for the above described CFA model of relationships between intelligence, aspiration and achievement is presented in Figure 9.1, where their respective nine indicators are consecutively denoted as Y_1 through Y_9.

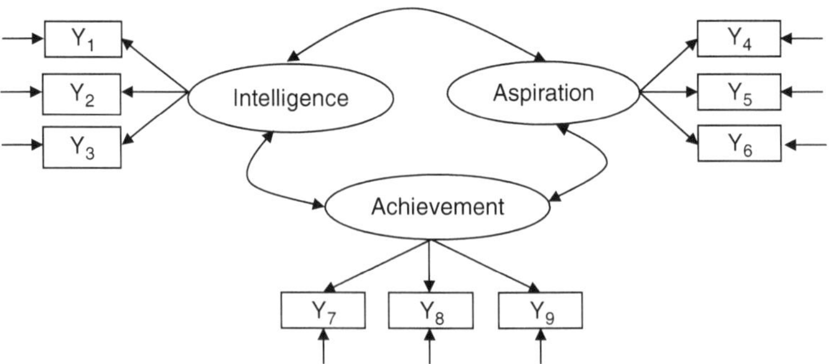

FIGURE 9.1
Proposed CFA model for nine observed measures.

With the proposed model pictorially presented in a path diagram, the next step of interest is to examine whether this model is consistent with the data collected in a study under consideration. Before we are able to deal with this question, however, we must discuss CFA in a more general framework that will allow us to readily address this query in subsequent sections.

9.3 Confirmatory Factor Analysis Model

The CFA model resembles that underlying EFA, which we discussed in detail in Chapter 8, and is formally based on the following equation:

$$\underline{y} = \Lambda\, \underline{f} + \underline{e}, \tag{9.1}$$

where

$\underline{y}$ is a $p \times 1$ vector comprising the p observed variables assumed with zero mean ($p > 1$)

$\underline{f}$ is the $q \times 1$ vector of factors similarly with means of zero ($q > 0$)

$\underline{e}$ is the $p \times 1$ vector of error terms assumed to be uncorrelated with $\underline{f}$ and with zero mean (typically in applications, q is considerably smaller than p)

Later in the chapter we generalize the model to the case of arbitrary means of the manifest variables $\underline{y}$.

In spite of the fact that the definitional equations of CFA and EFA seem identical, there are a number of deeper differences between the two modeling approaches. For one, unlike the case in EFA, there is no need for rotation in CFA. The reason is that in CFA one obtains by definition a unique solution (provided it exists). In order to obtain this solution, the problem of identification, that is, unique estimability of the unknown aspects of the model, needs to be resolved before a CFA model is further considered and evaluated in an application. To handle the issue of identification, one needs to be aware of the parameters of the model. Similar to the EFA model, a CFA model has as parameters the factor loadings, error term variances, and factor covariances. However, unlike EFA, additional parameters in a CFA model may be the factor variances (which are typically assumed to be equal to 1 in EFA) as well as error covariances, and therein lies another main distinction between these two modeling approaches. Furthermore, in a CFA model one may impose restrictions on factor loadings and/or factor or error term variances and covariances, if these constraints are consistent with available knowledge in a given substantive domain or represent hypotheses of interest to be tested with the model. These constraints cannot be introduced in EFA.

The feature of "additional" parameters possible in a CFA model comes at the price of a general issue that is not unrelated to that of lack of unique solution in EFA. This is the issue of model and parameter identification indicated above. To begin its discussion, we observe that like those of any statistical model, the CFA model parameters are meaningful and useful only if their unknown values can be uniquely deduced from the available data (i.e., if and only if they are "identified" based on the data). This means that the analyzed data need to contain information that is sufficient to find unique values for each model parameter. The problem of model identification is in general fairly complicated and cannot be treated in full detail within the confines of this chapter. We present a more formal definition of it below, and refer the reader to a number of instructive sources for further discussion (Bekker, Merckens, & Wansbeek, 1994; Bollen, 1989; see also Raykov & Marcoulides, 2006). We also mention here that most statistical software used to evaluate CFA models routinely examine numerically parameter identification, and nearly all of the time will correctly determine whether a model is identified, that is, whether all its parameters are identified. While there is a general algebraic procedure to ascertain model identification, which is indicated later in the chapter, we note that an easily and widely applicable rule for facilitating identification of CFA models is to fix at a constant value (usually 1) one loading per factor or alternatively to fix the variance of the factor (usually at 1). In addition, depending on model size—number of observed variables, number of factors, and nonzero loadings—it is possible that only up to a certain number of error covariances are identified.

This discussion also brings us to yet another distinction between EFA and CFA. It consists in the fact that unlike the case in EFA, in a CFA model it is in general not true that all factors may load on all observed variables. The reason is that if the latter were true, then the CFA model will in general not be identified as it will contain a number of unidentified parameters (in particular some of these factor loadings and/or factor variances or covariances). In special CFA models, depending on their size, it is possible that some factors may have more than a single indicator without compromising model identification. We return to the issue of identification in a later section, and close its current discussion by stressing that in a CFA model (a) one needs to fix either a loading for each factor or its variance, usually at 1, in order to facilitate model identification; and in addition, (b) one should carefully examine the output of software used to conduct CFA modeling for any indication (warning, error message, or condition code) with respect to lack of identification of any parameter.

Having addressed the issue of model identification, the next question is concerned with evaluating the meaningfulness and usefulness of a CFA model under consideration. To begin answering this query, it is instructive to make first an important observation. From Equation 9.1, using rules for covariances of linear functions of random vectors (see discussion in

Chapter 8), the following expression for the population covariance matrix Σ of the observed variables $\underline{y}$ is readily obtained:

$$\Sigma = \Lambda\Phi\Lambda' + \Psi, \tag{9.2}$$

where $\Phi = \text{Cov}(\underline{f})$ is the covariance matrix of the factors and Ψ is the covariance matrix of the error terms (both Φ and Ψ need not in general be diagonal). Equation 9.2 states that the assumption of validity of the CFA model in Equation 9.1 has as a logical consequence that the population covariance matrix becomes structured in terms of parameters, which are more fundamental and usually considerably fewer in number than its $p(p+1)/2$ nonduplicated elements. Therefore, if we denote by $\underline{\theta}$ the $s \times 1$ vector of all unknown model parameters (typically $s > 0$), Equation 9.2 can be formally rewritten as

$$\Sigma = \Sigma(\underline{\theta}). \tag{9.3}$$

Equation 9.3 states that as an implication of assuming a CFA model (i.e., Equation 9.1), each element of the population covariance matrix represents a possibly nonlinear function of model parameters. This matrix, once expressed as a function of model parameters as in the right-hand side of Equation 9.3, is referred to as implied or reproduced covariance matrix for the considered model, and plays a fundamental role in the rest of the chapter. Any CFA model has an implied covariance matrix, obtained as above in this section, which in general need not be reproducible only by that particular model.

Equation 9.3 allows us now to more formally define the notion of CFA model identification and to approach a general resolution of it. A model is called identified, if there are no two distinct parameter vectors, say $\underline{\theta}_1$ and $\underline{\theta}_2$ (with meaningful, i.e., admissible, values for their elements), which have the property that they yield the same implied covariance matrix. That is, a CFA model is identified if the fact that $\Sigma(\underline{\theta}_1) = \Sigma(\underline{\theta}_2)$ implies $\underline{\theta}_1 = \underline{\theta}_2$, for any pair of parameter vectors $\underline{\theta}_1$ and $\underline{\theta}_2$ (leading to the same implied covariance matrix). In that case, also each and every model parameter is identified. Alternatively, when two distinct vectors $\underline{\theta}_1$ and $\underline{\theta}_2$ can be found with the property that $\Sigma(\underline{\theta}_1) = \Sigma(\underline{\theta}_2)$, then the elements in which $\underline{\theta}_1$ and $\underline{\theta}_2$ differ represent model parameters that are not identified and the model is not identified either.

From this formal definition, starting with Equation 9.3, a generally applicable algebraic procedure for ascertaining model identification consists of trying to express each model parameter in terms of elements of the population covariance matrix. If for each parameter one finds that there are only a finite number of ways to express it symbolically as a function of elements of Σ, then the model will be identified; however, if the assumption of infinitely many distinct values for at least one model parameter being consistent with Σ cannot be rejected, then this parameter is not identified, and the model is unidentified as it contains at least one unidentified parameter (for additional details, see Marcoulides & Hershberger, 1997; Raykov & Marcoulides, 2006). This algebraic procedure is unfortunately quite tedious to use in practice, especially with large models. However, any time popular statistical software is used to evaluate a CFA model, a numerical examination of model identification is also carried out simultaneously by the program. For nearly all practical purposes, this examination can be relied on as a means of ascertaining model identification. Therefore, we will not discuss CFA model identification further in this chapter.

9.4 Fitting Confirmatory Factor Analysis Models

If a CFA model under consideration is identified, its unknown parameters can be uniquely estimated in an optimal way, and as a by-product one can also evaluate the model for consistency or fit with the analyzed data. To this end, in the case of multinormal observed variables one can employ the widely used maximum likelihood (ML) method, which we have already been concerned with and utilized earlier in the book.

To discuss important issues pertaining to an application of ML for the purpose of CFA model estimation and testing, let us recall that we have assumed the observed variables as having zero mean (e.g., they have been mean centered beforehand; we will relax this assumption later in the chapter). Then, as is well known, the empirical covariance matrix S represents a sufficient statistic (Johnson & Wichern, 2002). Hence, choosing values for the unknown parameters of a given CFA model so as to ensure that its implied covariance matrix $\Sigma(\theta)$ becomes as close as possible to S represents a method of evaluating (fitting) the model in a way rendering it consistent with the highest amount of information contained in the analyzed data. In fact, it can be shown that this strategy leads to the same results as the maximization of the raw data likelihood (''probability'' of observing the raw data), which is the essence of the ML method as mentioned earlier. Accordingly, fitting the model to the sample covariance matrix S—estimating its parameters so that the distance between implied and empirical covariance matrix be minimized—leads to ML estimates of the model parameters (Bollen, 1989).

Minimization of the distance between the implied covariance matrix $\Sigma(\theta)$ and the sample covariance matrix S is accomplished by minimizing the following fit function with respect to the model parameters:

$$F_{\mathrm{ML}}(\Sigma(\theta), S) = -\ln|S[\Sigma(\theta)]^{-1}| + \mathrm{tr}\{S[\Sigma(\theta)]^{-1}\} - p, \qquad (9.4)$$

where p is the number of observed variables. The minimization of this fit function is a problem in nonlinear optimization and utilizes widely available numerical algorithms that proceed typically in iterative steps, beginning with some start values (that most of the time used statistical software for CFA modeling can determine readily). Upon convergence of this iterative process, if achieved, the vector minimizer of the right-hand side of Equation 9.4 becomes available, and its elements are the sought ML estimates of the model parameters (Kaplan, 2000).

The fit function in Equation 9.4 is obtained using the likelihood ratio principle when comparing the saturated model (fully parameterized model, or full model) with a particular CFA model under consideration (restricted model). The saturated model has as many parameters as there are nonduplicated elements of the analyzed covariance matrix, and thus fits the data perfectly. Hence, this model provides a highly useful benchmark against which the goodness of fit of other identified models for the same set of observed variables can be evaluated (Bollen, 1989).

Once a CFA model is fitted to the analyzed data in this manner, the likelihood ratio test of the null hypothesis that the model perfectly reproduces the population covariance matrix is based on the statistic

$$T = (n-1)F_{\min}, \tag{9.5}$$

where n is sample size and $F_{\min}$ is the attained minimum value of the fit function in Equation 9.4. Accordingly, if this null hypothesis is correct and sample size is large, the test statistic T follows a chi-square distribution with degrees of freedom, d, equal to

$$d = p(p+1)/2 - s \tag{9.6}$$

(s is the number of model parameters, also explicitly provided by utilized software); for this reason, T is often referred to as chi-square index for the fitted model (chi-square value) (Bollen, 1989).

Furthermore, at the minimizer of the fit function given by Equation 9.4, the square-rooted main diagonal elements of the simultaneously obtained information matrix provide large-sample standard errors for the model parameters. In addition, parameter restrictions can also be tested with a likelihood ratio test, which has as statistic the difference of the chi-square values for the restricted (constrained) and unrestricted (full) model. This statistic, under validity of the restrictions and large samples, follows a chi-square distribution with degrees of freedom being the difference in the degrees of freedom of the restricted model and the full model, as follows from the likelihood ratio theory (Kaplan, 2000).

A main limitation of the chi-square index T is, however, that with large samples it exhibits a spurious tendency of being associated with significant values (i.e., with pertinent p-values that are significant), even if a tested model may be incorrect to a very minor and substantively meaningless degree. This has motivated the development of numerous descriptive and alternative goodness of fit indices over the past 30 years or so. Their detailed discussion goes beyond the confines of this book, and at this point we only mention that one of the currently most popular alternative fit indexes is the root mean square error of approximation (RMSEA) (Steiger & Lind, 1980; see also Browne & Cudeck, 1993). The RMSEA, denoted π, is defined as badness of fit per degree of freedom, using the applicable of the following two equations:

$$\pi = 0 \quad \text{if } T \leq d$$

or

$$\pi = \sqrt{(T - d)/[d(n - 1)]} \quad \text{if } T > d. \tag{9.7}$$

It has been suggested that model fit may be considered good if in the analyzed data set its associated RMSEA is lower than .05, or the left endpoint of its 90%-confidence interval (CI) is under .05 (Browne & Cudeck, 1993). Owing to its current widespread use and popularity, when evaluating fit of CFA models in this book we will be primarily concerned with the RMSEA in addition to the chi-square index T and its corresponding p-value.

In order to fit a CFA model to a given data set in practice, it is typically necessary to make use of specifically developed computer software. A number of excellent programs are available to date, such as AMOS, EQS, LISREL, M*plus*, Mx, RAMONA, and SEPATH. For a detailed discussion of how these specialized software can be used, we refer the reader to their manuals and various related textbooks (Byrne, 1994, 1998, 2001; Dunn, Everitt, & Pickles, 1993; Raykov & Marcoulides, 2006). Each program may be thought of as utilizing a particular language with special commands, syntax structure, and related rules that must be followed to communicate to the program all needed details for analyzing a CFA model. Thereby, a set of command lines, collectively referred to as input file (or command file), is submitted to the software, and model interpretation is based on the resulting output generated by the program. In the input file, description of the data and its location are provided as well as such of the model to be fitted, and possibly a list of desired output results to be returned to the analyst. Although as indicated above, there are currently a number of software that can be employed for fitting CFA models, for our purposes in the remainder of this book the program M*plus* (Muthén & Muthén, 2006) may be viewed as most appropriate. In the next section, we provide a short introduction to this software, and specifically to its application for purposes of CFA, which will suffice for our goals in this and the following chapters.

9.5 A Brief Introduction to M*plus*, and Fitting the Example Model

M*plus* has nine main commands, each equipped with several options, subcommands, and keywords. The program can fit a very wide spectrum of model classes of relevance in social and behavioral science research. For fitting CFA models in this book, we use only a fraction of these commands and options. In particular, we will start the input file with the command TITLE, which will contain a short description of the analytic session goal. We will then provide the location of the data, using the command DATA. (For simplicity, we make use of free format as data file format throughout the book.) Next, with the command VARIABLE, we will assign names to the variables in the data file, in the order they appear in the latter. The model will then be defined in a section starting with the command MODEL. In it, we will state the label (name) of each factor followed by its indicators, using the keyword "BY" to assign each observed variable to its corresponding factor. By default, all factor correlations are model parameters to be freely estimated, but if we wish to introduce also error covariances as parameters, we will do so by connecting through the keyword "WITH" the names of the indicators to which they pertain. Any particular output results beyond the default provided by the software will be requested with the command OUTPUT followed by stating the respective options.

To illustrate this discussion using a CFA model, let us return to the earlier considered example of the high school junior study of intelligence, aspiration, and achievement (see Figure 9.1 for the path-diagram of this model). Assume that for the available data from $n = 350$ students, the multinormality assumption is plausible, and that the covariance matrix of the nine involved measures is as given in Table 9.1; thereby, the first three measures are indicators of the factor intelligence (denoted Y_1 through Y_3),

TABLE 9.1

Covariance Matrix of Intelligence, Aspiration, and Achievement Measures

Y_1	Y_2	Y_3	Y_4	Y_5	Y_6	Y_7	Y_8	Y_9
125.37								
81.36	133.23							
87.25	79.45	129.55						
34.26	31.39	41.25	144.21					
45.31	42.37	51.33	101.35	152.62				
50.61	49.41	52.15	95.33	97.35	155.77			
61.58	55.71	57.36	62.36	59.87	61.44	151.98		
58.95	61.33	62.76	59.61	63.56	66.31	91.56	145.36	
64.67	62.55	59.47	63.76	58.45	62.33	90.54	101.33	144.53

the next three of aspiration (Y_4 through Y_6), and the last triple are proxies of achievement (Y_7 through Y_9). (This covariance matrix is provided in the file ch9ex1.dat available from www.psypress.com/applied-multivariate-analysis, with variables following the order in Table 9.1.)

To fit the CFA model in Figure 9.1 with M*plus* to these data, we use the following command file. (Detailed clarifications of its lines are provided subsequently.)

```
TITLE:      CFA MODEL OF INTELLIGENCE, ASPIRATION, AND
            ACHIEVEMENT
DATA:       FILE = CH9EX1.COV;
            TYPE = COVARIANCE;
            NOBSERVATIONS = 350;
VARIABLE:   NAMES = IQ1 IQ2 IQ3 ASP1 ASP2 ASP3 ACH1 ACH2
            ACH3;
MODEL:      INTELL BY IQ1-IQ3;
            ASPIRE BY ASP1-ASP3;
            ACHIEV BY ACH1-ACH3;
```

As indicated above, we start the command file with a title (which can be of any length; here we use only a couple of lines for the sake of space). We note that with the exception of the title lines, each of the following command lines terminates with a semicolon. The DATA command that appears next indicates the name of the file containing the covariance matrix (if it is not in the current directory, an appropriate path needs to be provided here as well). M*plus* needs also to be informed of the nature of the data, which happens with the subcommand TYPE, and similarly needs to be given the sample size—use the subcommand NOBSERVATIONS for this aim. (When analyzed are the raw data, which is the default option in this software, instead of these three lines only the name of the file containing the data needs to be provided with the FILE subcommand.) Then the VARIABLE command assigns names to the observed variables, which appear in order identical to that followed by the measures in the analyzed covariance matrix (or raw data file). The model command next, initiated with MODEL, states each factor as being measured by the observed variables whose names appear subsequently. To this end, the name of the factor is first given, and then the keyword ''BY'' (for ''measured by'') is followed by a list of that factor's indicators. Note that we can use the dash to represent all applicable variables between an initial and final one, in the sequence provided in the variable command. Thus, the last three lines of the above command file define the factor INTELL (for intelligence) as being measured by the variables labeled IQ1, IQ2, and IQ3; the factor ASPIRE (for aspiration) as being assessed by the measures labeled ASP1, ASP2, and ASP3; and the factor ACHIEV (for achievement) as being indicated by the variables labeled ACH1, ACH2, and ACH3. By default, the factors are assumed interrelated (i.e., the program will estimate their covariances), and the identification problem is resolved by assuming that for each factor, the loading of its first appearing indicator after the "BY" keyword is fixed at 1.

The discussed command file produces the following output. (Its sections are presented next in a font different from the main text, and clarifying comments are inserted after appropriate portions. We present the full output, but dispense in the rest of the book with the opening and final sections that display software version information and such regarding the execution timing of analysis.)

```
Mplus VERSION 4.0
MUTHEN & MUTHEN
12/03/2007  9:29 AM
```

This output part indicates the version used and date as well as time of carrying out the analysis. It may be relevant to the user for future reference.

```
INPUT INSTRUCTIONS

TITLE:     CFA MODEL OF INTELLIGENCE, ASPIRATION, AND
           ACHIEVEMENT
DATA:      FILE=IAA.COV.INP;
           TYPE=COVARIANCE;
           NOBS=350;
VARIABLE:  NAMES=IQ1 IQ2 IQ3 ASP1 ASP2 ASP3 ACH1 ACH2
           ACH3;
MODEL:     INTELL BY IQ1-IQ3;
           ASPIRE BY ASP1-ASP3;
           ACHIEV BY ACH1-ACH3;

INPUT READING TERMINATED NORMALLY
```

This section echoes the input file and is important to check since it lists the commands accepted and followed by M*plus*. The last line of the presented section states that no syntax errors were found by the program, and should be found at this location in the output of a fitted model in order to continue with interpreting its following sections. In case syntax errors are made by the analyst, warning messages will be issued here by the software that aim at guiding the input file revision effort.

```
CFA MODEL OF INTELLIGENCE, ASPIRATION, AND ACHIEVEMENT

SUMMARY OF ANALYSIS

Number of groups                                        1
Number of observations                                350
```

```
Number of dependent variables                                   9
Number of independent variables                                 0
Number of continuous latent variables                           3

Observed dependent variables

  Continuous
  IQ1      IQ2      IQ3      ASP1      ASP2      ASP3
  ACH1     ACH2     ACH3

Continuous latent variables
  INTELL    ASPIRE    ACHIEV

Estimator                                                      ML
Information matrix                                       EXPECTED
Maximum number of iterations                                 1000
Convergence criterion                                   0.500D-04
Maximum number of steepest descent iterations                  20

Input data file(s)
  IAA.COV.INP

Input data format FREE
```

This output part provides technical details regarding the internal software arrangements for conducting the requested analysis, and should also be monitored in order to ensure that the correct analysis is being carried out on the data of concern.

```
THE MODEL ESTIMATION TERMINATED NORMALLY
```

This line states that the iterative numerical process underlying minimization of the fit function in Equation 9.4 has terminated uneventfully, that is, it has converged. The statement does not say, however, whether the found solution (set of parameter estimates) makes sense or whether it is associated with a well-fitting model, but instead represents the first necessary condition for consideration of the remaining part of the output. If convergence of the minimization process has not been achieved, a warning will be issued by the software at this point, which will inform the analyst that no convergence was attained. In that case, no part of the remaining output should generally be trusted, and measures should be taken to deal with the problem of nonconvergence that may have multiple causes (Bollen, 1989). Furthermore, with lack of model identification, a warning that indicates it can also be found here, possibly singling out the parameter that is not identified. In the present case, no such warnings are found since the fitted model is identified. (In particular, we capitalized on the default feature of M*plus* to fix at 1 a loading for each factor.)

```
TESTS OF MODEL FIT

Chi-Square Test of Model Fit
  Value                          32.889
  Degrees of Freedom                 24
  P-Value                        0.1064
```

The reported model fit is the result of the likelihood ratio test comparing the model in Figure 9.1 with the saturated model. Accordingly, the model of interest seems to fit the data acceptably well, since the *p*-value of this test is not significant (assuming a conventional significance level of .05). We do not make a final decision with regard to its fit until we examine the RMSEA below, however. We stress that this is the first point in the output where one could find out the degrees of freedom explicitly stated (as calculated by the software, based on the submitted command file). A necessary, but not sufficient, condition for model identification is that the degrees of freedom be nonnegative (Raykov & Marcoulides, 2006), so here one can also readily perform a check of this condition.

```
Chi-Square Test of Model Fit for the Baseline Model

  Value                          1620.820
  Degrees of Freedom                   36
  P-Value                          0.0000
```

This section gives the likelihood ratio test statistic for the model stipulating that there are no interrelationships at all between the analyzed variables, which is sometimes also referred to as the "null model" (or baseline model). In this sense, this is the least interesting model that one could come up with, and in the present example is clearly rejected.

```
CFI/TLI

  CFI            0.994
  TLI            0.992
```

These are two descriptive fit indices that, unlike the inferential chi-square fit index examined above, tell only how well the model fits the sample at hand. There is a large body of literature on descriptive fit indices that we will not be concerned with in this book but instead refer to other sources discussing them (Bentler, 2004; Bollen, 1989; Marcoulides & Schumacker, 1996). We merely note that values for these two indices fairly close to 1 are indicative of good model fit, as is the case in this example.

```
Loglikelihood

  H0 Value            -11478.191
  H1 Value            -11461.746
```

These two lines provide the loglikelihood values associated with the fitted model (lower line) and the saturated model (upper line). Twice the difference between them, taken in this order, equals the chi-square goodness of fit index for the fitted model, which we already discussed above.

```
Information Criteria

  Number of Free Parameters                   21
  Akaike (AIC)                         22998.381
  Bayesian (BIC)                       23079.398
  Sample-Size Adjusted BIC             23012.778
    (n* = (n + 2) / 24)
```

These indices utilize the concept of information-based fit criteria, and become particularly important when comparing a set of competing or alternative models for given data. Under these circumstances, the model with the smallest of them may be considered desirable to select and use further.

```
RMSEA (Root Mean Square Error Of Approximation)

  Estimate                              0.033
  90 Percent C.I.                       0.000     0.058
  Probability RMSEA <= .05              0.859
```

The RMSEA index is well within the range of acceptable model fit (up to .05, as mentioned above), and the left endpoint of its 90%-CI is as good as it can be, that is, 0. These results further point to the fitted model being a plausible means of data description and explanation.

```
SRMR (Standardized Root Mean Square Residual)

  Value          0.027
```

Finally, in terms of fit criteria, if all variables were standardized prior to the analysis, the SRMR index could be considered to be the average covariance residual (i.e., the average unexplained by the fitted model element of the analyzed covariance matrix). A finding of this residual being fairly small, as in the present example, is another piece of fit information corroborating plausibility of a proposed model.

```
MODEL RESULTS
                       Estimates        S.E.      Est./S.E.

INTELL BY
  IQ1                    1.000         0.000        0.000
  IQ2                    0.926         0.063       14.630
  IQ3                    0.985         0.063       15.732
```

```
ASPIRE BY
  ASP1               1.000          0.000          0.000
  ASP2               1.025          0.066         15.647
  ASP3               0.987          0.066         15.015

ACHIEV BY
  ACH1               1.000          0.000          0.000
  ACH2               1.092          0.074         14.741
  ACH3               1.087          0.074         14.727

ASPIRE WITH
  INTELL            44.557          6.633          6.717

ACHIEV WITH
  INTELL            58.191          7.077          8.222
  ASPIRE            57.515          7.305          7.873

Variances
  INTELL            87.698          9.785          8.963
  ASPIRE            97.112         11.118          8.735
  ACHIEV            84.172         10.911          7.715
```

This output section lists, within respective rows, the model parameter estimates (in the first column, after factor and indicator names), their standard errors (second column), and finally their corresponding *t*-values that equal the ratio of estimate to standard error. We observe that all factor loadings are significant (see "BY" part of last presented output section), as are all factor covariances (see "WITH" part in it) and following factor variances. This is because each of their *t*-values lies outside of the nonsignificance interval (−1.96, 1.96), if one were to employ the conventional significance level of $\alpha = .05$. (Disregard the 0's in the third column for the fixed factor loadings, as these numbers are meaningless because the pertinent parameters were originally fixed at 1 to achieve model identification.)

```
Residual Variances
  IQ1            37.314            4.727            7.893
  IQ2            57.597            5.586           10.311
  IQ3            44.014            5.009            8.787
  ASP1           46.686            5.509            8.474
  ASP2           50.126            5.847            8.572
  ASP3           60.751            6.257            9.710
  ACH1           67.373            6.300           10.694
  ACH2           44.606            5.217            8.550
  ACH3           44.608            5.195            8.587
```

It is important to note here that a meaningful solution—a set of model parameter estimates—cannot contain negative variance estimates (or correlations that are larger than 1 or smaller than −1). In the residual variances part of the output, one could spot possible indications of an inadmissible solution; this would be the case if a variance were estimated at a negative value here. (We note that in such a case the program will also issue a pertinent warning, to be found in an early part of the output.) In the present example, we observe also that all error term variances are significant, a typical finding in social and behavioral research where most measures are fallible proxies of underlying factors.

```
  Beginning Time:  09:29:25
     Ending Time:  09:29:26
    Elapsed Time:  00:00:01

MUTHEN & MUTHEN
3463 Stoner Ave.
Los Angeles, CA 90066

Tel: (310) 391-9971
Fax: (310) 391-8971
Web: www.StatModel.com
Support: Support@StatModel.com

Copyright (c) 1998-2006 Muthen & Muthen
```

This final part of the output lists the timing of analysis execution and contact information for the vendor of the software. (As mentioned before, we will dispense with it and the opening section in future output presentations.)

Internal Estimation of Factor Correlations

The analysis carried out thus far quantified the relationships among the intelligence, aspiration, and achievement factors, and those between them and their measures. In addition, indication was found that the CFA model in Figure 9.1 was a plausible means of data description and explanation. That is, the hypothesis was tested—and could not be rejected—that there were three factors behind the nine observed variables of the empirical study under consideration, with successive triples of them loading correspondingly on the intelligence, aspiration, and achievement factors. With regard to the factor interrelations, however, the above analysis only estimated their variances and covariances. Even though from these estimates one can directly obtain estimates of the latent correlations, it would not be as simple to furnish standard errors and CIs for them. This is partly because we capitalized on the default feature of M*plus* to fix for each factor a loading at 1 and consider that factor's variance as an unknown parameter.

Direct point and interval estimation of factor correlations is possible with an alternative approach to model parameterization that as indicated earlier also ensures identification of this CFA model. Employing this approach, factor variances are set at 1 while all factor loadings are treated as parameters to be freely estimated. As a result of having unitary factor variances, the factor covariances become equal to their correlations, and thus can be straightforwardly point and interval estimated.

To accomplish this parameterization, all we need to do is override the default settings in M*plus* just mentioned while setting all factor variances equal to 1. This is achieved with the following altered and added command lines in the MODEL section of the last input file, without changing any other part of it:

```
MODEL:    INTELL BY IQ1* IQ2 IQ3;
          ASPIRE BY ASP1* ASP2 ASP3;
          ACHIEV BY ACH1* ACH2 ACH3;
          INTELL-ACHIEV@1;
OUTPUT:   CINTERVAL;
```

With the first three lines in this command file section, one defines as model parameters all three loadings per factor. This is accomplished by adding the asterisk immediately after the first indicator name (otherwise the software will automatically fix that factor loading to 1; we note that in the M*plus* syntax, asterisk is typically used to denote a freely estimated parameter). Further, with the fourth line, one achieves model identification in a different way than the one used in the last modeling session. This is done by alternatively setting the factor variances equal to 1—to this end, after listing the names of the factors in question, one uses the symbol @, for "fix at," followed by the constant value that these variances are being set equal to (viz. 1). The last line of this input file section requests CIs (at the 95%- and 99%-levels by default) for all model parameters, and hence also for the factor correlations. As mentioned before, nothing else needs to be changed in the input file, but for completeness of this discussion we present next the entire new command file.

```
TITLE:      CFA MODEL OF INTELLIGENCE, ASPIRATION, AND
            ACHIEVEMENT
DATA:       FILE = CH9EX1.DAT;
            TYPE = COVARIANCE;
            NOBS = 350;
VARIABLE:   NAMES = IQ1 IQ2 IQ3 ASP1 ASP2 ASP3 ACH1 ACH2
            ACH3;
MODEL:      INTELL BY IQ1* IQ2 IQ3;
            ASPIRE BY ASP1* ASP2 ASP3;
            ACHIEV BY ACH1* ACH2 ACH3;
            INTELL-ACHIEV@1;
OUTPUT:     CINTERVAL;
```

Since only the factor loading and variances are altered in this input file (in a way that compensates changes in the former for those in the latter), just their resulting estimates are changed, with no consequences for model fit. For this reason, all output sections are identical to the corresponding ones presented above, except the model result parts; in addition, a new section is provided, which presents the 95%- and 99%-CIs. For the sake of space, therefore, we present next only these two output sections.

```
MODEL RESULTS
                    Estimates        S.E.      Est./S.E.
INTELL BY
  IQ1                   9.365       0.522         17.926
  IQ2                   8.675       0.559         15.512
  IQ3                   9.228       0.537         17.179

ASPIRE BY
  ASP1                  9.855       0.564         17.470
  ASP2                 10.102       0.581         17.387
  ASP3                  9.725       0.597         16.290

ACHIEV BY
  ACH1                  9.174       0.595         15.429
  ACH2                 10.017       0.557         17.983
  ACH3                  9.975       0.556         17.950

ASPIRE WITH
  INTELL                0.483       0.051          9.503

ACHIEV WITH
  INTELL                0.677       0.039         17.271
  ASPIRE                0.636       0.042         15.196

Variances
  INTELL                1.000       0.000          0.000
  ASPIRE                1.000       0.000          0.000
  ACHIEV                1.000       0.000          0.000
```

According to these results, all factor correlations are significant—see their *t*-values in the last column that are all higher than the significance threshold of 1.96. (We recall that a test of factor correlation significance was not possible when carrying out EFA; see discussion in Chapter 8.) The (linear) relation between aspiration and intelligence is marked but not strong, estimated at 0.48. The achievement correlations with intelligence and aspiration are, however, more prominent, estimated at 0.68 and 0.64, respectively. To examine informally this difference in relationship strength, we can look at these correlations' CIs and see how they are positioned relative to one another. To this end, we need to examine the last portion of the output. (Alternatively, the 95%-CIs say can be directly obtained by subtracting and adding 1.96 times the standard errors to respective correlation estimates.)

```
CONFIDENCE INTERVALS OF MODEL RESULTS

              Lower .5%  Lower 2.5%  Estimates  Upper 2.5%  Upper .5%
 INTELL BY
   IQ1           8.019      8.341      9.365     10.389      10.710
   IQ2           7.234      7.579      8.675      9.771      10.115
   IQ3           7.845      8.176      9.228     10.281      10.612

 ASPIRE BY
   ASP1          8.402      8.749      9.855     10.960      11.308
   ASP2          8.606      8.963     10.102     11.241      11.599
   ASP3          8.187      8.555      9.725     10.895      11.263

 ACHIEV BY
   ACH1          7.643      8.009      9.174     10.340      10.706
   ACH2          8.582      8.925     10.017     11.109      11.452
   ACH3          8.544      8.886      9.975     11.065      11.407

 ASPIRE WITH
   INTELL        0.352      0.383      0.483      0.582       0.614

 ACHIEV WITH
   INTELL        0.576      0.600      0.677      0.754       0.778
   ASPIRE        0.528      0.554      0.636      0.718       0.744

 Variances
   INTELL        1.000      1.000      1.000      1.000       1.000
   ASPIRE        1.000      1.000      1.000      1.000       1.000
   ACHIEV        1.000      1.000      1.000      1.000       1.000

 Residual Variances
   IQ1          25.138     28.049     37.315     46.581      49.492
   IQ2          43.208     46.648     57.597     68.545      71.985
   IQ3          31.112     34.196     44.014     53.833      56.917
   ASP1         32.495     35.888     46.686     57.483      60.876
   ASP2         35.064     38.665     50.126     61.587      65.188
   ASP3         44.634     48.487     60.750     73.014      76.867
   ACH1         51.145     55.024     67.372     79.719      83.599
   ACH2         31.167     34.380     44.606     54.831      58.044
   ACH3         31.227     34.426     44.608     54.790      57.989
```

As we can readily find out from the CI table (see in particular its third and fifth columns for 95%-confidence level), the CIs for the correlations of aspiration with intelligence and that of intelligence with achievement do not overlap. Thereby, the 95%-CI for the aspiration correlation with intelligence is located completely to the left of that CI for the correlation of achievement and intelligence. One may, therefore, suggest that the degree of linear relationship between intelligence and achievement is considerably stronger than this relationship of intelligence and aspiration. In the same way, it is seen that the linear relationship of achievement with intelligence is similar in strength to that relationship between achievement and aspiration, since their CIs overlap nearly completely. Also, the relationship between aspiration and intelligence seems to be notably less strong than the relationship of intelligence with achievement because the CI of the latter correlation is almost entirely above the CI of the former correlation. (We stress that the three CI-based comparative statements regarding latent correlations, which were made in this paragraph, are informal and do not result from statistical tests.)

In conclusion of this section, we note that its method of fitting and testing CFA models and obtaining interval estimates of their parameters will prove very beneficial later in the book.

9.6 Testing Parameter Restrictions in Confirmatory Factor Analysis Models

Behavioral and social researchers are typically interested in developing models of studied phenomena that are parsimonious and, in addition, fit analyzed data acceptably well. These models provide best (most precise) estimation of their unknown parameters and lend themselves most readily to subject-matter interpretation. One can arrive at such models after imposing plausible and substantively meaningful parameter restrictions in an initially considered model. These restrictions usually represent theory-based relationships or hypotheses of interest in their own right, and when they involve parameters of CFA, models can be tested within the framework underlying the present chapter. To this end, as indicated earlier, one fits two models, one with and one without the restriction(s) of interest. With large samples, it follows from the likelihood ratio theory (Johnson & Wichern, 2002) that if the difference in their chi-square values is significant when compared to the chi-square distribution with degrees of freedom being the difference in these of the involved models, one can reject the null hypothesis of the restriction(s) being valid in the studied population; otherwise the restriction(s) may be interpreted as plausible.

In the context of the empirical example used throughout this chapter, a substantively interesting question refers to whether a given set of indicators evaluate the same factor (latent construct) in the same units of measurement. (In the psychometric and sociometric literature, such indicators are referred to as tau-equivalent or true-score equivalent measures.) For example, it would be of interest to find out whether the IQ tests evaluate intelligence with the same units of measurement. This query can be addressed by testing the restriction of equal loadings of the IQ tests on their common factor, intelligence. To evaluate this restriction, what we need to do is instruct the software to keep the three IQ test loadings on this factor the same while searching for the optimal estimates of all model parameters. This is accomplished by a minor modification of the last used M*plus* input file. Specifically, in the end of the command line defining the indicators of the intelligence factor, we assign an integer number placed within parentheses, for example, finalize that line with "(1)." For completeness of this discussion, we include next the so-modified command file (see end of its 6th line for this single change relative to the last command file).

```
TITLE:      CFA MODEL OF INTELLIGENCE, ASPIRATION, AND
            ACHIEVEMENT
DATA:       FILE=CH9EX1.DAT;
            TYPE=COVARIANCE;
            NOBS=350;
VARIABLE:   NAMES=IQ1 IQ2 IQ3 ASP1 ASP2 ASP3 ACH1 ACH2
            ACH3;
MODEL:      INTELL BY IQ1* IQ2 IQ3 (1);
            ASPIRE BY ASP1* ASP2 ASP3;
            ACHIEV BY ACH1* ACH2 ACH3;
            INTELL-ACHIEV@1;
OUTPUT:     CINTERVAL;
```

This leads to the following results, whereby only the chi-square value portion of the output is presented that is of relevance for testing parameter restrictions (after ensuring, of course, admissibility and substantive meaningfulness of the solution, as is the case here).

```
THE MODEL ESTIMATION TERMINATED NORMALLY

TESTS OF MODEL FIT

Chi-Square Test of Model Fit

  Value                          34.268
  Degrees of Freedom                 26
  P-Value                        0.1284
```

We observe that the chi-square value of this model increased relative to the full model fitted in the preceding section, since this is a restricted model. The term "nested model" is a widely used reference to models obtained from one another by imposing additional restrictions in one of them. In particular, a model is said to be nested in another model if the former can be obtained from the latter by introducing one or more parameter restrictions. Examples of such restrictions are fixing a parameter to zero or another constant, setting at least two parameters equal to each other, or involving them in a more general linear or nonlinear constraint (restriction, restraint). Owing to this feature, the nested model has a larger number of degrees of freedom, and (typically) a chi-square value that cannot be smaller than the value associated with the model in which it is nested.

With this in mind, we see that the presently considered model is nested in the last model fitted in the previous section. To test now the null hypothesis of the intelligence measures being true-score equivalent, we obtain the difference in the chi-square values of this model and the one in which it is nested (see preceding section for the latter's chi-square value and degrees of freedom): $34.268 - 32.889 = 1.379$. Since the difference in degrees of freedom is $26 - 24 = 2$, this test statistic is nonsignificant as it is smaller than the cutoff value of 5.99 for the relevant chi-square distribution with two degrees of freedom (at a conventional significance level of .05). We therefore conclude that there is no evidence in the analyzed data, which would warrant rejection of the claim that the IQ tests assess intelligence with the same units of measurement.

In the context of this discussion, a more general note is worth making. When considering a set of nested models, it is recommendable to retain in later analyses restrictions that are not rejected earlier, since they make a model more parsimonious. When ending up with a tenable model while following this recommendation, that model represents a parsimonious and plausible means of data description and explanation. (One would gain further confidence in the last model after a replication study; see next section. For additional discussion of important issues related to model parsimony, see Raykov & Marcoulides, 1999.)

9.7 Specification Search and Model Fit Improvement

In social and behavioral research, it is not infrequent that an initially considered model does not fit an analyzed data set well. For example, a tested theory that is embodied in the model may be poorly developed and possibly require changes or adjustments. The question that immediately comes to mind then is how to improve the model so as to fit the data better. That is, how can one modify the model in order to render it well fitting the data at hand? The process of answering this question is called specification search. It should best begin by a careful examination of two possible main sources of model misfit: (a) the analyzed data are not of good quality (e.g., the measures used have low validity and/or reliability), and (b) the model is deficient in some sense, that is, misspecified. In the general context of CFA, model misspecification frequently means omitting a loading of an observed variable on a factor other than the one that it is already assumed to be an indicator of, or perhaps omitting a latent variable covariance or error covariance. Further, it is possible that a restriction(s) imposed in the model is incorrect, for example, the assumption of equal loadings for several indicators, or that of some factor loadings being equal to a given constant(s) or involved in linear/nonlinear constraint(s).

In order to locate (a) omitted parameters that are meaningful from a substantive viewpoint and considerably improve model fit if added, or (b) potentially incorrect parameter restrictions, it is possible to use the so-called modification indices (MIs) that are also known as Lagrange multiplier test statistics (Bentler, 2004). There is an MI for each parameter involved in a constraint by itself or with one or more other parameters, or not included in a given model (i.e., fixed at 0 in the latter). In particular, there is an MI for each factor loading assumed to equal 0 in a model (e.g., cross-loading from another factor to an indicator of a given factor), as well as for each error covariance assumed to be 0. The MI approximates the improvement (drop) in the chi-square value that would be gained if the parameter to which it pertains is released from the constraint imposed on it in a current model. For this reason, an MI may be viewed as approximately following a chi-square distribution with 1 degree of freedom. Therefore, parameters that are involved in a constraint (i.e., either being fixed at 0 or another constant, or participating in a linear/nonlinear restriction with other model parameters), which do make substantive sense if freed and are associated with MIs higher than 5—a conservative cutoff for the chi-square distribution with 1 degree of freedom—may be considered for inclusion in a subsequently fitted model version. We emphasize, however, that only such parameters should be considered releasing from a constraint, which make substantive sense and may be only seen as having been unintentionally omitted in a given model. Recently, some automated specification search procedures (e.g., genetic algorithms, ant colony optimization, and Tabu search) have also been introduced in order to make the entire model modification process more manageable (Marcoulides & Drezner, 2001; Marcoulides, Drezner, & Schumacker, 1998). We warn, however, that use of MIs to improve model fit may lead to substantial capitalization on chance, and should therefore be conducted as rarely as possible and only on strong substantive grounds. Furthermore, repeated use of MI to improve model fit effectively moves the researcher into an exploratory position, and he or she is no longer able to claim they performed "pure" CFA across all modeling sessions.

To illustrate this discussion, consider the following example. In a study of $n = 300$ high school seniors, three motivation and three college aspiration measures were administered. Their empirical covariance matrix is given in Table 9.2. (The following covariance matrix is provided in the file ch9ex2.dat available from www.psypress.com/applied-multivariate-analysis.)

On the basis of the composition of this set of observed variables, a researcher hypothesizes that there are two underlying factors, which load each only on the corresponding of the two consecutive triples of measures. To test this hypothesis, a CFA is carried out by fitting a model

TABLE 9.2

Covariance Matrix of Motivation and College Aspiration Measures

Y_1	Y_2	Y_3	Y_4	Y_5	Y_6
2.605					
1.321	3.184				
1.433	1.716	3.559			
0.664	0.792	0.858	2.911		
2.158	2.351	2.545	2.480	6.782	
0.993	1.188	1.287	2.100	3.720	4.754

with two correlated factors, each measured by a triple of pertinent indicators, to the data in Table 9.2. (These factors are next named MOTIV and ASPIRE, respectively.) The following M*plus* input file is used for this purpose. (Note the shorthand used for assigning names to variables in the VARIABLE command line.)

```
TITLE:      CFA OF MOTIVATION AND COLLEGE ASPIRATION
DATA:       FILE=CH9EX2.DAT;
            TYPE=COVARIANCE;
            NOBS=300;
VARIABLE:   NAMES=MOT1-MOT3 ASP1-ASP3;
MODEL:      MOTIV BY MOT1-MOT3;
            ASPIRE BY ASP1-ASP3;
```

This leads to the following model fit and parameter estimate output sections.

```
THE MODEL ESTIMATION TERMINATED NORMALLY

TESTS OF MODEL FIT

Chi-Square Test of Model Fit

  Value                                          32.554
  Degrees of Freedom                                  8
  P-Value                                        0.0001

Chi-Square Test of Model Fit for the Baseline Model

  Value                                         660.623
  Degrees of Freedom                                 15
  P-Value                                        0.0000

CFI/TLI

  CFI                                             0.962
  TLI                                             0.929
```

```
Loglikelihood

  H0 Value                              -3426.071
  H1 Value                              -3409.794

Information Criteria

  Number of Free Parameters                    13
  Akaike (AIC)                           6878.142
  Bayesian (BIC)                         6926.291
  Sample-Size Adjusted BIC               6885.063
    (n* = (n + 2) / 24)

RMSEA (Root Mean Square Error Of Approximation)

  Estimate                                  0.101
  90 Percent C.I.                           0.066         0.139
  Probability RMSEA <= .05                  0.009

SRMR (Standardized Root Mean Square Residual)

  Value                                     0.047


MODEL RESULTS

                      Estimates      S.E.     Est./S.E.
MOTIV BY
  MOT1                   1.000      0.000        0.000
  MOT2                   1.147      0.121        9.455
  MOT3                   1.241      0.129        9.592

ASPIRE BY
  ASP1                   1.000      0.000        0.000
  ASP2                   2.434      0.230       10.584
  ASP3                   1.481      0.152        9.757

ASPIRE WITH
  MOTIV                  0.831      0.129        6.463

Variances
  MOTIV                  1.168      0.200        5.839
  ASPIRE                 1.037      0.192        5.404

Residual Variances
  MOT1                   1.428      0.149        9.565
  MOT2                   1.635      0.179        9.148
  MOT3                   1.746      0.198        8.840
  ASP1                   1.865      0.164       11.338
  ASP2                   0.616      0.324        1.899
  ASP3                   2.465      0.237       10.396
```

As can be seen from the respective output parts, model fit is not satisfactory. In particular, the chi-square value is much higher than model degrees of freedom, its associated p-value is very low, and especially the RMSEA is well above the commonly used threshold of .05 for acceptable fit; in addition, the left endpoint of its 90%-CI is markedly above the threshold as well. These results indicate that the model is deficient and hence in need of modification to improve its degree of consistency with the analyzed data. (This assumes of course that the analyzed data are not deficient themselves, in the sense indicated above, which is a presumption we follow throughout the rest of this chapter.) On the basis of substantive considerations, the researcher suspects that some of the college aspiration measures may have stronger relationships to motivation than may be warranted by the fitted model. She or he decides to examine this conjecture by inspecting the MIs associated with the loadings of the aspiration measures onto the motivation factor, which are fixed at 0 in the initially fitted model. To request MIs, one only needs to add the following line at the very end of the above input file:

```
OUTPUT:  MODINDICES(5);
```

The so-extended command file yields the same output as the one presented last, plus a section containing the MIs, which is as follows:

```
MODEL MODIFICATION INDICES

Minimum M.I. value for printing the modification index   5.000

                          M.I.     E.P.C.      Std      StdYX
                                             E.P.C.    E.P.C.

BY Statements

MOTIV    BY ASP1          5.573    -0.423    -0.457    -0.268
MOTIV    BY ASP2         32.381     2.330     2.518     0.969
MOTIV    BY ASP3         11.026    -0.790    -0.854    -0.392

WITH Statements

ASP2     WITH MOT1        5.777     0.339     0.339     0.081
ASP2     WITH ASP1       11.034    -0.816    -0.816    -0.184
ASP3     WITH ASP1       32.393     0.890     0.890     0.240
ASP3     WITH ASP2        5.576    -0.957    -0.957    -0.169
```

Examination of this output part suggests, based on the magnitude of pertinent MIs, considering inclusion of several cross-loadings (see section labeled "BY Statements") or alternatively error covariances (see section "WITH Statements"). From all these options, error covariances are hard to justify on substantive grounds. In particular, the error covariances pertaining to the aspiration measures do not make much sense since these measures' interrelationships are supposed to be already explained by their own factor, aspiration; thus, there is no reason to expect on subject-matter grounds any meaningful error covariances as indices of relationship among unexplained parts of these three observed variables. Further, the MI associated with the covariance between the errors of the first motivation measure and the second aspiration measure is not really remarkable in magnitude, relative to the others, and that covariance is, in addition, hard to interpret substantively. This is because the correlation between the motivation and aspiration factors would be expected to account at least to a marked degree for the correlation between their observed measures.

In difference to the error covariances, however, the cross-loading parameters make more substantive sense. They typically indicate commonality of indicators of different factors that cannot be accounted for by a single factor correlation. In the present case, the expectation of the researcher that aspiration measures may have some strong communality with motivation indicators, in addition to the high value of the MI of the second aspiration measure loading on the motivation factor, contributes to a suggestion to include this factor loading. Since this parameter is substantively motivated, we introduce it in the model version fitted next. We also note that the expected parameter change column (titled "E.P.C.") contains approximate values of the pertinent parameters if freely estimated (since they are fixed at 0 in the currently considered model), and their standardized values follow within respective row in the subsequent columns.

To include the loading of the motivation factor upon the second aspiration measure, we only need to correspondingly modify the line defining the motivation factor to reflect that, without changing anything else in the command file of the last fitted model. Specifically, that line now looks as follows (note in particular its final portion):

```
MOTIV BY MOT1-MOT3 ASP2;
```

The following output results when the so-modified command file is used. We insert comments at appropriate places.

```
THE MODEL ESTIMATION TERMINATED NORMALLY

TESTS OF MODEL FIT

Chi-Square Test of Model Fit
  Value                              1.635
  Degrees of Freedom                     7
  P-Value                           0.9773
```

Model fit is satisfactory now, after having been significantly improved (chi-square difference is $32.554 - 1.635 = 30.919$, which is well above 3.84, that is, the relevant cutoff for the pertinent chi-square distribution with 1 degree of freedom; see above in this section). This significant fit improvement can be seen as evidence in favor of a cross-loading of the second aspiration measure on the motivation factor.

```
Chi-Square Test of Model Fit for the Baseline Model

  Value                                  660.623
  Degrees of Freedom                          15
  P-Value                                 0.0000

CFI/TLI

  CFI                                      1.000
  TLI                                      1.018

Loglikelihood

  H0 Value                             -3410.611
  H1 Value                             -3409.794

Information Criteria

  Number of Free Parameters                   14
  Akaike (AIC)                          6849.223
  Bayesian (BIC)                        6901.076
  Sample-Size Adjusted BIC              6856.676
    (n* = (n + 2) / 24)

RMSEA (Root Mean Square Error Of Approximation)

  Estimate                                 0.000
  90 Percent C.I.                          0.000          0.000
  Probability RMSEA <= .05                 0.997
```

The RMSEA is now also very good, suggesting the conclusion that the model fits acceptably well.

```
SRMR (Standardized Root Mean Square Residual)
  Value                    0.007


MODEL RESULTS

                         Estimates      S.E.     Est./S.E.

MOTIV BY
  MOT1                     1.000       0.000       0.000
  MOT2                     1.133       0.118       9.572
  MOT3                     1.226       0.126       9.717
  ASP2                     1.032       0.154       6.682
```

```
ASPIRE BY
  ASP1                 1.000        0.000         0.000
  ASP2                 1.264        0.150         8.416
  ASP3                 1.500        0.135        11.099

ASPIRE WITH
  MOTIV                0.686        0.124         5.543

Variances
  MOTIV                1.188        0.201         5.927
  ASPIRE               1.395        0.224         6.236

Residual Variances
  MOT1                 1.408        0.147         9.560
  MOT2                 1.648        0.177         9.288
  MOT3                 1.762        0.196         8.999
  ASP1                 1.506        0.156         9.686
  ASP2                 1.474        0.237         6.216
  ASP3                 1.598        0.251         6.372
```

We note that the added cross-loading is significant (see last line of first part of the section "Model Result"). In fact, this finding could be expected due to the significant drop in the chi-square fit index, which followed introduction of this parameter in the model.

We conclude this section by stressing that in order to have more trust in a model arrived at through a specification search using MIs (or using results from earlier analyses of the same data set), one needs to examine whether it would replicate on an independent sample from the same studied population. Only when this happens can one place more confidence in that modified model. This procedure can be readily carried out if a large sample is initially available. Splitting it randomly into two halves (parts) and performing specification search (or other analyses) on one as outlined in this section, can lead to at least one model version of interest that can be tested on the other half. (See also above discussion on information criteria of model fit that can also be used for selection purposes with more than one models of interest resulting from the analyses on the first half.)

9.8 Fitting Confirmatory Factor Analysis Models to the Mean and Covariance Structure

So far in the chapter, we have fit models of interest to the sample covariance matrix, that is, analyzed the covariance structure of a set of observed variables. This is a correct decision on the assumption that the observed measures have zero means (e.g., have been mean centered beforehand). When this assumption has not been made or is incorrect, however, analysis of the covariance structure is associated with potentially serious loss of sample information that is contained in the means. The reason is the well-known fact that variances and covariances (like correlations) are insensitive to variable means; that is, the former can have essentially any values that they can take, regardless of what the means are. This is of particular relevance in empirical studies where origins of analyzed variables make substantive sense and are not arbitrary or meaningless. In those circumstances, the means need to be involved in the analysis rather than disregarded as would be the case if only the covariance structure of the manifest measures were analyzed. Well-known examples when analysis of the means is necessary are studies that involve multiple-populations and/or longitudinal (repeated measure) investigations, which we will deal with in later sections/chapters of the book. To prepare the ground for them, as well as to complete the CFA discussion of the present chapter, we extend here the CFA model used so far to the case where the earlier assumption of zero means on all observed variables cannot be made. That is, in this section we allow all analyzed variables to have means that need not be zero.

To permit inclusion of variable means into the process of CFA modeling, first it is noted that just like there is an implied covariance matrix $\Sigma(\underline{\theta})$ of any considered model for a given set of observed variables (implied covariance structure), there is also a vector of implied means for these variables, which is often referred to as implied mean structure. Denoting generically population mean by μ, the implied mean vector is symbolized by $\underline{\mu}(\underline{\theta})$ and represents the collection of observed variable means expressed as functions of model parameters (see below). We stress that when we make the assumption of zero means on all observed variables (in appropriately constructed models), all implied means should be zero as well. For this reason, when zero observed means are assumed, one does not need to be concerned with the implied mean structure (by those models).

To work out the implied mean vector, we use a well-known relationship for the mean of a linear combination of random variables. Accordingly, the mean of a linear combination of any number of random variables is the same linear combination of their means. That is, using $M(\cdot)$ to denote mean, the following equation is true:

$$M(a_1y_1 + a_2y_2 + \ldots + a_py_p) = a_1\mu_1 + a_2\mu_2 + \ldots + a_p\mu_p, \qquad (9.8)$$

where $\mu_i = M(y_i)$ are the means of the random variables y_i, and a_i are given constants ($i = 1, \ldots, p$). In other words, denoting by $\underline{\mu}$ the $p \times 1$ vector of observed variable means, Equation 9.8 states that

$$M(\underline{a}'\underline{y}) = \underline{a}'\underline{\mu}, \tag{9.9}$$

where $\underline{a} = (a_1, \ldots, a_p)'$.

With this mean relationship in mind, the extension of the CFA model in Equation 9.1 to the case when the assumption of zero means on all observed variables is not made is as follows:

$$\underline{y} = \underline{\alpha} + \Lambda\underline{f} + \underline{e}, \tag{9.10}$$

where the same assumptions are advanced as in the CFA model of Equation 9.1, and in addition to that earlier notation $\underline{\alpha}$ designates the $p \times 1$ vector of observed variable intercepts, typically referred to as mean intercepts (or variable intercepts). We note that the latter parameters in Equation 9.10 play the same role as the intercepts in conventional regression analysis models, which Equation 9.10 conceptually is, although this regression cannot be performed in practice since there are no available observations on the factors $\underline{f}$.

From Equation 9.10, the implied mean structure of the CFA model under consideration is obtained by using Equation 9.8 on the individual observed variables and compacting the result into corresponding vectors and matrices:

$$\underline{\mu} = \underline{\alpha} + \Lambda\underline{\nu}, \tag{9.11}$$

where $\underline{\nu} = M(\underline{f})$ is the vector of factor means. Equation 9.11 can be formally rewritten, by analogy to the implied covariance matrix, as

$$\underline{\mu} = \underline{\mu}(\underline{\theta}), \tag{9.12}$$

where now $\underline{\theta}$ contains not only the earlier discussed factor loadings, variances and covariances as well as error variances and possibly covariances but also the mean intercepts in $\underline{\alpha}$ and factor means in $\underline{\nu}$ (under the assumption of model identification, to be addressed below).

The p equalities contained in Equation 9.12 shows how each observed variable mean is structured in terms of model parameters. For this reason, Equation 9.12 is aptly called implied mean vector, or implied mean structure. The p equations in 9.12 represent each the way in which the pertinent variable first-order moment (i.e., mean) is a function of model parameters. Further, as mentioned earlier, the covariance structure $\Sigma(\underline{\theta})$, which consists of p^2 equations in nature similar to those in Equation 9.12, indicates for each observed variable variance or covariance with another variable (i.e., second-order moments) the way in which these represent functions of model parameters. This discussion demonstrates that for a given set of observed variables, each CFA model has (a) an implied mean structure, and (b) an implied covariance structure. Oftentimes in the literature, when considering these two structures simultaneously, they are referred to with the single term "mean and covariance structure" (MCS).

Equation 9.11 shows that if, say, the q factor means in the vector $\underline{\nu}$ $(q > 0)$ are unknown parameters in a model under consideration, along with all mean intercepts in $\underline{\alpha}$, then the model will not be identified. This will happen even if all factor loadings, variances and covariances, as well as error variances and allowed error covariances are identified. The reason is that the model would then have more than p parameters that only have implications for the p variable means—these parameters are the q factor means plus the p mean intercepts. (The model will not be identified then even in case there is a single nonzero factor mean as a model parameter.) Indeed, since the means have no implications for variances and covariances, the latter do not conversely contain any information about the means. Therefore, we need to identify factor mean and mean intercepts only from the observed variable means. Yet, the latter are p in number, and hence fewer than these unknown parameters that count in total $p + q > p$.

Hence, to identify these critical parameters and thus the model, we need to impose appropriate in number and nature restrictions on the former. An available choice is the well-known constraint, used routinely earlier in this book, that all factor means are 0. (We will identify a version of the model in a different way in the next section and in a later chapter.) In fact, the assumption that all observed variables have zero mean implies that we have to assume also all factors with zero means, as done thus far in the book. More precisely, if we do not make the assumption of observed variable zero means, then we cannot have as unknown parameters more than p in total factor means and mean intercepts. Only having up to p such parameters will render the model identified, as far as the mean structure is concerned. (The model will then have to be identified also with regard to its covariance structure, an issue we discussed at considerable length in an earlier section.) This will be an important observation that we will bear in mind in the rest of the chapter.

In addition to resolving appropriately these model identification issues, to accomplish analysis of the MCS—that is, to simultaneously analyze the means of a given set of observed variables along with their covariance matrix—a correspondingly extended fit function needs to be minimized. In case of multinormality (and nonzero means on all observed variables), both the empirical means and covariance matrix are sufficient statistics (Johnson & Wichern, 2002). Therefore, choosing parameter values in a model so that the distance is minimized between the sample means, variances, and covariances to those implied by the model ensures that the process of model fitting capitalizes on the maximal possible information contained in the available data. Then it can be shown that maximizing the

TABLE 9.3

Means of the Nine Analyzed Variables for the Model in Figure 9.1

35.67	53.34	77.54	53.66	47.88	56.52	59.76	63.46	67.33

likelihood of the raw data is equivalent to fitting the model such that a particular fit function is minimized. This fit function is obtained when one adds to that used when analyzing the covariance structure a special term accounting for the degree to which the model explains the variable means (Bollen, 1989). That is, formally, the fit function minimized when fitting a model to the MCS is

$$F_{ML}(\underline{m}, \mu(\underline{\theta}), S, \Sigma(\underline{\theta})) = -\ln|S\Sigma(\underline{\theta})^{-1}| + \mathrm{tr}(S\Sigma(\underline{\theta})^{-1}) - p + (\underline{m} - \mu(\underline{\theta}))'\Sigma(\underline{\theta})^{-1}(\underline{m} - \mu(\underline{\theta})), \tag{9.13}$$

where $\underline{m}$ denotes the vector of observed variable means. Parameter standard errors, model fit evaluation, and tests of restrictions are obtained or carried out just as in the earlier considered case of covariance structure analysis (i.e., the case of CFA under the assumption of zero mean observed variables; see preceding sections).

To exemplify the discussion of MCS analysis, let us assume that the means of the variables participating in the earlier considered CFA model in Figure 9.1 were as follows (in the order intelligence indicators, aspiration indicators, and, finally, achievement indicators):

To fit the CFA model in Figure 9.1 to the MCS of these variables (see Table 9.1 for their covariance matrix), we also need to provide to M*plus* the values of the means presented in Table 9.3. In addition, we need to instruct the program to carry out MCS analysis. We accomplish this with the following input file. (The relevant data are available in file ch9ex3.dat available from www.psypress.com/applied-multivariate-analysis.)

```
TITLE:      CFA MODEL OF INTELLIGENCE, ASPIRATION, AND
            ACHIEVEMENT, MEAN STRUCTURE ANALYSIS
DATA:       FILE = CH9EX3.DAT;
            TYPE = MEANS COVARIANCE;
            NOBS = 350;
VARIABLE:   NAMES = IQ1 IQ2 IQ3 ASP1 ASP2 ASP3 ACH1 ACH2 ACH3;
ANALYSIS:   TYPE = MEANSTRUCTURE;
MODEL:      INTELL BY IQ1* IQ2 IQ3;
            ASPIRE BY ASP1* ASP2 ASP3;
            ACHIEV BY ACH1* ACH2 ACH3;
            INTELL-ACHIEV@1;
OUTPUT:     CINTERVAL;
```

Before we move on to examine the output, a few comments are in order with regard to some features of this new command file. First, we note the extended title that now hints in its second row to MCS analysis. Second, the type of analyzed data is now defined as MEANS COVARIANCES. In this way, the software is instructed that in the file with name CH9EX3.DAT the first row contains the variable means and the remaining nine rows the covariance matrix of the analyzed variables. Third, in the middle section of this input file, a new command appears: ANALYSIS: TYPE = MEANSTRUCTURE. This command invokes MCS analysis, rather than the default covariance structure analysis that would be performed if no analysis command were included, as was the case earlier in the chapter. The rest of this input file is identical to that needed earlier for fitting the model in Figure 9.1 (see Section 9.5).

This M*plus* command file produces the following output:

```
THE MODEL ESTIMATION TERMINATED NORMALLY

TESTS OF MODEL FIT

Chi-Square Test of Model Fit

   Value                                        32.889
   Degrees of Freedom                               24
   P-Value                                      0.1064
```

We stress that the fit of the model is unchanged relative to the last fitted unrestricted model in the preceding section. This is because as seen next there are nine added model parameters—for just as many additionally included means—to the data points that the model is fitted to (the sample means and nonduplicated elements of the empirical covariance matrix); by default, M*plus* assumes that the factor means are 0. In fact, because the model is thus saturated in its mean structure, no part of the output is changed except that the following section of variable intercepts is added. These equal the variable means, for the same reason, and the gain of fitting the model to the MCS is that we now obtain standard errors for the mean of any measure in the context of the remaining eight measures that are all interrelated.

```
Intercepts

                  Estimate         S.E.     Est./S.E.
   IQ1             35.670         0.598       59.684
   IQ2             53.340         0.616       86.578
   IQ3             77.540         0.608      127.633
   ASP1            53.660         0.641       83.716
   ASP2            47.880         0.659       72.612
   ASP3            56.520         0.666       84.843
   ACH1            59.760         0.658       90.819
   ACH2            63.460         0.644       98.613
   ACH3            67.330         0.642      104.927
```

This analysis and preceding discussion naturally give rise to the following question, "When should one analyze the covariance structure (i.e., fit a model of interest to the covariance matrix only), or the MCS (i.e., fit the model to the variable means and covariance matrix)?" The answer to this question is twofold. If one has not made the assumption of observed variables having zero mean, and the origins of their scales are meaningful rather than arbitrary, then one should analyze the MCS. This is because otherwise the model generally will not be in a position to account (i.e., fit) part of the data, namely, the variable means. The latter omission will typically be the case in multiple-population and/or repeated measure studies (but not necessarily only then), when a considered model is fitted to the covariance structure only.

Conversely, if the assumption of observed variable zero means is made, which would be meaningful when the origins of the scales of observed variables are arbitrary (and the variables have been mean-centered beforehand), then one will need to additionally assume the factor means as zero. This is because otherwise there is in general no substantively interpretable possibility for the implied observed means to be 0—this follows from the assumption of zero observed variable means and the CFA model in Equation 9.1. Under the assumption of zero means on the observed variables, it will be essentially immaterial whether one analyzes the covariance structure or the MCS in case no restrictions are imposed on mean intercepts and/or factor means. Then one obtains in the mean intercept estimates such of the observed variable means.

In short, unless we assume zero observed variable means, we should analyze the MCS, that is, fit the model to the observed means and covariance matrix. If we assume zero means on all observed measures, then we can analyze the covariance structure (i.e., fit the model to the covariance matrix only), making the additional assumption of factor means being zero as well. From this perspective, any CFA model may be viewed as being fitted to the MCS. In the special case when the assumption of observed (and thus also factor) means of zero is made, and no parameter constraints are imposed that have implications for the implied moment structure, the same results—both with regard to fit and estimates of factor loadings, variances and covariances, as well as error variances and covariances, if any—follow as when the model is fitted to the covariance structure only.

In conclusion of this section, one does not need to ask the question of whether to fit the model only to the covariance matrix or to it and the variable means, if analyzing the raw data. (Then M*plus* will fit the model to the MCS by default.) Hence, also from this perspective, it is recommendable to always fit models to the raw data, if available, rather than to summary statistics, such as variances, covariances, and possibly means of a given set of analyzed manifest measures.

9.9 Examining Group Differences on Latent Variables

In Chapter 6, we familiarized ourselves with the widely used method of multivariate analysis of variance (MANOVA). That method is utilized whenever one is concerned with the question of whether two or more groups differ with regard to the means on several dependent variables that are typically interrelated. A main characteristic of settings with MANOVA applicability is that these variables are observed measures. Frequently, however, behavioral and social scientists are interested in examining group differences on studied factors or constructs, such as intelligence, ability, extraversion, neuroticism, depression, to name only a few. For example, a scholar may be concerned with gender differences in writing ability of third graders, or in socioeconomic class differences in depression in elderly adults, or with whether there are gender differences in neuroticism in middle aged persons. These constructs are not directly observable, as mentioned in Chapter 8, that is, they are latent (hidden) variables, at times perhaps hypothetical and/or not very well defined. For this reason, the factors are not really measurable themselves but instead only their manifestations in studied subjects can be evaluated under certain circumstances. This leads to data sets containing multiple observed variables that are fallible indicators (proxies) of the unobservable factors.

When group mean differences on such latent variables are of concern, MANOVA is not applicable since it is a statistical procedure that deals only with observed variables of main interest. Then it is desirable to have a method allowing the study of these group differences using data from observable indicators of the latent variables involved. The CFA framework underlying the present chapter provides an approach to accomplishing this aim, which is dealt with in this section.

To begin, suppose we have data from G groups (i.e., samples from G distinct populations, where $G > 1$), and are interested in finding out whether the means of several factors or constructs differ across them. We assume that the following CFA model holds in each population (for the purposes of this discussion, we restate Equation 9.10 next in each group):

$$\underline{y}_g = \underline{\alpha}_g + \Lambda_g \underline{f}_g + \underline{e}_g, \tag{9.14}$$

where

$\underline{y}_g$ is the vector of p observed variables in the gth group ($p \geq 2$)
$\underline{f}_g$ is the vector of q factors in it (usually, $q < p$, assuming of course $q \geq 1$)
$\underline{\alpha}_g$ is the $p \times 1$ vector of mean intercepts in that group
$\underline{e}_g$ is the pertinent vector of error terms there, which is uncorrelated with $\underline{f}_g$ ($g = 1, \ldots, G$)

From Equation 9.14 it follows (see, e.g., Chapter 8) that the implied covariance matrix in each population is

$$\Sigma(\underline{\theta}_g) = \Lambda_g \Phi_g \Lambda_g' + \Psi_g, \tag{9.15}$$

where $\underline{\theta}_g$ denotes the vector of model parameters while Φ_g and Ψ_g are the covariance matrices of the factors and error terms, respectively, which need not be diagonal ($g = 1, \ldots, G$). Further, the implied mean structure in each population is (see previous section)

$$\mu(\underline{\theta}_g) = \underline{\alpha}_g + \Lambda_g \underline{\nu}_g, \tag{9.16}$$

where $\underline{\nu}_g$ denotes the mean vector of the factors ($g = 1, \ldots, G$). The intercept parameters $\underline{\alpha}_g$ in Equation 9.14 are frequently referred to as location parameters, and the factor loadings in Λ_g as scale parameters; the reason is that the elements of $\underline{\alpha}_g$ govern the origin of the scale on which the factors are measured, while Λ_g have consequences for the units in which they are evaluated ($g = 1, \ldots, G$).

With these assumptions, we next observe that in order to be able to examine group differences on the latent factors $\underline{f}_g$, they need to be measured on the same scale in all G groups under consideration. This means that one must make sure that (a) the origin, and (b) the units of measurement of the scales of these factors are identical in all populations. Under these circumstances, a comparison across groups of the means of the factors would give an answer to the question which population is on average higher (lower) on which factor. We note that it is the mean intercepts that contain information about scale origins (being location parameters), and it is the factor loadings that contain information about units of measurement (being scale parameters).

The required scale identity, in order to be in a position to compare groups on latent means, will be ensured when the factor indicators have the same mean intercepts and the same loadings in all groups under consideration. More formally, examination of mean group differences on latent factors would be possible under the following two assumptions:

$$\Lambda_1 = \Lambda_2 = \ldots = \Lambda_G, \tag{9.17}$$

which is often referred to as factorial structure invariance, and

$$\underline{\alpha}_1 = \underline{\alpha}_2 = \ldots = \underline{\alpha}_G. \tag{9.18}$$

Then a direct cross-group comparison of the means on any factor would answer the question whether there are group differences on that latent dimension.

Since being able to test group differences on latent factors depends on the assumptions in Equations 9.17 and 9.18, it is important to be in a position to evaluate their plausibility. This can be accomplished using a special multistage modeling procedure (Muthén & Muthén, 2006). An essential part of it consists in conducting CFA simultaneously across several populations. This is achieved via single-session CFA modeling in all groups, which is concerned with minimizing a multiple-group fit function representing a weighted sum of the fit functions for the CFA model in each of the groups; the weights in this sum are n_g/n, where $n = n_1 + n_2 + \ldots + n_G$ is the total sample size while n_g denotes the sample size in the gth group ($g = 1, \ldots, G$). That is, multiple-group CFA is carried out using the same general principles as outlined earlier in this chapter, with the difference that minimized is now the multigroup fit function

$$F = \frac{n_1}{n} F_1 + \frac{n_2}{n} F_2 + \ldots + \frac{n_g}{n} F_g, \tag{9.19}$$

where F_g is the fit function used in each of the groups ($g = 1, \ldots, G$) (Bollen, 1989). Evaluation of parameter standard errors and testing of parameter constraints follow then the same procedure as in the single-group case already discussed.

The above-mentioned multistage procedure for evaluating the assumptions in Equation 9.18 consists of the following steps (Muthén & Muthén, 2006):

Step 1. Fit the CFA model of Equation 9.1 in each sample (with all factor means assumed 0), in order to ensure that it is plausible in each of the populations under consideration.

Step 2. Fit the CFA model of Equation 9.1 simultaneously in all samples, allowing all parameters to be freely estimated (all factor means being assumed 0). This will be a baseline model, or full model, for the first test of relevance carried out next, that of population identity in the factor loadings.

Step 3. Fit the CFA model of Equation 9.1 to the covariance structure in all samples, whereby $\Lambda_1 = \Lambda_2 = \ldots = \Lambda_G$ is set (and in all groups all latent means are assumed 0). This is a nested model in the one fitted in step 2. With the present model, the factor loading equality restriction is tested using the likelihood ratio procedure, that is, by comparing the chi-square value obtained here with that of the model in step 2. When the restriction of group identity in the factor loading matrix is not rejected, factorial structure invariance is plausible and one can proceed to the next step.

Step 4. Fit the CFA model of Equation 9.1 as in step 3 but to the MCS, and with the restrictions $\Lambda_1 = \Lambda_2 = \ldots = \Lambda_G$ as well as $\underline{\alpha}_1 = \underline{\alpha}_2 = \ldots = \underline{\alpha}_G$; thereby, in order to facilitate testing subsequently group mean differences on the latent factors, fix their means in one of the groups to be all equal to 0 and free these factor means in all remaining groups. If the chi-square goodness of fit index of this model changes, relative to that index of the model fitted in step 3, to a degree comparable to the difference in degrees of freedom associated with these two models, one may view the restriction of identical intercepts across groups as plausible.

In the last considered model with group restrictions on the loading matrices and mean intercepts, a useful default in M*plus* is that the first sample listed in the DATA command has its mean automatically set equal to 0 (i.e., $\underline{\nu}_1 = \underline{0}$) while the latent variable means are freely estimated parameters in all remaining samples. Therefore, in this model version, significance of any factor mean in the gth sample ($g = 2, \ldots, G$) implies that the latter population is on average different on that factor from the first population (with fixed at 0 latent means). To test then differences on any given factor between any two other populations, one simply imposes the restraint of their pertinent factor means being the same and uses as above the likelihood ratio test to examine the null hypothesis of their identity.

We illustrate this discussion with the following example. In a study of reading and mathematics ability of elementary school children, three measures of each of these abilities were used. In the girls and boys groups, the following means and covariance matrices were obtained, respectively (see Table 9.4; the following data are provided in file ch9ex4.dat available from www.psypress.com/applied-multivariate-analysis, along with its relevant subsets needed for the following analyses, which are correspondingly named):

TABLE 9.4

Means (Top Row) and Covariance Matrices for Reading and Mathematics Ability Measures (Denoted RA1-RA3 and MA1-MA3, Respectively) in a Study of 230 Girls (Top Part) and 215 Boys (Bottom Part)

RA1	RA2	RA3	MA1	MA2	MA3
33.60	45.65	56.27	34.51	67.47	72.50
112.35					
77.22	103.42				
69.33	74.44	109.16			
21.22	23.46	29.51	111.32		
20.35	28.23	31.43	78.54	104.49	
21.53	34.33	29.23	72.75	87.97	114.62
32.69	45.94	56.32	35.61	68.43	73.57
113.38					
78.23	105.49				
69.92	74.43	108.15			
22.25	22.45	29.71	112.38		
21.32	27.26	31.63	78.65	105.49	
22.53	33.37	29.43	72.95	88.43	113.60

To follow the above multistage procedure for examining gender differences in reading and mathematics ability, we start with a two-factor CFA model (reading and mathematics ability being the underlying factors, with the above-mentioned pertinent indicators). This model is fitted separately to the data of each sample in turn. For the girls sample, the M*plus* input file needed is as follows:

```
TITLE:       FITTING A CFA MODEL FOR GIRLS (STEP 1)
DATA:        FILE = CH9EX4-GIRLS.DAT;
             TYPE = COVARIANCE;
             NOBS = 230;
VARIABLE:    NAMES = RA1-RA3 MA1-MA3;
MODEL:       READING BY RA1-RA3;
             MATHABIL BY MA1-MA3;
```

This renders the following output. (We note that model fit is acceptable—see in particular the left endpoint of the 90%-CI of RMSEA.)

```
THE MODEL ESTIMATION TERMINATED NORMALLY

TESTS OF MODEL FIT

Chi-Square Test of Model Fit

  Value                                  14.455
  Degrees of Freedom                          8
  P-Value                                0.0706

Chi-Square Test of Model Fit for the Baseline Model

  Value                                 789.522
  Degrees of Freedom                         15
  P-Value                                0.0000

CFI/TLI

  CFI                                     0.992
  TLI                                     0.984

Loglikelihood

  H0 Value                            -4805.576
  H1 Value                            -4798.348

Information Criteria

  Number of Free Parameters                  13
  Akaike (AIC)                         9637.151
  Bayesian (BIC)                       9681.846
  Sample-Size Adjusted BIC             9640.644
    (n* = (n + 2) / 24)
```

```
RMSEA (Root Mean Square Error Of Approximation)

  Estimate                                  0.059
  90 Percent C.I.                           0.000        0.107
  Probability RMSEA <= .05                  0.328

SRMR (Standardized Root Mean Square Residual)

  Value                                     0.029

MODEL RESULTS
                         Estimates        S.E.     Est./S.E.
READING BY
  RA1                      1.000        0.000        0.000
  RA2                      1.083        0.081       13.310
  RA3                      0.975        0.078       12.442

MATHABIL BY
  MA1                      1.000        0.000        0.000
  MA2                      1.196        0.083       14.403
  MA3                      1.123        0.082       13.781

MATHABIL WITH
  READING                 22.722        5.472        4.152

Variances
  READING                 70.636       10.366        6.814
  MATHABIL                65.209        9.785        6.664

Residual Variances
  RA1                     41.230        5.310        7.765
  RA2                     20.142        4.637        4.343
  RA3                     41.510        5.212        7.965
  MA1                     45.621        5.013        9.101
  MA2                     10.830        3.850        2.813
  MA3                     31.845        4.438        7.175
```

For the boys group, we only change the name of the reference data file (containing that sample's covariance matrix) and sample size. The output follows. (We note that also here the model fits the data reasonably well—see, in particular, the left endpoint of the 90%-CI of RMSEA.)

```
THE MODEL ESTIMATION TERMINATED NORMALLY

TESTS OF MODEL FIT

Chi-Square Test of Model Fit

  Value                                    13.211
  Degrees of Freedom                            8
  P-Value                                  0.1047
```

```
Chi-Square Test of Model Fit for the Baseline Model

  Value                                    736.784
  Degrees of Freedom                            15
  P-Value                                   0.0000

CFI/TLI

  CFI                                        0.993
  TLI                                        0.986

Loglikelihood

  H0 Value                               -4495.640
  H1 Value                               -4489.034

Information Criteria

  Number of Free Parameters                     13
  Akaike (AIC)                            9017.279
  Bayesian (BIC)                          9061.098
  Sample-Size Adjusted BIC                9019.903
    (n* = (n+2) / 24)

RMSEA (Root Mean Square Error Of Approximation)

  Estimate                                   0.055
  90 Percent C.I.                            0.000          0.106
  Probability RMSEA <= .05                   0.382

SRMR (Standardized Root Mean Square Residual)

  Value                                      0.028


MODEL RESULTS
                          Estimates       S.E.     Est./S.E.
READING BY
  RA1                         1.000      0.000         0.000
  RA2                         1.069      0.083        12.898
  RA3                         0.963      0.080        12.106

MATHABIL BY
  MA1                         1.000      0.000         0.000
  MA2                         1.199      0.087        13.808
  MA3                         1.128      0.085        13.301

MATHABIL WITH
  READING                    22.745      5.721         3.976
```

```
Variances
  READING                  72.377        10.876         6.654
  MATHABIL                 65.081        10.176         6.396

Residual Variances
  RA1                      40.474         5.505         7.352
  RA2                      22.227         4.886         4.549
  RA3                      40.556         5.321         7.622
  MA1                      46.775         5.278         8.862
  MA2                      11.413         4.011         2.846
  MA3                      30.323         4.506         6.730
```

As a next step, we fit this CFA model simultaneously in both samples, without any parameter constraints. The needed input file for this baseline model is as follows:

```
TITLE:       SIMULTANEOUS CFA IN TWO GROUPS (STEP 2)
DATA:        FILE=CH9EX4.DAT;
             TYPE=COVARIANCE;
             NGROUPS=2;
             NOBS=230 215;
VARIABLE:    NAMES=RA1-RA3 MA1-MA3;
MODEL:       READING BY RA1-RA3;
             MATHABIL BY MA1-MA3;
MODEL G2:    READING BY RA2-RA3;
             MATHABIL BY MA2-MA3;
```

With multiple group MCS analysis, the default in M*plus* is to have the same factor loadings in both groups. This is why the last two lines of this input file redefine the model in the second listed group in the initial data file (i.e., boys), which has the effect of requesting their estimation in the boys group freely from that of the factor loadings in the girls group. (Formally, this redefinition of the model in the last two command lines frees the factor loadings in the second group—referred to in the input file as G2—from being set equal by default to the same loadings in the first group.)

The following output results. We note that model fit is acceptable, and that the chi-square fit index and degrees of freedom are each equal to the sum of these indexes from the above analyses within each separate group (see also section titled "Chi-Square Contributions from Each Group"). Further, the parameter estimates and standard errors are the same in each group as they were found to be in the above single-group analyses. Last but not least, the girls group is next referred to as "Group G1" or simply "G1," while boys group as "Group G2" or "G2."

```
THE MODEL ESTIMATION TERMINATED NORMALLY

TESTS OF MODEL FIT

Chi-Square Test of Model Fit

  Value                                    27.666
  Degrees of Freedom                           16
  P-Value                                  0.0346

Chi-Square Contributions From Each Group

  G1                                       14.455
  G2                                       13.211

Chi-Square Test of Model Fit for the Baseline Model

  Value                                  1526.306
  Degrees of Freedom                           30
  P-Value                                  0.0000

CFI/TLI

  CFI                                       0.992
  TLI                                       0.985

Loglikelihood

  H0 Value                              -9301.215
  H1 Value                              -9287.382

Information Criteria

  Number of Free Parameters                    26
  Akaike (AIC)                          18654.431
  Bayesian (BIC)                        18760.981
  Sample-Size Adjusted BIC              18678.468
    (n* = (n + 2) / 24)

RMSEA (Root Mean Square Error Of Approximation)

  Estimate                                  0.057
  90 Percent C.I.                           0.015          0.092

SRMR (Standardized Root Mean Square Residual)

  Value                                     0.028

MODEL RESULTS
                              Estimates       S.E.      Est./S.E.
Group G1

READING BY
  RA1                            1.000       0.000        0.000
```

```
  RA2                  1.083       0.081      13.310
  RA3                  0.975       0.078      12.442

MATHABIL BY
  MA1                  1.000       0.000       0.000
  MA2                  1.196       0.083      14.404
  MA3                  1.123       0.081      13.782

MATHABIL WITH
  READING             22.724       5.472       4.153

Variances
  READING             70.638      10.366       6.815
  MATHABIL            65.216       9.786       6.664

Residual Variances
  RA1                 41.225       5.310       7.764
  RA2                 20.145       4.637       4.344
  RA3                 41.511       5.212       7.965
  MA1                 45.623       5.013       9.101
  MA2                 10.824       3.850       2.812
  MA3                 31.849       4.438       7.176


Group G2

READING BY
  RA1                  1.000       0.000       0.000
  RA2                  1.069       0.083      12.899
  RA3                  0.963       0.080      12.106

MATHABIL BY
  MA1                  1.000       0.000       0.000
  MA2                  1.199       0.087      13.807
  MA3                  1.128       0.085      13.300

MATHABIL WITH
  READING             22.741       5.721       3.975

Variances
  READING             72.381      10.877       6.655
  MATHABIL            65.078      10.175       6.396

Residual Variances
  RA1                 40.470       5.505       7.351
  RA2                 22.222       4.886       4.548
  RA3                 40.564       5.321       7.623
  MA1                 46.779       5.278       8.863
  MA2                 11.406       4.011       2.844
  MA3                 30.330       4.506       6.731
```

In step 3, we fit the last model, yet with factor loadings set equal across groups. Because of the default features of the software mentioned above, the command file needed is identical to the last used input file after deleting its final two lines, but we present it next because of its relevance in its own right and for the sake of completeness of this discussion. (Note that we analyze the same covariance matrices, so there is no need to change the file reference name in the second command line.)

```
TITLE:      TESTING FACTOR LOADING IDENTITY (STEP 3)
DATA:       FILE = CH9EX4.DAT;
            TYPE = COVARIANCE;
            NGROUPS = 2;
            NOBS = 230 215;
VARIABLE:   NAMES = RA1-RA3 MA1-MA3;
MODEL:      READING BY RA1-RA3;
            MATHABIL BY MA1-MA3;
```

The following output results:

```
THE MODEL ESTIMATION TERMINATED NORMALLY

TESTS OF MODEL FIT

Chi-Square Test of Model Fit

  Value                                 27.685
  Degrees of Freedom                        20
  P-Value                               0.1171

Chi-Square Contributions From Each Group

  G1                                    14.464
  G2                                    13.221

Chi-Square Test of Model Fit for the Baseline Model

  Value                               1526.306
  Degrees of Freedom                        30
  P-Value                               0.0000

CFI/TLI

  CFI                                    0.995
  TLI                                    0.992

Loglikelihood

  H0 Value                           -9301.225
  H1 Value                           -9287.382
```

```
Information Criteria

  Number of Free Parameters                    22
  Akaike (AIC)                          18646.450
  Bayesian (BIC)                        18736.607
  Sample-Size Adjusted BIC              18666.789
   (n* = (n + 2) / 24)

RMSEA (Root Mean Square Error Of Approximation)

  Estimate                                  0.042
  90 Percent C.I.                           0.000        0.076

SRMR (Standardized Root Mean Square Residual)

  Value                                     0.029


MODEL RESULTS

                            Estimates       S.E.     Est./S.E.

Group G1

READING BY
  RA1                          1.000        0.000        0.000
  RA2                          1.076        0.058       18.536
  RA3                          0.969        0.056       17.360

MATHABIL BY
  MA1                          1.000        0.000        0.000
  MA2                          1.197        0.060       19.953
  MA3                          1.125        0.059       19.154

MATHABIL WITH
  READING                     22.780        5.346        4.261

Variances
  READING                     71.265        9.170        7.771
  MATHABIL                    65.032        8.402        7.740

Residual Variances
  RA1                         41.076        5.141        7.990
  RA2                         20.259        4.260        4.755
  RA3                         41.516        5.051        8.220
  MA1                         45.640        4.934        9.250
  MA2                         10.821        3.480        3.109
  MA3                         31.839        4.224        7.537
```

```
Group G2

READING BY
  RA1                1.000       0.000       0.000
  RA2                1.076       0.058      18.536
  RA3                0.969       0.056      17.360

MATHABIL BY
  MA1                1.000       0.000       0.000
  MA2                1.197       0.060      19.953
  MA3                1.125       0.059      19.154

MATHABIL WITH
  READING           22.686       5.551       4.087

Variances
  READING           71.695       9.433       7.600
  MATHABIL          65.282       8.628       7.567

Residual Variances
  RA1               40.645       5.299       7.670
  RA2               22.099       4.505       4.905
  RA3               40.541       5.162       7.853
  MA1               46.756       5.200       8.992
  MA2               11.423       3.602       3.171
  MA3               30.328       4.257       7.124
```

To formally test now the restriction of identical factor loadings across the two gender groups, that is, the first critical assumption in Equation 9.18, as mentioned we obtain the pertinent likelihood ratio test statistic as the difference in chi-square values of this model and the last fitted one in which this is nested: $27.685 - 27.666 = .019$. This statistic is nonsignificant when referred to the relevant chi-square distribution with $20 - 16 = 4$ degrees of freedom, since the respective chi-square distribution cutoff is 9.48. We conclude that there is no evidence for gender differences in the loadings of the reading and mathematics tests correspondingly on the reading and mathematics ability factors. That is, the factorial structure of the mathematics and readings indicators is gender invariant.

As a final step before testing group differences in factor means that are of actual interest, we fit the last model to the MCS, whereby we impose, in addition, the restraint of identical intercepts across groups (which is the default in M*plus* as soon as one signals to the program that the mean structure is being analyzed; see below). To this end, we need to analyze the means and covariance matrices in the two groups, and use the following input file. (Note the changed name of the file reference for analyzed data in the second input line, as well as the request for analysis of the MCS.)

```
TITLE:       TESTING MEAN INTERCEPT IDENTITY (STEP 4)
DATA:        FILE = CH9EX4-MCS.DAT;
             TYPE = MEANS COVARIANCE;
             NGROUPS = 2;
             NOBS = 230 215;
VARIABLE:    NAMES = RA1-RA3 MA1-MA3;
ANALYSIS:    TYPE = MEANSTRUCTURE;
MODEL:       READING BY RA1-RA3;
             MATHABIL BY MA1-MA3;
```

The following output results:

```
THE MODEL ESTIMATION TERMINATED NORMALLY

TESTS OF MODEL FIT

Chi-Square Test of Model Fit

  Value                                  30.555
  Degrees of Freedom                         24
  P-Value                                0.1670

Chi-Square Contributions From Each Group

  G1                                     15.837
  G2                                     14.717

Chi-Square Test of Model Fit for the Baseline Model

  Value                                1526.306
  Degrees of Freedom                         30
  P-Value                                0.0000

CFI/TLI

  CFI                                     0.996
  TLI                                     0.995

Loglikelihood

  H0 Value                            -9302.660
  H1 Value                            -9287.382

Information Criteria

  Number of Free Parameters                  30
  Akaike (AIC)                        18665.319
  Bayesian (BIC)                      18788.261
  Sample-Size Adjusted BIC            18693.054
    (n* = (n + 2) / 24)
```

```
RMSEA (Root Mean Square Error Of Approximation)

  Estimate                                0.035
  90 Percent C.I.                         0.000        0.068

SRMR (Standardized Root Mean Square Residual)

  Value                                   0.027


MODEL RESULTS
                           Estimates      S.E.     Est./S.E.
Group G1

READING BY
  RA1                         1.000      0.000        0.000
  RA2                         1.076      0.058       18.480
  RA3                         0.970      0.056       17.329

MATHABIL BY
  MA1                         1.000      0.000        0.000
  MA2                         1.195      0.060       20.009
  MA3                         1.125      0.059       19.201

MATHABIL WITH
  READING                    22.818      5.352        4.263

Means
  READING                     0.000      0.000        0.000
  MATHABIL                    0.000      0.000        0.000

Intercepts
  RA1                        33.188      0.650       51.033
  RA2                        45.815      0.654       70.013
  RA3                        56.324      0.635       88.683
  MA1                        34.620      0.631       54.868
  MA2                        67.433      0.667      101.093
  MA3                        72.546      0.666      108.973

Variances
  READING                    71.211      9.178        7.759
  MATHABIL                   65.166      8.410        7.749

Residual Variances
  RA1                        41.331      5.164        8.003
  RA2                        20.343      4.271        4.763
  RA3                        41.466      5.054        8.204
  MA1                        45.624      4.935        9.246
  MA2                        10.883      3.475        3.132
  MA3                        31.800      4.222        7.532
```

```
Group G2

READING BY
  RA1                 1.000        0.000        0.000
  RA2                 1.076        0.058       18.480
  RA3                 0.970        0.056       17.329

MATHABIL BY
  MA1                 1.000        0.000        0.000
  MA2                 1.195        0.060       20.009
  MA3                 1.125        0.059       19.201

MATHABIL WITH
  READING            22.714        5.555        4.089

Means
  READING            -0.062        0.855       -0.072
  MATHABIL            0.869        0.798        1.089

Intercepts
  RA1                33.188        0.650       51.033
  RA2                45.815        0.654       70.013
  RA3                56.324        0.635       88.683
  MA1                34.620        0.631       54.868
  MA2                67.433        0.667      101.093
  MA3                72.546        0.666      108.973

Variances
  READING            71.616        9.438        7.588
  MATHABIL           65.402        8.635        7.574

Residual Variances
  RA1                40.939        5.327        7.685
  RA2                22.205        4.519        4.914
  RA3                40.478        5.166        7.836
  MA1                46.744        5.201        8.988
  MA2                11.486        3.597        3.193
  MA3                30.288        4.255        7.119
```

On the basis of the fit indices of this model, we gauge the plausibility of the restriction of identical mean intercepts across groups by looking at the difference in the chi-square values of this model and the last one fitted, namely, $30.555 - 27.685 = 2.870$, in relation to the difference of their degrees of freedom: $24 - 20 = 4$. Since the difference in fit indices is comparable to that in degrees of freedom, we conclude that there appear to be no gender differences in the mean intercepts as well. Together with the last finding of factorial invariance (gender invariance of the factor loading matrix), this result suggests that the assumptions in Equation 9.18 are plausible, which we stipulated as a prerequisite for testing factor mean differences across the two groups.

With these assumptions not being violated, we can now move on to testing for gender differences in the factors reading and mathematics ability. To this end, we recall first that their means are set by default at 0 in the first listed group in the data file analyzed, that is, girls. Therefore, the test of gender differences on the reading ability is tantamount to simply checking if this factor mean is significant in the boys group (Group G2 in the last presented output). Accordingly, boys show only slightly lower average on reading ability, namely, −.062. This difference is however not significant—see its standard error of .855 and, in particular, its nonsignificant *t*-value of −.072. In the same way, we find that boys are only marginally better on mathematics ability than girls, but also this difference of .869 is not significant—see its standard error of .798 and in particular nonsignificant *t*-value of 1.089. We, therefore, conclude that the analyzed data does not contain evidence warranting rejection of the null hypothesis of boys and girls having the same means on the reading and mathematics abilities.

We stress that here we tested group mean differences on unobserved factors, reading and mathematics ability, rather than on their observed indicators. When examination of the latter is of concern, a MANOVA would be employed on either of the two triples of reading and mathematics tests (or on all six of them, if that would be of substantive interest; see Chapter 6).

In conclusion of this chapter, we emphasize that the method used in it is applicable with any number of latent factors $q(q > 0)$, including the case of a single construct, that is, when $q = 1$. We also stress that this CFA method will prove very useful in subsequent chapters where we will be concerned with specific types of models that can be viewed as subsumed under its general framework.

10

Discriminant Function Analysis

10.1 Introduction

In Chapter 4, we discussed multivariate analysis of variance (MANOVA) and saw how it could be used to tackle research questions concerned with examining group differences when at least two dependent variables (DVs) are considered. If the underlying null hypothesis of mean vector equality across several groups is not rejected, one can conclude that the groups do not differ from one another with respect to the set of studied response measures. We also indicated in Chapter 4 that in the case of nonsignificant mean differences, it would not be appropriate in general to examine group differences on any of the dependent measures considered separately from the others (assuming of course that the research question was of a multivariate nature to begin with).

While this is a fairly straightforward approach to follow when there are no mean differences, how to handle the alternative situation when the above mentioned null hypothesis is rejected appears to be less clear-cut. As one possibility, we could examine separately for group differences each of the dependent measures, but since they are interrelated in general we will be using in our respective statistical conclusions partly overlapping pieces of sample information (because these univariate tests are as a consequence also related to one another). In addition, it may actually be the case that substantively meaningful combinations of the DVs may exhibit interesting group differences, and we will miss them if we remain concerned only with univariate follow-up analyses after a significant MANOVA finding. The method of discriminant function analysis (sometimes for short referred to as discriminant analysis, DA) allows us to examine linear combinations of dependent measures that exhibit potentially very important group differences. In addition, DA will enable us to utilize these variable combinations subsequently to classify other individuals into appropriate groups. In fact, when the intent is to describe group differences on the basis of observed variables, DA may be used as a primarily descriptive approach; alternatively, when the intent is to classify individuals on the basis of studied dependent variables, DA may be used predominantly as a predictive technique. For example, consider the situation in which a university admission officer is trying to determine what criteria to use in the process of admitting applicants to graduate school. Discriminant analysis could be used then in its descriptive capacity to find the most appropriate criterion to utilize from a large set of variables under consideration. Once the criterion is decided upon, DA could be used in its predictive capacity to classify new applicants into admit or do not admit groups.

To formally begin the discussion of DA, let us suppose that a set of DVs is given, denoted as $\underset{\sim}{x} = (x_1, x_2, \ldots, x_p)'$ $(p > 1)$, which have been observed across several samples (or groups) from multiple populations of interest. Assume also that in a MANOVA we have found that there were significant group differences, that is, the respective mean vectors differed more than what could be explained by chance factors only. The question that we will be mainly interested in the rest of this chapter is, ''What is the meaning of these differences?'' In other words, how can we interpret the finding of significant mean differences in a way that combines appropriately the group difference information across the DVs, and at the same time highlights important subject-matter aspects of the phenomenon under investigation? We stress that the finding of groups differing overall, which is all a significant MANOVA result tells us, may not be especially revealing. Instead, one would typically wish to be in a position of coming up with more substantively interpretable ways for making sense of the group differences. In particular, a social or behavioral researcher would usually want to know how the groups differ with regard to the dependent measures while accounting for all available information, including that about their interrelationships. He or she would then be especially interested in obtaining insightful combined information about group differences that is contained in the response variables.

10.2 What Is Discriminant Function Analysis?

Discriminant analysis (DA) is a method that can be used in a multigroup setting to find out how a set of dependent variables (DVs; or explanatory variables) is related to group membership, and in particular how they may be combined so as to enhance one's understanding of group differences. In more formal terms, DA aims at developing a rule for describing—and subsequently predicting if need be—group membership, based on a given set of variables (DVs). Once this rule becomes available, one may use it to make differential diagnosis, that is, group-membership prediction for particular subjects.

To illustrate this discussion, let us consider the following example. Suppose we are given three groups of (a) normally aging elderly,

(b) older adults with some memory problems, and (c) elderly exhibiting considerable cognitive decline. Assume that the members of the three groups are all measured on a set of psychological tests. The research question is, "Can a differential diagnosis be made with respect to these three groups of aged adults using the test results?" This aim could be attained by first appropriately combining group difference information contained in the available psychological measures. Then, depending on this combination's value for a given aged adult (that typically has not participated in the initial study), one may make a diagnosis, that is, prediction for his/her group membership. Obviously, group discrimination would become possible when combined information from the DVs is considerably different across groups. (In the context of DA, and in the rest of this chapter, to make this goal more readily attainable, we will interpret the notion of group diagnosis as meaning group-membership prediction.)

To be more precise, the initial goal of discriminant function analysis is to find a linear combination, denoted y, of p observed variables, $x_1, \ldots, x_p$ $(p > 1)$, which best discriminates between k given groups $(k \geq 2)$. That is, the individual scores on the new variable $y = \underline{a}'\underline{x} = a_1x_1 + a_2x_2 + \ldots + a_px_p$ in any one of the groups should be maximally dissimilar to the scores on this new, derived variable y in any other group. This linear combination, $y = \underline{a}'\underline{x}$, is called a discriminant function.

From the preceding discussion, in the case of $k = 2$ groups, it would be intuitively clear that we would be seeking a single discriminant function, y, that would differentiate as well as possible between the groups. That is, the groups should exhibit maximally different scores on this function. The case of $k > 2$ groups is not as transparent, however, and as we will see in a later section, is in fact considerably more complicated. Since a data set on p DVs may reflect quite complex variable interrelationships, for $k > 2$ groups there may be more than one way of reliably discriminating between the groups based on the p available measures.

To return for a moment to the above empirical example with $k = 3$ groups, it is possible that (a) one linear combination may discriminate well between the normally aging elderly on the one hand and those with memory deficits as well as those with considerable cognitive decline on the other hand, while (b) another linear combination may differentiate well between the latter two groups. With this in mind, we realize that the picture may be much more complicated with a larger number of groups, that is, when $k > 3$. This brings us to the specific, formal goal of discriminant function analysis. When p DVs are studied—with $p > k$ as would commonly be the case in most social and behavioral research, for it would be unusual to have more groups than variables under consideration—it can be shown that the aim of DA would be achieved if one could find $k - 1$ discriminant functions that are unrelated to one another, and which have the property that they best discriminate between the groups in the sense discussed so far.

10.3 Relationship of Discriminant Function Analysis to Other Multivariate Statistical Methods

As already mentioned, DA aims at explaining group differences in terms of a given set of dependent measures. Hence, DA can be conceptually thought of as a regression analysis (RA) where (a) group membership is the dependent variable, and (b) that set of measures play the role of independent variables (IVs, predictors). Such a formal approach would not be consistent with one of the main assumptions of conventional RA, namely that the response variable should be at least approximately continuous. Nonetheless, the conceptual consideration of DA as a regression of group membership on earlier considered DVs (e.g., in a MANOVA with a significant group difference finding) is very useful to keep in mind. In fact, this idea represents the relationship between DA and RA, and as indicated in a subsequent section, for the case of $k = 2$ groups DA and RA lead then to essentially the same results in a number of aspects.

This view of DA reveals, upon some reflection, that DA is actually a MANOVA but one that proceeds in a sense reversely. Indeed, as we will consider in more detail later in the chapter, in MANOVA one first searches for the linear combination that best discriminates between a given set of groups, and then compares its means across groups as in a univariate ANOVA. In more specific terms, in MANOVA the independent variable is "Groups," while in DA this is the dependent variable, i.e., just the opposite. Furthermore, in MANOVA the measured variables are the DVs, whereas in DA they are the predictors (IVs).

In fact, in many applications of DA there is no new mathematical technique involved other than the one underlying MANOVA, yet with a somewhat different focus. Indeed, the MANOVA question is whether group membership is associated with reliable mean differences on combined DV scores (viz. those of the maximally group differentiating linear combination). Specifically, the question is if one can predict the variable means based on knowledge of group membership. Conversely, the DA question is whether a given set of explanatory variables (e.g., DVs in a MANOVA) can be combined linearly with one another so as to allow prediction of group membership in a reliable manner. From this perspective, in conceptual terms there is no essential difference between MANOVA and DA, only that the direction of inference is the reverse of one another, as can be seen from the following figure (cf., e.g., Tabachnick & Fidell, 2007):

Schematic diagram of the relationship between MANOVA and discriminant function analysis:

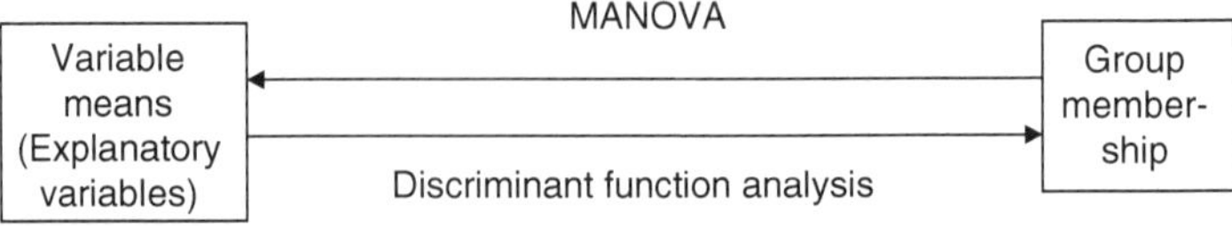

This discussion of similarities and differences between MANOVA and DA would be incomplete unless one mentions that once a DA is carried out, a subsequent related procedure may be of interest for a researcher to engage in, which is classification or group-membership prediction. In this procedure, the goal is to classify a newly made available observation(s) into the earlier given groups in a way that is associated with least possible error. As indicated previously, this application is commonly referred to as predictive DA. Furthermore, as we will see in the rest of this chapter, there is a distinct effort in DA that aims at understanding, and finding out, the dimensions (linear combinations of the DVs) along which the groups differ. There is no counterpart to this aim in MANOVA where one is interested mostly in whether there are group differences when all DVs are considered together.

This brings us to a more refined statement of the specific questions of DA. These deal with which, and how many, dimensions there are along which the groups differ reliably from one another. In addition, since these dimensions are represented by linear combinations of the DVs, the next query would be concerned with the substantive interpretation of the dimensions. We emphasize that when answering these questions, DA utilizes all information about the DVs and their interrelationships. This is quite different from the objective of MANOVA. Even though after finding with MANOVA a significant group difference one may pursue testing group differences on each of the DVs, those analyses would not be the same as DA since they would not use the information about dependent measure interrelationships. This information is utilized however in DA, and therein lies one of its main characteristic features as well as difference from MANOVA.

In conclusion of this section, we stress that in DA there are two facets: (a) description of group membership (i.e., finding the discriminant function(s)), and (b) classification or prediction of new observations to available groups (studied populations). Which of them will be pursued by a researcher in a particular study depends on the question(s) he/she is trying to answer. In fact, different studies may use one or the other of these two aspects of DA. In this chapter, we will be primarily interested in the first facet, but will also mention how DA can be used for purposes of classification. Last but not least, a question that may be lurking behind the context of the preceding discussion is, "When should one apply MANOVA and when DA?" As it turns out, its answer really depends on the research question of interest. If a scientist is dealing with a factorial design, MANOVA is likely to be preferred, as it would also be when using a within-subject design. In other cases, a DA is likely to be the preferred method of analysis, especially when one is concerned with combining the DVs in a MANOVA so as to make a substantively meaningful interpretation of found significant group mean differences on them.

10.4 Discriminant Function Analysis With Two Groups

The essence of DA in the case of $k=2$ groups is directly related to the intuitive notion of means' separation and multivariate distance. Suppose we have data on p variables, $\underline{x}=(x_1, x_2, \ldots, x_p)'$, from two random samples (groups) that have been drawn out of two studied populations denoted P_1 and P_2 (with $p>1$). We only assume that the respective population covariance matrices of these p variables are the same, and symbolize this common matrix as Σ. We wish to utilize the information available in $\underline{x}$ in such a way, which would allow us to discriminate as well as possible between the two groups, referred to as Group 1 and Group 2 below. This signifies that we seek a linear combination $y=\underline{a}'\ \underline{x}$, such that the resulting y scores in Group 1 are maximally different from the y scores furnished this way in Group 2. Our goal will be accomplished when we find the linear combination weights $\underline{a}=(a_1, a_2, \ldots, a_p)'$ leading to maximal group differentiation.

What does it really mean to have the y scores in Group 1 maximally dissimilar from the y scores in Group 2? This will be the case when the separation between the two groups is maximal. Group separation, denoted *sep* below, can be defined as the distance between the two group means, accounting for the units of measurement. Under the above assumption of same covariance matrices in both populations, it follows using the rules for variance in Chapter 2 (see also Chapter 8) that the variance of any linear combination of the variables under consideration, $x_1, x_2, \ldots, x_p$, will be the same in both groups. Hence, formally we can define group separation as follows:

$$sep = \frac{|\bar{y}_1 - \bar{y}_2|}{s_y},$$

that is,

$$sep^2 = \frac{(\bar{y}_1 - \bar{y}_2)^2}{s_y^2} = (\bar{y}_1 - \bar{y}_2)(s_y^2)^{-1}(\bar{y}_1 - \bar{y}_2), \tag{10.1}$$

where $s_y^2 = \frac{\sum_{i=1}^{n_1}(y_{1i}-\bar{y}_1)^2 + \sum_{i=1}^{n_2}(y_{2i}-\bar{y}_2)^2}{n_1+n_2-2}$ is the pooled variance estimate of the linear combination $y=\underline{a}'\ \underline{x}$, with n_1 and n_2 denoting the respective group (sample) sizes, while $\bar{y}_1$ and $\bar{y}_2$ denote the means of y in Groups 1 and 2, respectively, with y_{1i} and y_{2i} symbolizing the individual scores of y in these groups. This estimate is obtained using data from both groups that as mentioned have the same variance on the new measure y.

Now, we need to search for the weights $\underline{a}$ in the linear combination $y=\underline{a}'\ \underline{x}$, such that sep^2 is maximal. Note that sep^2 in Equation 10.1 is the ratio of the squared distance between the sample means of y, to the sample variance of y. Hence, from the right-hand side of Equation 10.1 is seen that sep^2 can be viewed as a special case of the concept of multivariate distance (in fact, when a single variable is under consideration, viz. y). We also notice that the separation measure is invariant to change of the units of measurement underlying the studied variables. This is because both its numerator and denominator are then affected to the same degree—being either multiplied or divided by the square of the scaling factor used. The next statement, credited to Fisher (1936), resolves this search for optimal weights.

Proposition 1 (R.A. Fisher): The maximal separation of the two groups under consideration is accomplished with that linear combination of the measures in $\underline{x}$, whose weights are given by the elements of the vector $\underline{a}$ defined as follows:

$$\underline{a}' = (\underline{\bar{x}}_1 - \underline{\bar{x}}_2)' S_{pooled}^{-1}, \tag{10.2}$$

where S_{pooled} is the pooled covariance matrix:

$$S_{pooled} = (\text{SSCP}_{(1)} + \text{SSCP}_{(2)})/(n_1 + n_2 - 2),$$

with $\text{SSCP}_{(i)}$ being the SSCP matrix in the ith group ($i = 1, 2$). Thereby, the attained maximum of the group separation measure, sep^2, is

$$D^2 = (\underline{\bar{x}}_1 - \underline{\bar{x}}_2)' S_{pooled}{}^{-1} (\underline{\bar{x}}_1 - \underline{\bar{x}}_2), \tag{10.3}$$

which is the Mahalanobis distance of any of the two group means to the other mean (with respect to the pooled covariance matrix).

The linear combination $y = \underline{a}'\ \underline{x}$, with weights in $\underline{a}$ as determined by Proposition 1 (Equation 10.2), is called Fisher's discriminant function, in honor of one of the most influential statisticians of the 20th century. Note that in order to obtain it, one only requires equality of the covariance matrix of the studied variables in both populations. We also observe that the weight vector $\underline{a}$ is determined up to a multiplicative constant. Indeed, while its components are not unique, their ratios (i.e., relative values) are -as could be expected, the vector $\underline{a}$ can provide us only with a direction in the multivariate space, along which the best group separation is possible.

If in addition we assume that the two studied variables follow a multinormal distribution in each population, then it can be shown that one can test the null hypothesis H_0: "The two populations are separable" using the test statistic (Johnson & Wichern, 2002)

$$\frac{n_1 + n_2 - p - 1}{p(n_1 + n_2 - 2)} \cdot \frac{n_1 n_2}{n_1 + n_2} D^2, \tag{10.4}$$

where D^2 is the above Mahalanobis distance (see Equation 10.3). If this hypothesis H_0 is rejected, we can conclude that the separation between the two populations is significant. The statistic in Equation 10.4 is identical to the one used to test the null hypothesis H_0: $\underline{\mu}_1 = \underline{\mu}_2$, as can be seen by comparing Equation 10.4 with the corresponding MANOVA test statistic in Chapter 4.

We stress that as seen from the expression in Equation 10.4, group separability is directly proportional to the Mahalanobis distance between the two group means: the larger this distance, the higher the group separability and vice versa. Conversely, one can conceptualize "overlap" between two multivariate distributions by the lack of their separability. From Equation 10.4, this is readily seen as another way of saying that the two groups are close to one another, because overlap between two groups is a function of the distance between their means, in the metric of the underlying covariance matrix. This statement is obviously the multivariate analog of the relationship between separability of two univariate distributions, on the one hand, and the standardized distance between their means, on the other hand (see Equation 10.1).

The preceding discussion shows that separation is essentially the same as group mean difference. Furthermore, the identity between the two discussed null hypotheses—viz. that of no separability and the one of no mean differences—illustrates that what underlies testing for group mean differences is finding a linear combination along which the two groups differ mostly, and then testing if the separation of their means is significant. This is precisely what is involved in MANOVA (or Hotelling's T^2 for the two sample case—see discussion in Chapter 4), when testing its null hypothesis of no group differences in the mean vectors for a set of DVs.

In addition to being instrumental for purposes of group discrimination, Proposition 1 can be shown to be useful for classification purposes employing the following rule for assignment of a new observation to one of two groups under consideration (e.g., Tatsuoka, 1988):

Assign a new observation with scores $\underline{x}_0$ to population P_1 if

$$y_0 = (\bar{\underline{x}}_1 - \bar{\underline{x}}_2)' S_{pooled}^{-1} \underline{x}_0 > (\bar{\underline{x}}_1 - \bar{\underline{x}}_2)' S_{pooled}^{-1} (\underline{x}_1 + \underline{x}_2) = (\bar{y}_1 + \bar{y}_2)/2,$$

and assign it to population P_2 otherwise.

We note that what counts for classification purposes is the location of Fisher's discriminant function value y_0 for the new observation, with respect to the average of the two group means on this function, i.e., the midpoint of the segment with ends $\bar{y}_1$ and $\bar{y}_2$. Indeed, if y_0 is closer to the mean of group 1 (i.e., $\bar{y}_1$), then y_0 is allocated to this group according to the above rule. Conversely, if y_0 is closer to the mean of group 2 (i.e., $\bar{y}_2$), then it is classified as a member of that group. We observe that the two group means, $\bar{y}_1$ and $\bar{y}_2$, their average or midpoint, $(\bar{y}_1 + \bar{y}_2)/2$, and the new observation's value of Fisher's discriminant function, y_0, all lie on a straight line. This collinearity feature follows from their definition.

To demonstrate, let us consider the following contrived example study of intelligence in university freshmen. In it, two groups of students, consisting of $n_1 = 85$ males and $n_2 = 155$ females, were tested on a battery of $p = 4$ intelligence tests. (The data are provided in the file "ch10ex1.dat" available from www.psypress.com/applied-multivariate-analysis, with variable names being "test.1" through "test.4" and "group.") For purposes of illustration, let us assume that the research question asks about the dimension along which there are most pronounced group differences (if there were such) in intelligence.

This query is resolved by carrying out a DA for $k = 2$ groups with a total of $n = 240$ subjects that have been measured on $p = 4$ variables (DVs, or x-variables in the notation used throughout this chapter). That is, the research question requires finding Fisher's discriminant function, which we can accomplish in SPSS using the following sequence of menu options:

Analyze → Classify → Discriminant (IV = tests; DV = group; Statistics: Means, Univariate ANOVA's, Box's M, Within-group covariance matrix).

To carry out the same activity in SAS, the following command statements within the PROC CANDISC (which is short for CANonical DIS-Criminant analysis) and the PROC DISCRIM procedures will be employed. It is important to note the use of the term *canonical* along with discriminant analysis in the CANDISC procedure, as well as throughout the output. This is because, as will be made clearer in the next chapter, in general DA may be thought of as a canonical correlation analysis between group membership and a set of considered variables. For now and the purposes of this chapter, we will just interpret pragmatically "canonical" to mean "discriminant."

```
DATA DA;
INFILE 'ch10ex1.dat';
INPUT test1  test2  test3  test4  group;
PROC CANDISC SIMPLE ANOVA PCOV;
CLASS group;
RUN;
PROC DISCRIM POOL = TEST;
CLASS group;
RUN;
```

In this set of SAS commands, the keyword SIMPLE is used with PROC CANDISC to request simple descriptive statistics for the total sample and within each group (class). The next following keyword ANOVA leads to a display of univariate statistics for testing hypotheses about the group means for each variable, and the keyword PCOV yields the pooled within-group covariance matrix. (Alternatively, one can use the keyword ALL instead, which activates all available options.) To request the test of homogeneity of the dependent variable covariance matrix, the POOL = TEST option is specified with PROC DISCRIM. We also note that the above SAS command file uses more than one procedure to accomplish DA, which is why we mentioned earlier that we make use of both PROC CANDISC and PROC DISCRIM for this purpose. In a later section, we will refer to yet another DA procedure, namely PROC STEPDISC.

The resulting outputs produced by SPSS and SAS follow next. For ease of presentation, they are organized into sections, and clarifying comments are accordingly inserted after each of them (some sections are self-explanatory and do not require such comments).

SPSS output

Discriminant

Group Statistics

group		Mean	Std. Deviation	Valid N (listwise)	
				Unweighted	Weighted
.00	test.1	44.7025	18.01038	85	85.000
	test.2	66.2758	15.16527	85	85.000
	test.3	47.9976	20.06096	85	85.000
	test.4	56.1810	11.99756	85	85.000
1.00	test.1	51.2596	18.51101	155	155.000
	test.2	73.1197	14.77488	155	155.000
	test.3	54.2651	19.90630	155	155.000
	test.4	59.4671	11.34053	155	155.000
Total	test.1	48.9373	18.56538	240	240.000
	test.2	70.6959	15.23976	240	240.000
	test.3	52.0454	20.14443	240	240.000
	test.4	58.3033	11.65930	240	240.000

Tests of Equality of Group Means

	Wilks' Lambda	F	df1	df2	Sig.
test.1	.971	7.020	1	238	.009
test.2	.954	11.560	1	238	.001
test.3	.978	5.412	1	238	.021
test.4	.982	4.423	1	238	.037

SAS output

The CANDISC Procedure

Observations	240	DF Total	239
Variables	4	DF Within Classes	238
Classes	2	DF Between Classes	1

Class Level Information

group	Variable Name	Frequency	Weight	Proportion
0	_0	85	85.0000	0.354167
1	_1	155	155.0000	0.645833

Simple Statistics

Total-Sample

Variable	N	Sum	Mean	Variance	Standard Deviation
test1	240	11745	48.93730	344.67342	18.5654
test2	240	16967	70.69586	232.25019	15.2398
test3	240	12491	52.04537	405.79794	20.1444
test4	240	13993	58.30328	135.93923	11.6593

group = 0

Variable	N	Sum	Mean	Variance	Standard Deviation
test1	85	3800	44.70248	324.37392	18.0104
test2	85	5633	66.27584	229.98532	15.1653
test3	85	4080	47.99759	402.44224	20.0610
test4	85	4775	56.18096	143.94140	11.9976

group = 1

Variable	N	Sum	Mean	Variance	Standard Deviation
test1	155	7945	51.25961	342.65739	18.5110
test2	155	11334	73.11974	218.29714	14.7749
test3	155	8411	54.26512	396.26067	19.9063
test4	155	9217	59.46712	128.60771	11.3405

```
                    The CANDISC Procedure

                 Univariate Test Statistics

          F Statistics,    Num DF=1,    Den DF=238

          Total    Pooled   Between
Vari-  Standard  Standard  Standard       R-  R-Square       F
able  Deviation Deviation Deviation  Square  /(1-RSq)   Value   Pr>F

test1   18.5654   18.3359    4.4350  0.0287    0.0295    7.02 0.0086
test2   15.2398   14.9138    4.6290  0.0463    0.0486   11.56 0.0008
test3   20.1444   19.9610    4.2391  0.0222    0.0227    5.41 0.0208
test4   11.6593   11.5767    2.2226  0.0182    0.0186    4.42 0.0365
```

The above displayed tables provide the entire sample and separate group descriptive statistics along with the univariate ANOVA findings that result from testing for group differences on each of the intelligence tests under consideration (i.e., on each dependent variable in turn), without taking into account its relationships to the other DVs. In simple terms, each of these *F* values gives a sense of the capability of the respective test measure by itself to predict group membership. Caution should be exercised however in interpreting these *F* tests because they do not account for the correlations among the DVs or any potential increase in Type I error that results from multiple univariate analyses being carried out on the same data set. These *F* ratios are not of main interest to us here, and we present them solely for the sake of completeness of this illustration.

SPSS output

Pooled Within-Groups Matrices[a]

		test.1	test.2	test.3	test.4
Covariance	test.1	336.204	198.059	331.226	157.527
	test.2	198.059	222.422	226.950	139.026
	test.3	331.226	226.950	398.442	177.196
	test.4	157.527	139.026	177.196	134.020

a. The covariance matrix has 238 degrees of freedom.

SAS output

```
                    The CANDISC Procedure

     Pooled Within-Class Covariance Matrix,    DF=238

Variable        test1         test2         test3         test4

test1     336.2044006   198.0589364   331.2256244   157.5272338
test2     198.0589364   222.4223763   226.9501247   139.0256098
test3     331.2256244   226.9501247   398.4423965   177.1956533
test4     157.5272338   139.0256098   177.1956533   134.0196020
```

The above tables each provide the pooled covariance matrix, S_{pooled}, whose inverse plays a major role in Proposition 1 regarding Fisher's discriminant function (see Equation 10.2).

SPSS output

Box's Test of Equality of Covariance Matrices

Log Determinants

group	Rank	Log Determinant
.00	4	18.642
1.00	4	17.995
Pooled within-groups	4	18.277

The ranks and natural logarithms of determinants printed are those of the group covariance matrices.

Test Results

Box's M		12.766
F	Approx.	1.251
	df1	10
	df2	141767.8
	Sig.	.253

Tests null hypothesis of equal population covariance matrices.

SAS output

```
            The DISCRIM Procedure
     Within Covariance Matrix Information

                                  Natural Log of the
              Covariance          Determinant of the
group         Matrix Rank         Covariance Matrix

     0                  4                  18.64232
     1                  4                  17.99498
Pooled                  4                  18.27709

            Test of Homogeneity
       of Within Covariance Matrices

    Chi-Square          DF          Pr > ChiSq
     12.505941          10              0.2526

Since the Chi-Square value is not significant at
the 0.1 level, a pooled covariance matrix will be
used in the discriminant function.
```

The above values of the group covariance matrix determinants are the building blocks of the overall test of the assumption we made in order to arrive at Fisher's discriminant function, namely, equality of the covariance matrices of the studied variables across groups. Note that the logarithms of these determinants, which are presented in the final columns, are numerically quite close to one another. We look next for the actual Box's M test or Bartlett's test statistic that is nonsignificant, as could be expected given the proximity of these determinant logarithms (see Equation 4.33 in Chapter 4). Consistent with this observation, Box's test or Bartlett's test indicates that the assumption of equal covariance matrices is plausible.

SPSS output

Summary of Canonical Discriminant Functions

Eigenvalues

Function	Eigenvalue	% of Variance	Cumulative %	Canonical Correlation
1	.059[a]	100.0	100.0	.235

a. First 1 canonical discriminant functions were used in the analysis.

Wilks' Lambda

Test of Function(s)	Wilks' Lambda	Chi-square	df	Sig.
1	.945	13.436	4	.009

SAS output

```
                         The CANDISC Procedure

                           Adjusted   Approximate        Squared
              Canonical   Canonical      Standard      Canonical
            Correlation  Correlation        Error    Correlation

            1  0.235252     0.210515     0.061105       0.055344

           Eigenvalues of Inv(E)*H
               = CanRsq/(1-CanRsq)

                                        Likelihood Approximate
Eigenvalue     Proportion Cumulative     Ratio       F  Num  Den   Pr>F
  Difference                                      Value   DF   DF
1  0.0586          1.0000      1.0000 0.94465   3.44    4  235 0.0093
```

```
         Multivariate Statistics and Exact F Statistics

Statistic                 Value    F Value    Num DF    Den DF     Pr>F

Wilks' Lambda        0.94465641       3.44         4       235   0.0093
Pillai's Trace       0.05534359       3.44         4       235   0.0093
Hotelling-Lawley     0.05858594       3.44         4       235   0.0093
  Trace
Roy's Greatest Root  0.05858594       3.44         4       235   0.0093
```

As indicated previously, we need not pay attention to the use of the term "canonical" (e.g., in SPSS it is used immediately before the heading "discriminant function"), whenever it appears in output from DA. This qualifier is commonly added by the software because as mentioned there are close connections between discriminant function analysis and canonical correlation analysis that is the topic of the next chapter. We will discuss this issue in detail there, but for now only reiterate that for the purposes of this chapter we can safely interpret "canonical" to mean "discriminant."

The multivariate test statistics provided in the above displayed results (i.e., Wilk's Lambda for SPSS, while SAS provides four test statistics—see also discussion in Chapter 4) pertain to the hypothesis of no group mean differences, i.e., of no group separability. These statistics incorporate empirical information stemming from the four interrelated intelligence tests, and suggest rejection of this hypothesis (see last column). Preempting some of the developments in a later section of this chapter, testing MANOVA's null hypothesis of no group mean differences is actually the same as testing the DA's null hypothesis of no significant discriminant functions (in this case, a single discriminant function); this relationship may well be intuitively clear based on the discussion in earlier sections of this chapter. Hence, the above output results mean that there is a significant discriminant function (that is, a linear combination of the DVs), which provides the direction of pronounced group differences.

SPSS output

Standardized Canonical Discriminant Function Coefficients

	Function
	1
test.1	.712
test.2	1.276
test.3	−.565
test.4	−.559

Structure Matrix

	Function
	1
test.2	.911
test.1	.710
test.3	.623
test.4	.563

Pooled within-groups correlations between discriminating variables and standardized canonical discriminant functions Variables ordered by absolute size of correlation within function.

SAS output

Pooled Within-Class Standardized Canonical Coefficients

Variable	Can1
test1	0.711800241
test2	1.276263324
test3	−0.565322068
test4	−0.559198028

Pooled Within Canonical Structure

Variable	Can1
test1	0.709572
test2	0.910541
test3	0.623015
test4	0.563234

The above tables display the standardized and structure coefficients, or weights for the single discriminant function here (as we are dealing with only two groups—males and females). There has been considerable discussion in the literature as to whether one should primarily examine the standardized discriminant function coefficients, or the structure coefficients (which SAS refers to as "Pooled Within Canonical Structure" coefficients). Each of these two views has its proponents, but it seems that the majority favor the structure coefficients. In our opinion, the choice depends on a researcher's purpose. The standardized coefficients convey the unique relationship between the variable and the discriminant function, much in the same way that a standardized partial regression coefficient (sometimes called beta weight) does in multiple regression. With the standardized coefficients, therefore, one may address the issue of relative importance of considered variables. In contrast, the structure coefficients reflect the correlation between the DVs and the respective discriminant function, thereby giving an indication of which variables are more closely associated with the function.

Based on these results, the measure or variable apparently mostly contributing to group differences on the discriminant function in question is the second intelligence test, namely, Test 2—see its highest (in magnitude) standardized coefficient. This observation is also corroborated by looking at the structure coefficients table which, as indicated, contains each measure's correlation with that discriminant function. We note that usually identifying a contributing variable as aligned with a discriminant function depends on both the relative and absolute value of its corresponding coefficient in this table. In particular, whenever variables with very low values are encountered in the standardized coefficient table, a researcher should be hesitant about concluding that these variables define the discriminant function, especially when their structure coefficients are small as well.

When many DVs are under consideration, a possible analytic strategy is to conduct a stepwise selection DA procedure to find out which variables could be used in the discriminant function. For example, in SAS this can be accomplished employing the procedure PROC STEPDISC in the above provided program statements. We note of course that when using stepwise DA one should take into account the same caveats as in stepwise multiple regression (Mills, Olejnik, & Marcoulides, 2005).

SPSS output

Canonical Discriminant Function Coefficients

	Function
	1
test.1	.039
test.2	.086
test.3	−.028
test.4	−.048
(Constant)	−3.659

Unstandardized coefficients

SAS output

Raw Canonical Coefficients

Variable	Can1
test1	0.0388200807
test2	0.0855758164
test3	−.0283212989
test4	−.0483038191
(Constant)	−3.6591000312

Based on the results presented in the last pair of tables, the optimal linear combination of the four intelligence tests administered to all 240 freshmen, which exhibits maximal dissimilarity in its values across both gender groups, is

$$y = -3.7 + .04 \times \text{Test.1} + .09 \times \text{Test.2} - .03 \times \text{Test.3} - .05 \times \text{Test.4}$$

(with "×" used to denote multiplication). That is, if one obtains the value of this function for each of the students, using his/her scores on the four behavioral measures used, then the gender means will be as distinct as possible. In other words, on this linear combination, the two gender groups are maximally separated. Their means on the combination in question are presented in the next output section.

SPSS output

Functions at Group Centroids

	Function
group	**1**
.00	−.325
1.00	.178

Unstandardized canonical discriminant functions evaluated at group means

SAS output

```
Class Means on Canonical Variables

     group              Can1

         0      -.3254887278
         1      0.1784938185
```

Up to this point in the present example, we have been exclusively concerned with group differentiation. As mentioned earlier in the chapter, however, discriminant function analysis can also be used to make classification, and we turn to this issue next. Let us suppose that a new observation becomes available after we have carried out the above DA on an initially provided data set. That is, in the currently considered empirical setting, assume that now the results on the four tests become available for a student who was not included in the originally analyzed sample of 240 freshmen. We have no information, however, on his/her gender. To make a well-educated guess of it, i.e., a prediction, one can work out the values of the following two linear combinations (one per column in either of the tables displayed next), and then classify that person as "male" or "female" according to the group with the higher of these functional values.

SPSS output

Classification Function Coefficients

	group	
	.00	**1.00**
test.1	−.037	−.017
test.2	.224	.267
test.3	−.191	−.206
test.4	.483	.459
(Constant)	−15.580	−17.387

Fisher's linear discriminant functions

SAS output

The DISCRIM Procedure

Linear Discriminant Function for group

Variable	0	1
Constant	−15.58008	−17.38729
test1	−0.03685	−0.01729
test2	0.22415	0.26728
test3	−0.19140	−0.20567
test4	0.48306	0.45871

That is, in order to carry out group assignment of a new subject (observation), we compare the values of the following two functions after substituting his/her scores on the four intelligence tests:

$$f_1 = -15.580 - .037 \times \text{test.1} + .224 \times \text{test.2} - .191 \times \text{test.3} + .483 \times \text{test.4}$$

and

$$f_2 = -7.387 - .017 \times \text{test.1} + .267 \times \text{test.2} - .206 \times \text{test.3} + .459 \times \text{test.4}.$$

We then predict their gender as being male if $f_1 > f_2$, or as female if $f_1 < f_2$. We note that while we have no guarantee that this prediction is identical with their actual gender, the possible error made with it has been minimized by following the DA procedure outlined so far in the chapter. (If $f_1 = f_2$, use all digits after decimal point in the pertinent output section. If even then $f_1 = f_2$, random assignment can be made.)

We stress that it is the pooled within-group covariance matrix (S_{pooled}) that was used to determine the discriminant function and carry out subsequent group-membership prediction. In other words, the above method of group assignment for new observations only required equality of the covariance matrix of the studied variables in both populations (groups) under consideration. When this assumption is fulfilled, we in fact employed a so-called *linear classification* method to assign a person to one of the groups (see Equation 10.4 and definition of discriminant function).

Alternatively, when there is evidence that the covariance matrices of the studied variables are not identical (i.e., $\Sigma_1 \neq \Sigma_2$ holds for their population covariance matrices), one needs to utilize each group-specific covariance matrix. Then a so-called *quadratic classification* rule is employed for the purpose of new observation assignment. We note that the quadratic classification method is automatically invoked within the SAS procedure PROC DISCRIM whenever the test for homogeneity of covariance matrices is found to be significant.

Another issue that is also important to consider when aiming at subject classification concerns the fact that we usually assume the prior probability of a person's membership in each of the groups to be the same. In other words, if there are just two groups, so far we were assuming that there is a prior probability of .5 for membership in either of them. It is likely, however, that this assumption may be incorrect in some empirical settings. If there are known differences in the prior probabilities of membership in the populations under study, it is essential that these be taken into account when assigning individuals to groups. Thus, if a researcher has knowledge that the prior probabilities of group membership are say .35 for one of these groups and .65 for the other, this information should be utilized in the classification procedure. We note that the default in SPSS and SAS is to assign an equal prior probability to each group (i.e., .5), unless the user specifies other prior probabilities. For example, in SAS one can use the PRIORS option to assign selected prior probability values; specifically, including the option line "PRIORS "0" = .35 "1" = .65"; after the PROC DISCRIM statement would assign these specified prior probabilities to each group.

One can also readily create a classification table for an originally collected data set, to see how well the DA may predict group membership on that data. The classification table presented below was generated on the previously analyzed sample of 240 freshmen (using a linear classification procedure with equal prior probabilities). We caution that this classification is related to possible capitalization on chance since one would utilize an available data set more than once—for deriving the discrimination function and for classifying the same subjects used thereby. For this reason, the results of this classification should not be overinterpreted.

```
                    The DISCRIM Procedure

Number of Observations and Percent Classified into group

    From group            0             1          Total

             0           46            39             85
                      54.12         45.88         100.00

             1           63            92            155
                      40.65         59.35         100.00

         Total          109           131            240
                      45.42         54.58         100.00

        Priors          0.5           0.5

               Error Count Estimates for group

                          0             1          Total

          Rate       0.4588        0.4065         0.4326
        Priors       0.5000        0.5000
```

10.5 Relationship Between Discriminant Function and Regression Analysis With Two Groups

As we indicated earlier in the chapter, it has been known for more than a half century that with two groups a DA yields essentially the same results as a RA with the dependent variable being group membership and the initially given set of x-variables (DVs, or explanatory variables) as predictors. In particular, the optimal linear combination weights (discriminant function weights) are proportional to the partial regression coefficients in a multiple regression equation of dichotomous group-membership response upon all predictors of interest (e.g., Tatsuoka, 1988). In other words, with two groups, DA reduces to a nonconventional RA with a dichotomous response measure. This fact led some early writers to claim that DA is nothing more than a particular RA, also when $k > 2$ groups are involved, which is however not correct. As we will see in the next chapter, in cases with more than two groups, DA does not reduce to an RA but instead to what is called canonical correlation analysis (which is the subject of that chapter).

To demonstrate the relationship between DA and RA in the case of two groups, let us reconsider the previous example study but now using this RA approach to analyze the data. The results follow below (we only provide the output generated by invoking the RA approach in SPSS, since identical results would be obtained with a similarly specified SAS regression procedure).

Regression

Variables Entered/Removed[b]

Model	Variables Entered	Variables Removed	Method
1	test.4, test.1, test.2, test.3[a]		Enter

a. All requested variables entered.
b. Dependent Variable: group

Model Summary

Model	R	R Square	Adjusted R Square	Std. Error of the Estimate
1	.235[a]	.055	.039	.46976

a. Predictors: (Constant), test.4, test.1, test.2, test.3

Note that the R^2 index here is the same as the squared correlation provided in the "conventional" DA output (referred to as "canonical correlation" there), which we examined in the preceding section.

ANOVA[b]

Model		Sum of Squares	df	Mean Square	F	Sig.
1	Regression	3.038	4	.760	3.442	.009[a]
	Residual	51.858	235	.221		
	Total	54.896	239			

a. Predictors: (Constant), test.4, test.1, test.2, test.3
b. Dependent Variable: group

In this table, the *p*-value of the overall regression equation, .009, is identical to the significance (*p*-value) associated with Wilk's lambda in the "conventional" DA output discussed above, since the pertinent statistical tests are the same. This identity should not be surprising because the test of MANOVA's null hypothesis of no group differences is the same as the test of whether group membership is related to the studied variables (means). The latter is however the test of significance of the R^2 index, i.e., the test of overall regression significance presented in the last table.

Coefficients[a]

Model		Unstandardized Coefficients		Standardized Coefficients	t	Sig.
		B	Std. Error	Beta		
1	(Constant)	.244	.177		1.381	.169
	test.1	.004	.004	.165	1.077	.283
	test.2	.009	.004	.299	2.557	.011
	test.3	−.003	.004	−.131	−.798	.426
	test.4	−.005	.005	−.129	−1.090	.277

a. Dependent Variable: group

The partial regression coefficients for each predictor are all obtained as a constant times the pertinent discriminant weights (discriminant function coefficients). This multiplier, as can be readily found out, is 9.75 for the analyzed data set; that is, when the partial regression coefficients are multiplied with this number, one obtains the corresponding DA weights.

We stress that this example was merely used here to illustrate the relationship between RA and DA in case of two groups. The example was not meant to suggest however that this RA-based strategy should be generally utilized instead of DA. In addition, we emphasize that a conceptually similar but formally more complicated relationship holds in case of multiple groups that we next turn to.

10.6 Discriminant Function Analysis With More Than Two Groups

When data from more than two groups are available, there is added complexity relative to the case of only two groups. In order to reduce this complexity, we will be interested in finding and studying the directions or dimensions along which the major group differences occur. These are the directions (dimensions) along which large group differences exist, i.e., the groups can be discernibly differentiated from one another. Equivalently, we need to find linear combinations of the original DVs that show large mean differences across groups. We note that once we find one such linear combination, denoted y, we have notably reduced the complexity stemming from considering multiple variables (predictors, explanatory variables, or original DVs) to differences on that derived variable y. We stress that with more than two groups, it is possible that there are more than just a single dimension along which the groups show pronounced differences. How to find them is the subject of this section.

To develop a strategy of addressing the group discrimination question in this context, we begin by considering the issue of group differences in the univariate case, specifically with respect to a linear combination $y = \underline{a}' \underline{x}$ of the originally considered measures in $\underline{x}$. For this case, we recall that univariate group differences are evaluated by the following F ratio in the pertinent ANOVA setting:

$$F = \frac{SS_{between}}{SS_{within}} \cdot \frac{n-k}{k-1}, \qquad (10.5)$$

where the numerator contains the between sum of squares on the new variable y, the denominator the within sum of squares for y, and n is the total number of studied subjects across all k groups ($k > 2$). We note that the "post-factor," $(n-k)/(k-1)$, in Equation 10.5 serves the role of insuring reference to a known distribution, in this case the F distribution. This factor is a constant for a particular study design, and thus the group differentiability on the linear combination y of the original measures depends only on the ratio of the between-group to within-group sum of squares. To handle this ratio further we need the following statement, whose validity can be shown directly using rules for variances and covariances of linear combinations of random variables, which we discussed in Chapters 8 and 9.

Proposition 2: For the F ratio in Equation 10.5, with regard to the linear combination $y = \underline{a}' \underline{x}$, the following representation holds:

$$\frac{SS_{between}(y)}{SS_{within}(y)} = \frac{\underline{a}'B\underline{a}}{\underline{a}'W\underline{a}}, \qquad (10.6)$$

where B is the SSCP matrix for the means of the original variables $\underline{x}$ between groups (which is the analog to the univariate between-group sum of squares), and W is the within-group sum of squares and cross-product matrix for those variables, $W = \text{SSCP}_{(1)} + \ldots + \text{SSCP}_{(k)}$ (which is the analog of the univariate within-group sum of squares).

The ratio in the right-hand side of Equation 10.6, $\lambda = \frac{\underline{a}' B \underline{a}}{\underline{a}' W \underline{a}}$, is called the discriminant criterion. As we will see, this criterion will play a major role in the rest of the chapter. This is because it quantifies the degree to which groups differ on the linear combination $y = \underline{a}' \, \underline{x}$, relative to the variability of the y scores within each of the groups. That is, the discriminant criterion is a measure of dissimilarity of the groups. The above Proposition 2 demonstrates that group differentiability on y will be maximal when the discriminant criterion will be maximal across all choices of $\underline{a}$, the vector of linear combination weights. A method for insuring this optimal choice of the critical vector a is given by the following statement (see Proposition 2 for the definition of the two involved matrices B and W).

Proposition 3: The discriminant criterion λ is maximized when the vector of linear combination weights $\underline{a}$ is taken as the eigenvector pertaining to the largest eigenvalue of the matrix $A = W^{-1}B$.

This vector, denoted $\underline{a}_{(1)}$, yields what is called the first discriminant function, $y_{(1)} = \underline{a}'_{(1)} \, \underline{x}$. The latter gives the dimension along which the k groups differ maximally. As we recall from Chapter 4, an eigenvector is determined up to a constant multiplier; therefore, the eigenvector in question is also chosen to have unit length. Once having found the first discrimination function, $y_{(1)}$, the natural question is how many similar functions are out there. It can be shown that there are as many discriminant functions, which represent the directions (dimensions) along which large group differences exist, as is the number of positive eigenvalues of the matrix A, viz. $r = \min(p, k - 1)$, which is the smaller of the number of studied variables and the number of groups less one (e.g., Tatsuoka, 1988). That is, there are r dimensions along which there are group differences. Each one of them is a linear combination of the original predictors (DVs, i.e., the x-variables), which is determined by the pertinent eigenvector of A with unit length. In much of social and behavioral research, the number of explanatory variables (DVs in a MANOVA) is larger than the number of groups; then the number of discriminant functions, r, is one fewer than the number of groups analyzed, i.e., $k - 1$. These r dimensions of pronounced (large) group differences have the additional property that they are uncorrelated but are not orthogonal to each other, and in fact can be viewed as an oblique rotation of an appropriate subset of the original system of p axes in the multivariate space. To interpret these discriminant functions, or dimensions of large group differences, as in the case of $k = 2$ groups one can use the standardized weights in their pertinent linear combination, as well as their correlations with the original p variables in $\underline{x}$ (see example below). However, as will be oftentimes the case in empirical research, not all of these r dimensions will be relevant, and in fact one could be concerned only with the first few. How to find out if this is the case, is the topic of the next section.

10.7 Tests in Discriminant Function Analysis

We begin here by recalling from Chapter 4 that the null hypothesis in a one-way MANOVA, H_0: $\underline{\mu}_1 = \underline{\mu}_2 = \ldots = \underline{\mu}_k$, can be tested with Wilk's lambda, $\Lambda = |W|/|T|$, where T is the total-sample SSCP matrix. It can be shown that one may also rewrite the inverse of this ratio of determinants as follows:

$$\begin{aligned} 1/\Lambda &= |T|/|W| = |W^{-1}T| = |W^{-1}(W+B)| = |I_p + W^{-1}B| \\ &= (1+\lambda_1)(1+\lambda_2)\ldots(1+\lambda_r), \end{aligned} \tag{10.7}$$

where $\lambda_1, \ldots, \lambda_r$ denote the nonzero eigenvalues of the matrix $A = W^{-1}B$ (e.g., Johnson & Wichern, 2002).

From Equation 10.7 it follows that testing the MANOVA null hypothesis, H_0, is in fact equivalent to testing whether all population eigenvalues of the matrix A are 0. This means that rejecting H_0 is the same as stating that there is at least one nonzero eigenvalue of the matrix product $W^{-1}B$, and thus that there is at least one significant discriminant function, that is, a direction along which there are significant group differences (see Proposition 3). Hence, the MANOVA concern is resolved by finding a linear combination of the given DVs (the initial x-variables in the notation used in this chapter), such that the group differences on it are maximal, and testing then if there are significant mean group differences on it. That is, testing the MANOVA null hypothesis is the same as testing the univariate ANOVA null hypothesis of group differences on this linear combination, namely, the first discriminant function.

Once the null hypothesis of no significant discriminant functions is rejected—and if it is not, then the MANOVA's H_0 is not rejected either (see Equation 10.7)—one can use a similar argument to test for significance the second discriminant function, etc. In this way, one can test for significance all discriminant functions in succession, until the first nonsignificant result is found. If this occurs at the qth step say ($1 \le q \le r$), then this means that there are $q - 1$ significant discriminant functions.

After the group discrimination issue is dealt with in this manner, classification questions can be addressed. In the multigroup case ($k > 2$), however, classification is more involved than in the two-group case. On the one hand, the same principle holds—for a given new observation with scores on all dependent measures (the x-variables in this chapter), work out the value of the classification function supplied by the software for the jth group ($j = 1, \ldots, k$), and compare the resulting k functional values across all groups. The group for which the classification function value is highest, is the one to which the new observation in question is assigned. The basic classification function (equation) for the jth group is $C_j = c_{j0} + c_{j1}x_1 + \ldots + c_{jp}x_p$, where the $p \times 1$ vector consisting of all c-coefficients but c_{j0} is $\underline{c}_j = W^{-1}\underline{\bar{x}}_j$, and $c_{j0} = \underline{c}_j'\underline{\bar{x}}_j$ ($\underline{\bar{x}}_j$ denotes the mean vector or centroid of the jth group, $j = 1, \ldots, p$).

To exemplify this discussion, let us consider a study in which $k = 3$ groups of aged adults were measured on $p = 4$ psychological tests. The first group includes 96 elderly with considerable cognitive decline, the second group comprises 114 older adults diagnosed with memory problems, and the third group consists of 30 normally aging elderly. The research question is whether there are significant dimensions along which the three groups differ, and if so, how to obtain them? (The data are provided in the file "ch10ex2.dat" available from www.psypress.com/applied-multivariate-analysis, where the respective variable names are "pt.1" through "pt.4" and "group.")

To respond to this question, with SPSS we use the following sequence of menu options:

Analyze → Classify → Discriminant (IVs: the four tests; grouping variable: Group, with values 0, 1, 2; Statistics: means, ANOVA, Box's M, Within-group covariance matrix).

To carry out the same analysis with SAS, the following command statements within the PROC CANDISC and PROC DISCRIM procedures can be used:

```
DATA DA;
INFILE 'ch10ex2.dat';
INPUT pt1 pt2 pt3 pt4 group;
PROC CANDISC SIMPLE ANOVA;
CLASS group;
RUN;
PROC DISCRIM POOL = TEST;
CLASS group;
RUN;
```

The resulting output produced by SPSS and SAS is provided next with clarifying comments inserted accordingly after each section.

SPSS output

Discriminant

Group Statistics

disorder		Mean	Std. Deviation	Valid N (listwise)	
				Unweighted	Weighted
.00	pt.1	33.8346	12.29708	96	96.000
	pt.2	61.0627	14.29844	96	96.000
	pt.3	36.3800	15.01888	96	96.000
	pt.4	51.2194	10.69525	96	96.000
1.00	pt.1	55.9692	13.26007	114	114.000
	pt.2	75.7116	12.41931	114	114.000
	pt.3	59.4790	14.48892	114	114.000
	pt.4	61.8836	9.81840	114	114.000
2.00	pt.1	70.5450	15.08808	30	30.000
	pt.2	82.4620	9.69656	30	30.000
	pt.3	73.9269	16.20181	30	30.000
	pt.4	67.3665	8.27592	30	30.000
Total	pt.1	48.9373	18.56538	240	240.000
	pt.2	70.6959	15.23976	240	240.000
	pt.3	52.0454	20.14443	240	240.000
	pt.4	58.3033	11.65930	240	240.000

Tests of Equality of Group Means

	Wilks' Lambda	F	df1	df2	Sig.
pt.1	.496	120.544	2	237	.000
pt.2	.713	47.695	2	237	.000
pt.3	.544	99.318	2	237	.000
pt.4	.731	43.630	2	237	.000

SAS output

```
                    The CANDISC Procedure

 Observations     240       DF Total                  239
 Variables          4       DF Within Classes         237
 Classes            3       DF Between Classes          2

                   Class Level Information

           Variable
group          Name    Frequency       Weight    Proportion

  0              _0           96      96.0000      0.400000
  1              _1          114     114.0000      0.475000
  2              _2           30      30.0000      0.125000
```

Simple Statistics

Total-Sample

Variable	N	Sum	Mean	Variance	Standard Deviation
pt1	240	11745	48.93730	344.67342	18.5654
pt2	240	16967	70.69586	232.25019	15.2398
pt3	240	12491	52.04537	405.79794	20.1444
pt4	240	13993	58.30328	135.93923	11.6593

group = 0

Variable	N	Sum	Mean	Variance	Standard Deviation
pt1	96	3248	33.83455	151.21829	12.2971
pt2	96	5862	61.06273	204.44536	14.2984
pt3	96	3492	36.37996	225.56671	15.0189
pt4	96	4917	51.21938	114.38843	10.6953

group = 1

Variable	N	Sum	Mean	Variance	Standard Deviation
pt1	114	6380	55.96917	175.82937	13.2601
pt2	114	8631	75.71162	154.23929	12.4193
pt3	114	6781	59.47899	209.92869	14.4889
pt4	114	7055	61.88361	96.40090	9.8184

group = 2

Variable	N	Sum	Mean	Variance	Standard Deviation
pt1	30	2116	70.54497	227.65002	15.0881
pt2	30	2474	82.46197	94.02336	9.6966
pt3	30	2218	73.92693	262.49850	16.2018
pt4	30	2021	67.36650	68.49090	8.2759

The CANDISC Procedure

Univariate Test Statistics

F Statistics, Num DF = 2, Den DF = 237

Variable	Total Standard Deviation	Pooled Standard Deviation	Between Standard Deviation	R-Square	R-Square/ (1-RSq)	F Value	Pr > F
pt1	18.5654	13.1265	16.1130	0.5043	1.0172	120.54	<.0001
pt2	15.2398	12.9227	9.9780	0.2870	0.4025	47.70	<.0001
pt3	20.1444	14.9208	16.6250	0.4560	0.8381	99.32	<.0001
pt4	11.6593	10.0098	7.3922	0.2691	0.3682	43.63	<.0001

Similarly to the two-group case, the above tables provide the univariate ANOVA test results that focus on each DV in isolation, and thus do not account for the interrelationships between the four tests. For this reason, these results are not of concern to us here, but are provided again for completeness of the discussion.

SPSS output

Pooled Within-Groups Matrices[a]

		pt.1	pt.2	pt.3	pt.4
Covariance	pt.1	172.305	101.210	161.315	82.947
	pt.2	101.210	166.996	126.192	95.043
	pt.3	161.315	126.192	222.630	99.858
	pt.4	82.947	95.043	99.858	100.196

a. The covariance matrix has 237 degrees of freedom.

SAS output

The CANDISC Procedure

Pooled Within-Class Covariance Matrix, DF = 237

Variable	pt1	pt2	pt3	pt4
pt1	172.3050896	101.2104754	161.3152795	82.9472976
pt2	101.2104754	166.9958931	126.1917306	95.0433331
pt3	161.3152795	126.1917306	222.6296919	99.8576308
pt4	82.9472976	95.0433331	99.8576308	100.1959421

The above tables give the pooled within-group covariance matrix that is an estimate of the assumed common covariance matrix of the four explanatory variables in the three groups (see Equation 10.2).

SPSS output

Box's Test of Equality of Covariance Matrices

Log Determinants

disorder	Rank	Log Determinant
.00	4	17.877
1.00	4	17.303
2.00	4	17.137
Pooled within-groups	4	17.627

The ranks and natural logarithms of determinants printed are those of the group covariance matrices.

Test Results

Box's M		27.085
F	Approx.	1.305
	df1	20
	df2	28387.165
	Sig.	.163

Tests null hypothesis of equal population covariance matrices.

SAS output

```
The DISCRIM Procedure

Within Covariance Matrix Information

                                      Natural Log of the
                 Covariance           Determinant of the
group            Matrix Rank          Covariance Matrix

     0                     4                   17.87733
     1                     4                   17.30342
     2                     4                   17.13726
Pooled                     4                   17.62742

Test of Homogeneity of Within Covariance Matrices

      Chi-Square           DF            Pr > ChiSq
       26.121774           20                0.1618

Since the Chi-Square value is not significant at the
0.1 level, a pooled covariance matrix will be used in
the discriminant function.
```

As indicated in the previous section, the determinants listed in the last pair of tables are the building blocks of the overall test of the assumption of covariance matrix equality across groups (e.g., Equation 4.33 in Chapter 4). We notice that the logarithms of the determinants are quite close to each other and that Box's M or Bartlett's test statistic value is not significant. Thus, a main assumption of discriminant function analysis is plausible, that of covariance matrix homogeneity.

SPSS output

Summary of Canonical Discriminant Functions

Eigenvalues

Function	Eigenvalue	% of Variance	Cumulative %	Canonical Correlation
1	1.050[a]	99.5	99.5	.716
2	.005[a]	.5	100.0	.074

a. First 2 canonical discriminant functions were used in the analysis.

Wilks' Lambda

Test of Function(s)	Wilks' Lambda	Chi-square	df	Sig.
1 through 2	.485	170.371	8	.000
2	.995	1.286	3	.732

SAS output

The CANDISC Procedure

	Canonical Correlation	Adjusted Canonical Correlation	Approximate Standard Error	Squared Canonical Correlation
1	0.715727	0.710740	0.031549	0.512265
2	0.073802	0.017865	0.064332	0.005447

Eigenvalues of Inv(E)*H
= CanRsq/(1-CanRsq)

	Eigen-value	Differ-ence	Propor-tion	Cumula-tive	Likelihood Ratio	Approximate F Value	Num DF	Den DF	Pr > F
1	1.0503	1.0448	0.9948	0.9948	0.48507811	25.49	8	468	<.0001
2	0.0055		0.0052	1.0000	0.99455333	0.43	3	235	0.7324

Multivariate Statistics and F Approximations

S=2 M=0.5 N=116

Statistic	Value	F Value	Num DF	Den DF	Pr > F
Wilks' Lambda	0.48507811	25.49	8	468	<.0001
Pillai's Trace	0.51771203	20.52	8	470	<.0001
Hotelling-Lawley Trace	1.05577172	30.80	8	331.97	<.0001
Roy's Greatest Root	1.05029523	61.70	4	235	<.0001

The entries labeled in the above tables as "Eigenvalue" (in the second column from left in the pertinent panels) are the eigenvalues of the critical matrix $A = W^{-1}B$, and are presented in descending order. These results are followed by the multivariate statistical tests of group differences. As can be seen, the three groups do differ significantly when the four psychological tests are considered simultaneously—see, e.g., the first row of the last presented table per software that as mentioned displays the results of the MANOVA null hypothesis of mean vector equality across groups. However, as seen from the above tables, only the first discriminant function is significant, unlike the second one that is nonsignificant (see last column in second row of the corresponding SPSS and SAS output part). That is, there is only one dimension along which there are significant group differences. In order to see which of the initial four psychological tests contribute mostly to this direction, we move on to the next output sections containing the standardized and structure coefficients for the discriminant functions.

SPSS output

Standardized Canonical Discriminant Function Coefficients

	Function	
	1	2
pt.1	.802	−.739
pt.2	.058	1.106
pt.3	.301	−.131
pt.4	−.160	.276

Structure Matrix

	Function	
	1	2
pt.1	.984*	−.013
pt.3	.893*	.168
pt.4	.591*	.534
pt.2	.616	.782*

Pooled within-groups correlations between discriminating variables and standardized canonical discriminant functions Variables ordered by absolute size of correlation within function.

*. Largest absolute correlation between each variable and any discriminant function

SAS output

```
Pooled Within-Class Standardized Canonical
                Coefficients

Variable              Can1           Can2

pt1            0.802115879   -0.739484471
pt2            0.057895852    1.105905097
pt3            0.301387665   -0.131255286
pt4           -0.159597192    0.276305266
```

Pooled Within Canonical Structure

Variable	Can1	Can2
pt1	0.984141	−0.013318
pt2	0.616465	0.781803
pt3	0.893223	0.168192
pt4	0.590816	0.534291

As can be seen by examining the standardized and structure coefficients, the variables that are mostly contributing to group differences on the first discriminant function are Test 1 (pt.1) and Test 3 (pt.3). We note that only entries in the first column should be examined here (i.e., the first discriminant function), since the other function was found not to be significant in this example.

SPSS output

Functions at Group Centroids

	Function	
disorder	**1**	**2**
.00	−1.170	−.031
1.00	.545	.067
2.00	1.671	−.153

Unstandardized canonical discriminant functions evaluated at group means

SAS output

Class Means on Canonical Variables

group	Can1	Can2
0	−1.169517430	−0.031307626
1	0.545233077	0.066537371
2	1.670570084	−0.152657606

Finally, in the last presented pair of tables we note how distant the group means are on the first discrimimant function, which is to be expected given the earlier finding of it being significant. We do not observe similar group discrepancies on the second discriminant function, but this is due to the fact that it is not significant.

As indicated in the previous section, in order to carry out assignment of a new subject, we could consult the values of the following three functions (one for each group), after substituting his/her observed scores on the four psychological tests. The function providing the highest value then, indicates the group to which we would assign the person in question. For completeness, we also present below the classification table of the original analyzed sample of 240 elderly (using a linear classification with equal prior probabilities).

Linear Discriminant Function for group			
Variable	0	1	2
Constant	−15.87920	−21.66011	−25.80061
pt1	0.00432	0.10359	0.18470
pt2	0.23571	0.25177	0.23805
pt3	−0.18215	−0.14837	−0.12371
pt4	0.46556	0.44092	0.41693

The DISCRIM Procedure

Number of Observations and Percent Classified into group

From group	0	1	2	Total
0	75	20	1	96
	78.13	20.83	1.04	100.00
1	20	61	33	114
	17.54	53.51	28.95	100.00
2	2	4	24	30
	6.67	13.33	80.00	100.00
Total	97	85	58	240
	40.42	35.42	24.17	100.00
Priors	0.33333	0.33333	0.33333	

Error Count Estimates for group

	0	1	2	Total
Rate	0.2188	0.4649	0.2000	0.2946
Priors	0.3333	0.3333	0.3333	

10.8 Limitations of Discriminant Function Analysis

We conclude this chapter by pointing out some important limitations of DA that need to be kept in mind when applying it in empirical research. We commence by noting that since DA is a "reversed" MANOVA, the same limitations as those mentioned in Chapter 4 concerning MANOVA apply to DA as well. Hence, DA results are most trustworthy under the same conditions under which MANOVA results are so. Due to their special relevance, next we would like to stress in the context of DA some of these conditions.

First, with regard to the normality and covariance matrix homogeneity assumptions, DA has been found to be somewhat robust against violations of the former. When sample sizes are equal, DA has some robustness also against up to moderate violations of covariance matrix homogeneity. Alternatively, at least for classification purposes, one could proceed by using a quadratic classification function, as indicated earlier in the chapter. As we mentioned in the concluding section of Chapter 4, Box's M test of the homogeneity assumption has been known to be quite sensitive to nonnormality. In that case, and with large samples, one can use instead latent variable modeling to evaluate group homogeneity for the covariance matrix of the variables in question (e.g., Raykov, 2001).

Second, DA is sensitive to multicollinearity of the dependent measures within groups. Lack of multicollinearity is thus an important requirement, since if it is not fulfilled DA cannot proceed mathematically. Hence, one needs to examine the DVs for multicollinearly beforehand. In cases where multicollinearity is indicated, appropriately limiting the number of analyzed variables may be considered, or alternatively application of principal component analysis or factor analysis with the goal of obtaining fewer, more fundamental dimensions that are related with one another only to a weak degree if at all. On these new dimensions one can obtain individual subject scores—component scores or factor scores—that can be used subsequently in a discriminant function analysis (or other analyses) as explanatory variables with respect to group membership.

We conclude this chapter by emphasizing that the DVs (or explanatory, predictor variables) used in DA should exhibit (stochastic) linear relationships. This is an implication of normality, as mentioned before, but is specifically stressed here again. Hence, when considering application of DA, the initial set of x-variables (the vector $\underline{x}$) needs to comprise only measures that exhibit linear relationships among themselves. Nonlinear relations between dependent measures indicate lack of normality, in which case one may consider use of logistic regression in lieu of DA (Hosmer & Lemeshow, 2000; see also Hastie, Tibshiriani, & Friedman, 2001). This alternative analysis can also be worthwhile employing when the assumption of covariance matrix homogeneity is violated to a considerable degree and sample sizes are not similar across groups.

11

Canonical Correlation Analysis

11.1 Introduction

Thus far in the book we have discussed a number of multivariate statistical analysis methods. Quite naturally, a question may arise at this point as to whether any part of them may be unified under a more general framework from which they may be considered in some sense special cases. As it turns out, canonical correlation analysis (CCA), which we deal with in the present chapter, represents this general framework. Specifically, CCA can be viewed as one of the most general multivariate analysis methods, from which regression analysis (RA), multivariate analysis of variance (MANOVA), discriminant function analysis, and other closely related techniques are all particular cases.

To introduce the topic of CCA, consider two sets of variables, which for ease of presentation are simply called set A and set B. Assume that set A consists of p members that make up the vector $\underline{x}$, while set B consists of q members that give rise to the vector $\underline{y}$ (with $p > 1$, $q > 1$). The variables in set A may or may not be considered dependent variables (DVs) in a given study, and similarly those in set B may or may not be considered independent variables (IVs), or vice versa. That is, whether we have some special focus on variables in A or B is immaterial for the following discussion. In fact, CCA treats the variables in both sets A and B in a completely symmetric fashion. For example, set A may correspond to a number of variables that have to do with socioeconomic status (SES) (e.g., income level, educational level, etc.), whereas set B may comprise a number of cognitive related variables (e.g., verbal ability, spatial ability, etc.). Either of these sets might be considered as IVs or DVs, according to particular research questions, but this type of distinction is not necessary in order for CCA to proceed.

Once these two sets of variables are given, let us take a look at the correlation matrix R of them all together. This matrix is of size $(p+q) \times (p+q)$, that is, contains in total $(p+q) \cdot (p+q-1)/2$ nonredundant (nonduplicated) correlations. Even if p and q are not large numbers on their own, it is easy to see that there are many nonredundant elements of R.

It may, therefore, be of interest to know whether there is some way of reducing the multitude of this potentially quite large number of correlations to a more manageable group of interrelationship indices that represent the way in which variables in set A covary with variables in B. In other words, our focus will be on examining the interrelationships among the two sets of variables.

This is precisely the aim of CCA, which as a method is specifically concerned with studying the interset correlations across A and B. Thereby, the purpose of CCA is to obtain a small number of derived variables from those in A and those in B, which show high correlations across the two sets. That is, CCA aims at "summarizing" the correlations between variables in set A and those in set B into a much smaller number of interrelationship measures that in some sense are representative of those correlations. In this capacity, CCA can essentially be used as a method for (a) examining independence of two sets of variables, as well as (b) data reduction.

Specifically, the pragmatic goal of CCA is to reduce the information on variable interrelationships contained in these $(p+q)\cdot(p+q-1)/2$ correlations among the variables in A and B to as few as possible indices (correlations) that characterize nearly as well their interrelations. Accomplishing this goal is made feasible through the following steps. First, a linear combination Z_1 of the variables $\underline{x}$ in A is found, and a linear combination W_1 of the variables $\underline{y}$ in B, such that their correlation $\rho_{1,1}=\text{Corr}(Z_1, W_1)$ is the highest possible across all choices of combination weights for W_1 and Z_1. We call these linear combinations or composites Z_1 and W_1 the first pair of canonical variates, and their correlation $\rho_{1,1}$ the first canonical correlation. Once this is done, in the next step another linear combination of variables in A is found, denoted Z_2, and a linear combination of variables in B, designated W_2, with the following property: their correlation $\rho_{2,2}=\text{Corr}(Z_2, W_2)$ is the highest possible under the assumption of Z_2 and W_2 being uncorrelated with the variables in the first combination pair, Z_1 and W_1. These new combinations Z_2 and W_2 are referred to as the second pair of canonical variates, and their correlation $\rho_{2,2}$ is called the second canonical correlation. This process can be continued until as many pairs of canonical variates are obtained in this way, as is the smaller of the numbers p and q (i.e., the smaller of the numbers of variables in the sets A and B). That is, the process of canonical variate construction yields the pairs $(Z_1, W_1), \ldots, (Z_t, W_t)$, where $t=\min(p, q)$ (with min(.,.) denoting the smaller of the numbers p and q). While in many social and behavioral studies this number t may be fairly large, it is oftentimes the case that only up to the first two or three pairs of canonical variates are really informative.

When the pairs of canonical variates and pertinent canonical correlations become available, one can make conclusions as to whether the two initial sets of variables, A and B, can be considered largely (linearly) unrelated. The latter will be the case if these pairs are not notably correlated, that is, if all canonical correlations are uniformly weak and close to zero. Otherwise, one could claim that there is some (linear) interrelationship between variables in A with those in B, which cannot be explained by chance factors only. Furthermore, once the canonical variates are determined, individual scores on them can be computed and used as values on new variables in subsequent analyses. These scores may be attractive for the latter purposes because they capture the essence of the cross-set variable interrelationships.

To illustrate this discussion using a simple example, consider the case of $p = q = 2$ variables comprising the sets A and B, respectively. Suppose that set A consists of the variables measuring arithmetic speed (denoted x_1) and arithmetic power (designated x_2). Assume set B comprises the measures reading speed (denoted y_1) and reading power (symbolized as y_2). These four variables give rise to six correlations among themselves, and in particular four cross-set correlations of a variable in set A with one in set B. The question now is whether these correlations could be reduced to one or two correlations between one or two derived measure(s) of arithmetic ability (linear combinations of the two arithmetic ability scores), on the one hand, and one or two derived measure(s) of reading ability, on the other hand. If this would be possible, the interrelationship information contained in the six correlations, and especially the four cross-set correlations, will be reduced to one or two canonical correlations. This is an example that the pioneer of canonical correlation, Harold Hotelling, used to illustrate this method in a paper introducing it in 1935.

To demonstrate some additional important features of CCA, let us look at one more example that is a little more complicated. Consider the case where the set A consists of $p = 5$ personality measures, $\underline{x}$, while the set B comprises $q = 4$ measures, $\underline{y}$, of success as a senior in high school. The substantive question of concern in this example study is, "what sort of personality profile tends to be associated with what pattern of academic achievement?" Accordingly, the formal question that can be answered with CCA is, "are there linear combinations of personality measures (i.e., Zs in the above notation) that correlate highly with linear combinations of academic achievement measures (i.e., Ws in the above notation)?" If so, what is the minimum number of such pairs of within-set combinations that can be found to nearly completely represent the cross-set correlations of personality with achievement measures?

The objective of CCA becomes particularly relevant in empirical research with an even larger number of variables in at least one of the sets A and B. For example, if one uses the 16 PF Questionnaire by Cattell and Eber to assess personality (i.e., $p = 16$), and takes $q = 8$ measures of academic performance for freshmen in college, there will be over 120 cross-set correlations (and many additional ones if one were to count the interset correlations; see Cattell, Eber, & Tatsuoka, 1970).

More than 120 cross-correlations might be meaningfully reduced to a handful of canonical correlations using CCA, in which case the power of CCA—also as a data reduction technique—can be seen very transparently.

11.2 How Does Canonical Correlation Analysis Proceed?

As indicated in the preceding section, to begin conducting CCA we seek two linear combinations $Z_1 = \underline{a}_1'\underline{x}$ and $W_1 = \underline{b}_1'\underline{y}$ correspondingly from each of two given variable sets A and B, such that $\rho_{1,1} = \text{Corr}(Z_1, W_1)$ is at the maximal possible value. To accomplish this goal, let us first look at the covariance matrix S of the entire set of $p+q$ variables in A and B:

$$S = \begin{bmatrix} S_{11} & S_{12} \\ S_{21} & S_{22} \end{bmatrix},$$

where S_{11} is the covariance matrix of the p variables in A, S_{22} that of the q variables in B, S_{21} is the covariance matrix of the q variables in B with the p in A, and S_{12} is the covariance matrix of the p variables in A with the q measures in B. We note in passing that even though S_{12} and S_{21} consist of the same elements, they are of different size and hence are not identical as matrices—specifically, S_{12} is of size $p \times q$ while S_{21} is a $q \times p$ matrix.

It can be shown (Johnson & Wichern, 2002) that this maximum correlation $\rho_{1,1}$ will be achieved if the following hold:

1. $\underline{a}_1$ is taken to be the (generalized) eigenvector pertaining to the largest solution ρ^2 of the following equation:

$$|S_{12}S_{22}^{-1}S_{21} - \rho^2 S_{11}| = 0 \tag{11.1}$$

(where $|.|$ denotes determinant; e.g., Chapter 2), that is, $\underline{a}_1$ fulfills the equation

$$\left(S_{12}S_{22}^{-1}S_{21} - \rho^2 S_{11}\right) \underline{a}_1 = \underline{0}, \tag{11.2}$$

with ρ^2 being the largest solution of Equation 11.1, and

2. $\underline{b}_1$ is the (generalized) eigenvector pertaining to the largest root of the following equation:

$$|S_{21}S_{11}^{-1}S_{12} - \pi^2 S_{22}| = 0, \tag{11.3}$$

that is, $\underline{b}_1$ fulfills the equation

$$\left(S_{21}S_{11}^{-1}S_{12} - \pi^2 S_{22}\right)\underline{b}_1 = \underline{0}, \tag{11.4}$$

with the largest π^2 satisfying Equation 11.3.

Thereby, the solutions of these two determinantal equations (Equations 11.1 and 11.3) are identical, that is, $\rho^2 = \pi^2$. Moreover, the positive square root of the largest of them equals $\rho_{(1)} = \pi_{(1)} = \rho_{1,1} = \text{Corr}(Z_1, W_1)$, the maximal possible correlation between a linear combination of variables in A with a linear combination of variables in B. These two optimal linear combinations, $Z_1 = \underline{a}_1'\underline{x}$ and $W_1 = \underline{b}_1'\underline{y}$, are the first canonical variate pair, while this maximal correlation—that is, their correlation $\text{Corr}(Z_1, W_1)$— is the first canonical correlation.

After the first canonical variate pair is obtained, the second canonical variate pair is furnished as a linear combination of the A variables using the eigenvector pertaining to the second largest solution of Equation 11.1, on the one hand, and a linear combination of the B variables using the second largest solution of Equation 11.3, on the other hand. Their correlation is the second canonical correlation. One continues in the same manner until $t = \min(p, q)$ canonical variate pairs are obtained and the corresponding canonical correlations are calculated as their interrelationship indices.

An important feature that follows from the construction of the canonical variates is that they are uncorrelated with one another:

$$\text{Cov}(Z_i, Z_j) = \text{Cov}(W_i, W_j) = \text{Cov}(Z_i, W_j) = 0 \quad (\text{for all } i \neq j; \; i, j = 1, \ldots, t). \tag{11.5}$$

That is, the canonical variates are uncorrelated within as well as across the sets of variables A and B they pertain to. In other words, if we look at the correlations of all Zs with all Ws, only the Zs and Ws with the same subindex will be nonzero (i.e., when $i = j$ for the corresponding covariances in Equation 11.5).

To demonstrate this discussion further, let us look at the following example. Consider again the above mentioned study of arithmetic speed and power (measures in set A) and reading speed and power (measures in set B). Denote by Z_1 and Z_2 the canonical variates for set A and by W_1 and W_2 those for set B. Then the correlation matrix of these variates, that is, the canonical correlation matrix, would be

$$R = \begin{bmatrix} 1 & 0 & \rho_1 & 0 \\ 0 & 1 & 0 & \rho_2 \\ \rho_1 & 0 & 1 & 0 \\ 0 & \rho_2 & 0 & 1 \end{bmatrix}, \tag{11.6}$$

where the first two rows (and columns) pertain to Z_1 and Z_2, and the last two rows (and columns) to W_1 and W_2, and the only off-diagonal elements that are nonzero are the two canonical correlations ρ_1 and ρ_2.

11.3 Tests and Interpretation of Canonical Variates

Despite the fact that there are $t = \min(p, q)$ canonical variate pairs and canonical correlations, not all of them will always be important when one aims at understanding the relationships across two prespecified sets of observed variables. To find out which of these pairs are of relevance for this purpose, statistical tests are available which evaluate the importance of canonical variate pairs and thus aid a researcher in determining how many of them should be retained for further analysis. These tests are based on the assumption of multivariate normality (MVN), follow the same logic as the tests for significance of discriminant functions, and specifically examine the significance of the canonical correlations.

Typically, the first test performed evaluates the null hypothesis that all canonical correlations are 0. If this test is rejected, the conclusion is that (a) at least the first canonical variate pair is of relevance when one aims at understanding the interrelationship between the variable sets A and B, and in particular (b) at least the first canonical correlation is not zero in the studied population. Subsequently, the second test examines the null hypothesis that apart from the first canonical correlation, all remaining ones are 0; and so on. If the first test is, however, not rejected, this means that the two initial variable sets, A and B, are (linearly) unrelated to one another. This is because in CCA, one is effectively interested in ''giving the best chance'' to the sets A and B to demonstrate their pattern of (linear) interrelationships, by building optimal, in this sense, linear combinations, namely Z_1 and W_1.

After completing these tests, and in case at least the first canonical correlation is significant, the next question is how to interpret the canonical variates. To this end, as with discriminant functions, one can use the correlations of each canonical variate with variables within its pertinent set, in order to facilitate its interpretation. That is, when trying to interpret Z_1, one would look at its correlations with the x-variables in the initial set A. Similarly, when trying to interpret W_1, one needs to examine its correlations with the variables in set B; and so on for the following canonical variate pairs and their members. Thereby, one interprets Z_1, W_1, and all remaining canonical variates as representing the common features of initial variables correlated highly with the canonical variate in question. Two types of canonical variate coefficients can then be considered, standardized and structure coefficients, and both are comparable in meaning and interpretation to their discriminant analysis counterparts.

The properties of standardized and structure coefficients discussed in Chapter 10 apply here also, as does our cautious preference for interpreting canonical variates mainly based on the structure coefficients.

Furthermore, for any given canonical correlation $\rho_i = \pi_i$, its square ρ_i^2 can be interpreted as a squared multiple correlation coefficient or the regression relating the *i*th canonical variate for any of the sets A or B, with the variables of the other set (B or A, respectively) ($i = 1, \ldots, t$). For this reason, ρ_i^2 can be viewed as the proportion shared variance between A and B, as captured by the *i*th canonical variate pair ($i = 1, \ldots, t$). In particular, the square of the first canonical correlation is, therefore, also interpretable as a measure of "set overlap." Another value that is also sometimes examined is the so-called redundancy index (Stewart & Love, 1968), which is sometimes used to assess how the variability of any variable in one of the sets is accounted for by the other set of variables. There have been a number of criticisms concerning the use of this index, however (Cramer & Nicewander, 1979), which is perhaps the reason why it is not very popular in the more recent literature.

Like principal component scores and factor scores, canonical variates can be used to work out individual subject scores on them. These derived scores can be used in subsequent analyses as say predictor variables. We also note that like principal components, the units of any canonical variate may not be meaningful. Further, we emphasize that canonical variates are not latent variables, but instead share the same observed status as manifest variables, since they are simply linear combinations or composites of the latter.

To exemplify these developments, let us consider a study in which a sample of $n = 161$ older adults were tested with an intelligence test battery consisting of eight tests. The first five tests, called set A, were related to fluid intelligence (i.e., our innate ability to solve problems seen for the first time), while the remaining three tests—denoted set B—tapped into crystallized intelligence (i.e., knowledge accumulated through one's life experiences; cf. Baltes, Dittmann-Kohli, & Kliegl, 1986). The set A variables were inductive reasoning for symbols, figural relations, inductive reasoning for letters, culture-fair, and Raven matrices tests. The set B variables were perceptual speed I, perceptual speed II, and vocabulary. The research question is, "can these 15 cross-set correlations be reduced to meaningful summary indices of set interrelationship?" (The data are provided in the set ch11ex1.dat available from www.psypress.com/applied-multivariate-analysis, with respective variable names corresponding to those mentioned above.)

To respond to this question, with SPSS we use the following series of menu options:

File → New → Syntax

and enter the following command file, to be submitted to the software (cf. Tabachnick & Fidell, 2007):

```
MANOVA ir.symb to raven WITH pers.sp1 to vocab
/DISCRIM = (RAW, STAN, ESTIM, COR)
/PRINT = SIGNIF(EIGEN, DIMENR)
/DESIGN.
```

To accomplish the same activity with SAS, the following command statements within PROC CANCORR can be used:

```
DATA cca;
INFILE 'ch11ex1.dat';
INPUT irsymb figrel irlett cultfr raven persspl perssp2 vocab;
CARDS;
PROC CANCORR ALL;
VAR irsymb figrel irlett cultfr raven;
  WITH persspl perssp2 vocab;
RUN;
```

We note that when utilizing PROC CANCORR, one set of variables must be arbitrarily designated as the "VAR" set (i.e., in this case the set of variables related to fluid intelligence), whereas the other is then designated as the "WITH" set of variables (i.e., the variables related to crystallized intelligence).

The resulting outputs produced by SPSS and SAS follow next, organized into sections, and clarifying comments are accordingly inserted after output portions.

SPSS output

```
* * * * * * * * * * * * * * * * * * * *Analysis of Variance* * * * * * * * * * * * * * * * * * *

161 cases accepted.
  0 cases rejected because of out-of-range factor values.
  0 cases rejected because of missing data.
  1 non-empty cell.
  1 design will be processed.
-------------------------------------------------------------------------

* * * * * * * * * * * * * * * * *Analysis of Variance - design 1* * * * * * * * * * * * * * * *

EFFECT .. WITHIN CELLS Regression
Multivariate Tests of Significance (S=3, M=1/2, N=75 1/2)

Test Name       Value     Approx. F    Hypoth. DF    Error DF    Sig. of F

Pillais         .50433     6.26450       15.00        465.00        .000
Hotellings      .80634     8.15295       15.00        455.00        .000
Wilks           .53310     7.21308       15.00        422.77        .000
Roys            .41554
-------------------------------------------------------------------------
```

SAS output

The CANCORR Procedure

VAR Variables	5
WITH Variables	3
Observations	161

Multivariate Statistics and F Approximations

S=3 M=0.5 N=75.5

Statistic	Value	F Value	Num DF	Den DF	Pr > F
Wilks' Lambda	0.53309665	7.21	15	422.77	<.0001
Pillai's Trace	0.50432725	6.26	15	465	<.0001
Hotelling-Lawley Trace	0.80633525	8.17	15	283.81	<.0001
Roy's Greatest Root	0.71096714	22.04	5	155	<.0001

NOTE: F Statistic for Roy's Greatest Root is an upper bound.

The first three multivariate tests in the above output from SPSS and the four multivariate tests from SAS evaluate whether all three possible canonical correlations are significant. In contrast, the displayed value of Roy's test statistic provided by SPSS pertains only to the first canonical correlation. According to the presented results, the null hypothesis of all three canonical correlations being simultaneously zero is rejected, and thus we can conclude that at least the first canonical correlation is not zero in the studied elderly population.

SPSS output

Eigenvalues and Canonical Correlations

Root No.	Eigenvalue	Pct.	Cum. Pct.	Canon Cor.	Sq. Cor
1	.711	88.173	88.173	.645	.416
2	.084	10.357	98.530	.278	.077
3	.012	1.470	100.000	.108	.012

Dimension Reduction Analysis

Roots	Wilks L.	F	Hypoth. DF	Error DF	Sig. of F
1 TO 3	.53310	7.21308	15.00	422.77	.000
2 TO 3	.91211	1.81224	8.00	308.00	.074
3 TO 3	.98829	.61237	3.00	155.00	.608

SAS output

The CANCORR Procedure

Canonical Correlation Analysis

	Canonical Correlation	Adjusted Canonical Correlation	Approximate Standard Error	Squared Canonical Correlation
1	0.644620	0.627947	0.046206	0.415535
2	0.277630	0.236207	0.072963	0.077078
3	0.108229	0.061898	0.078131	0.011714

Eigenvalues of Inv(E)*H
= CanRsq/(1-CanRsq)

Test of H0: The canonical correlations in the current row and all that follow are zero

	Eigenvalue	Difference	Proportion	Cumulative	Likelihood Ratio	Approximate F Value	Num DF	Den DF	Pr > F
1	0.7110	0.6275	0.8817	0.8817	0.53309665	7.21	15	422.77	<.0001
2	0.0835	0.0717	0.1036	0.9853	0.91211085	1.81	8	308	0.0742
3	0.0119		0.0147	1.0000	0.98828649	0.61	3	155	0.6080

As can be seen by examining the columns labeled "Canon. Cor." and "Canonical Correlation" in the SPSS and SAS outputs, respectively, their entries are the values of the canonical variate pair correlations (i.e., the canonical correlations themselves). The columns labeled "Sq. Cor." and "Squared Canonical Correlation," respectively, pertain to their squared values. For example, the value in the first row of these tables corresponds to the proportion of variance common to the first canonical variate pair, which is the "set overlap" measure mentioned above. In the second row, the proportion common to the second canonical variate pair follows, and in the last row the one associated with the third canonical variate pair.

The columns labeled "Wilks L." and "Likelihood Ratio" (LR) in the SPSS and SAS outputs, respectively, which are located in the second segments of the above sections, pertain to the tests of significance of the canonical correlations. In particular, the first rows of these segments contain the results of the test of the null hypothesis H_0: $\rho_{1,1}=\rho_{2,2}=\rho_{3,3}=0$. This hypothesis asserts that there is no linear relationship at all between the two sets of variables, that is, between the used fluid and crystallized intelligence measures. Not rejecting this H_0 means that carrying out a CCA is going to be a fruitless exercise, unlike in the alternative case. If this test is significant, however, just like it is here, one may conclude that there is at least one significant canonical variate pair, in which case one would want to proceed with further interpretation of this output table. Its second row presents the test of the hypothesis that after eliminating the effect of the first canonical variate pair, the remaining interrelationships between the two sets of variables can be explained with chance effects only. If this hypothesis is rejected, there are at least two significant canonical variate pairs. Similarly for the third row and so forth in the general case. In the present example, from the second row of the output panels obtained with SPSS and SAS, we see that the test of there being only one nonzero canonical correlation in the studied elderly population cannot be rejected. That is, there is only one optimal linear combination of fluid intelligence measures and one of crystallized intelligence tests, which is associated with the highest possible correlation that is not zero in that population.

SPSS output

```
EFFECT .. WITHIN CELLS Regression (Cont.)
Univariate F-tests with (3,157) D. F.

Variable   Sq. Mul. R   Adj. R-sq.   Hypoth. MS   Error MS          F

IR.SYMB      .38560       .37386     6951.38925   211.64400   32.84473
FIG.REL      .28801       .27440     3276.63066   154.78205   21.16932
IR.LET       .38152       .36970     7985.77236   247.37389   32.28220
CULT.FR      .33192       .31915     2222.19964    85.46873   26.00015
RAVEN        .26828       .25429     6204.35383   323.35882   19.18721
```

```
Variable  Sig. of F

IR.SYMB    .000
FIG.REL    .000
IR.LET     .000
CULT.      .000
FR RAVEN   .000
```

SAS output

The CANCORR Procedure

Canonical Redundancy Analysis

Squared Multiple Correlations Between the VAR Variables and the First M Canonical Variables of the WITH Variables

M	1	2	3
irsymb	0.3848	0.3854	0.3856
figrel	0.2792	0.2864	0.2880
irlett	0.3812	0.3814	0.3815
cultfr	0.3231	0.3318	0.3319
raven	0.2628	0.2682	0.2683

The last output parts, labeled in SPSS as univariate F tests and as canonical redundancy analysis in SAS, provide the R^2 indexes that correspond to the squared value of what we referred to as the so-called redundancy index. Herewith each variable in set A is treated in turn as a DV and the variables in set B are used as predictors (we note that in SPSS, all three variables are used at once, whereas in SAS they are given incrementally—so the last column in the SAS output refers to using the three B variables as predictors in a single regression model). We observe that the strongest (linear) relationship between crystallized intelligence measures and a fluid intelligence test is that predicting Inductive reasoning with symbols from the former measures. We also notice that there are discernible linear relationships between each fluid measure and all crystallized tests considered together.

SPSS output

```
* * * * * * * * Analysis of Variance -- design 1 * * * * * * * *


Raw canonical coefficients for DEPENDENT variables
                       Function No.

Variable             1                2

IR.SYMB            .023             .065
FIG.REL            .000           -.042
IR.LET             .015           -.023
CULT.FR            .017           -.107
RAVEN              .008             .051
------------------------------------------------

Standardized  canonical  coefficients  for  DEPENDENT
variables
                       Function No.

Variable             1                2

IR.SYMB            .426            1.204
FIG.REL           -.002            -.620
IR.LET             .305            -.456
CULT.FR            .192           -1.196
RAVEN              .164            1.059
------------------------------------------------
```

SAS output

```
                      The CANCORR Procedure

                 Canonical Correlation Analysis

        Raw Canonical Coefficients for the VAR Variables

                     V1                 V2                 V3

irsymb         0.0231583623       -0.065460894       -0.079782093
figrel         -0.000114338       0.0424762808       -0.099793827
irlett           0.01538746       0.0230226172       0.0943550511
cultfr         0.0171027787       0.1067854492       0.0697619955
raven          0.0078925649       -0.050858014       0.0076579995
```

The CANCORR Procedure

Canonical Correlation Analysis

Standardized Canonical Coefficients for the VAR Variables

	V1	V2	V3
irsymb	0.4258	−1.2035	−1.4668
figrel	−0.0017	0.6204	−1.4575
irlett	0.3048	0.4561	1.8693
cultfr	0.1916	1.1964	0.7816
raven	0.1644	−1.0591	0.1595

The first segments of this output section contain the unstandardized weights for the linear combination (canonical variate) Z_1 of the fluid tests, which as observed previously is the only significant canonical variate and thus of interest. From the second segment of this output section, which contains the pertinent standardized coefficients, it is suggested that the induction tests (i.e., "IRSYMB" and "IRLET") are the ones most saliently contributing to this canonical variate. (We note that this is an informal observation.)

SPSS output

Correlations between DEPENDENT and canonical variables

Variable	Function No. 1	2
IR.SYMB	.962	.086
FIG.REL	.820	−.305
IR.LET	.958	−.052
CULT.FR	.882	−.337
RAVEN	.795	.265

Variance in dependent variables explained by canonical variables

CAN. VAR.	Pct Var DE	Cum Pct DE	Pct Var CO	Cum Pct CO
1	78.508	78.508	32.623	32.623
2	5.738	84.246	.442	33.065

SAS output

```
                        The CANCORR Procedure

                         Canonical Structure

 Correlations Between the VAR Variables and Their Canonical Variables

                         V1                V2                 V3

 irsymb              0.9623           -0.0858            -0.1376
 figrel              0.8198            0.3053            -0.3670
 irlett              0.9578            0.0525             0.1022
 cultfr              0.8818            0.3369             0.0775
 raven               0.7953           -0.2647             0.0729

                    Canonical Redundancy Analysis

       Standardized Variance of the VAR Variables Explained by

                 Their Own                        The Opposite
             Canonical Variables               Canonical Variables

Canonical
Variable                 Cumulative  Canonical              Cumulative
Number       Proportion  Proportion   R-Square  Proportion  Proportion

      1          0.7851      0.7851     0.4155      0.3262      0.3262
      2          0.0574      0.8425     0.0771      0.0044      0.3307
      3          0.0351      0.8775     0.0117      0.0004      0.3311
```

The previous suggestion with respect to the contribution of the two inductive reasoning tests is corroborated in the last presented section as well (see the magnitude of the correlations provided in columns titled "1" and "V1" in the SPSS and SAS outputs, respectively). We also see from the magnitude of variance values that 78.5% of the variance in the fluid measures is explained in terms of the first canonical variate Z_1 (see entries in first row and second column of the SPSS and SAS outputs), which again is the only one of those presented that is of interest to interpret. At the same time, this variate Z_1 explains 32.6% of variance in the crystallized intelligence tests. As indicated earlier, this index is also called in general redundancy but now it reflects the contribution of the first canonical variate for the entire set A, in this case all five fluid tests.

The final parts of the generated output pertain to set B, here the crystallized intelligence tests, which are referred to by SPSS throughout as "covariates" and by SAS as the "WITH" variables. They are interpreted in complete analogy to their corresponding preceding output sections that are associated with set A (the five fluid intelligence measures in this example), and for this reason are left without any additional comments.

SPSS output

```
* * * * * * * * * * Analysis of Variance -- design 1 * * * * * * * * * *

Raw canonical coefficients for COVARIATES

                         Function No.

COVARIATE               1                  2

PERS.SP1             .047               .108
PERS.SP2             .032              -.075
VOCAB                .039               .000

---------------------------------------------------------------

Standardized canonical coefficients for COVARIATES

                         CAN. VAR.

COVARIATE               1                  2

PERS.SP1             .437              1.014
PERS.SP2             .462             -1.074
VOCAB                .439              -.004

---------------------------------------------------------------

Correlations between COVARIATES and canonical variables

                         CAN. VAR.

Covariate               1                  2

PERS.SP1             .821               .430
PERS.SP2             .772              -.525
VOCAB                .649               .124

---------------------------------------------------------------

Variance in covariates explained by canonical variables

CAN. VAR. Pct Var DE Cum Pct DE Pct Var CO Cum Pct CO

  1         23.420     23.420     56.362     56.362
  2          1.224     24.644     15.882     72.244

---------------------------------------------------------------
```

SAS output

The CANCORR Procedure

Canonical Correlation Analysis

Raw Canonical Coefficients for the WITH Variables

	W1	W2	W3
perssp1	0.0465942578	−0.108116279	0.0581828945
perssp2	0.032383677	0.0753262832	0.0155114688
vocab	0.0392325106	0.0003812354	−0.085252118

Canonical Correlation Analysis

Standardized Canonical Coefficients for the WITH Variables

	W1	W2	W3
perssp1	0.4370	−1.0140	0.5457
perssp2	0.4619	1.0743	0.2212
vocab	0.4388	0.0043	−0.9535

Canonical Structure

Correlations Between the WITH Variables and Their Canonical Variables

	W1	W2	W3
perssp1	0.8208	−0.4302	0.3758
perssp2	0.7721	0.5252	0.3577
vocab	0.6488	−0.1244	−0.7507

Canonical Redundancy Analysis

Standardized Variance of the WITH Variables Explained by

	Their Own Canonical Variables			The Opposite Canonical Variables	
Canonical Variable Number	Proportion	Cumulative Proportion	Canonical R-Square	Proportion	Cumulative Proportion
1	0.5636	0.5636	0.4155	0.2342	0.2342
2	0.1588	0.7224	0.0771	0.0122	0.2464
3	0.2776	1.0000	0.0117	0.0033	0.2497

The preceding discussion in this chapter lets us now represent schematically the essence of CCA in the empirical example under consideration as follows (cf. Tabachnick & Fidell, 2007; Marcoulides & Hershberger, 1997).

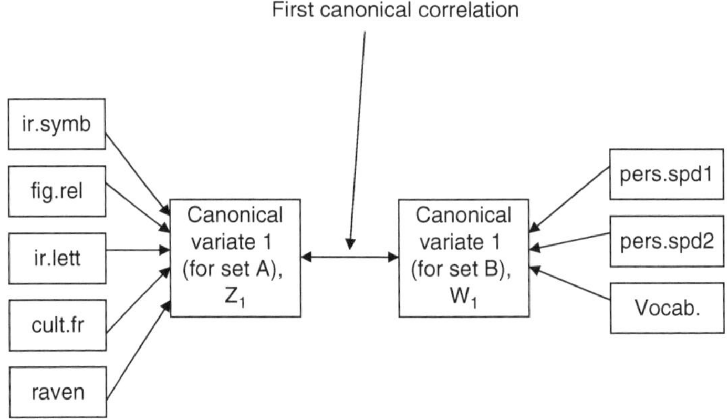

Graphical representation of CCA example results
(using path-diagrammatic notation)*

Set A ($x_1, x_2, \ldots, x_5$)
Fluid intelligence measures

Set B (y_1, y_2, y_3)
Crystallized intelligence measures

11.4 Canonical Correlation Approach to Discriminant Analysis

In the previous chapter, we mentioned that in case of $k = 2$ groups DA is essentially equivalent to a RA of the group membership variable treated as a dependent measure and the remaining variables used as predictors (IVs). This equivalency, however, cannot be extended to the general case with $k > 2$ groups. Specifically, it can be shown (Tatsuoka, 1971) that with $k > 2$ groups DA is identical to CCA using artificially defined variables D_1, $D_2, \ldots, D_{k-1}$ as comprising set A, while the original explanatory (predictor) variables $\underline{x} = (x_1, x_2, \ldots, x_p)'$ are treated as set B ($p > 1$). In particular, these dummy variables $D_1, D_2, \ldots, D_{k-1}$ are defined in the same way they would be for purposes of RA with categorical predictors (Pedhazur, 1997), which is also explicated next for completeness of the present discussion:

* In this notation, a one-way arrow means an explanatory relationship, a two-way arrow stands for an unspecified relationship (correlation), and a square signals an observed variable. We note that unlike factors in a factor analysis (but much like components in principal component analysis), here we are dealing only with observed variables and linear combinations of them, which can therefore be treated as derived observed variables as well.

	D_1	D_2	D_3	...	D_{k-1}
All subjects in Group 1 get	1	0	0	...	0
All subjects in Group 2 get	0	1	0	...	0
All subjects in Group 3 get	0	0	1	...	0
...					
All subjects in Group $k-1$ get	0	0	0	...	1
All subjects in Group k get	0	0	0	...	0

If one now performs a CCA with sets A and B as outlined in this section, the results will be identical to those obtained when carrying out a discriminant function analysis on the original variables $\underline{x}$. Specifically, the first canonical variate on the side of B will be the same as the first discriminant function; the second canonical variate on B's side will be the same as the second discriminant function and so forth. The test for significance of the canonical correlations is then a test of significance of discriminant functions, and the number of significant such functions and canonical correlations is the same.

Furthermore, each eigenvalue for the discriminant criterion, v_i, is related to a corresponding generalized eigenvalue (determinantal equation root) ρ_i (i.e., π_i) as follows:

$$v_i = \frac{\rho_i^2}{1-\rho_i^2} (i = 1,2,\ldots,r). \tag{11.7}$$

That is, each eigenvalue for the discriminant criterion is a nonlinear function of a corresponding squared canonical correlation. Thereby, this relationship is such that one is 0 if and only if the other is so (i.e., $v_i = 0$ if and only if $\pi_i = \rho_i = 0$; $i = 1, \ldots, r$). In particular, Wilks' lambda statistic can be expressed in terms of these two quantities as follows:

$$\Lambda = \frac{1}{(1+v_1)(1+v_2)\ldots(1+v_r)} = (1-\rho_1^2)(1-\rho_2^2)\ldots(1-\rho_r^2)$$
$$= (1-\pi_1^2)(1-\pi_2^2)\ldots(1-\pi_r^2). \tag{11.8}$$

Hence, testing the significance of discriminant functions, which as mentioned in Chapter 10 is the same as testing for significance the eigenvalues v_i, is equivalent to testing significance of canonical correlations. Indeed if $v_s = 0$, where v_s is the sth largest eigenvalue of the discriminant criterion $(1 \leq s \leq r)$, then all subsequent eigenvalues will be 0 as well, and thus from Equation 11.7 all ρs (or πs for the same matter) with the same indices will be 0 too. Then, the right-hand side of Equation 11.8 will not contain the factors pertaining to these ρs (πs) leaving the remaining ρs (πs) as nonzero since that would be the only way that $\Lambda = 1$ if $v_1 = 0$, or $\Lambda < 1$ when the first vanishing eigenvalue is v_2 or one after it.

To demonstrate this discussion, consider a study of $n = 161$ freshmen from three SES groups who were each given $p = 4$ college aspiration measures. It is of interest to find out whether there are any group differences on these four variables, and if so, what the dimensions of group differences are. To respond to this question, first we carry out a DA (see Chapter 10 for software instructions). Since DA output was discussed extensively in the previous chapter, for ease of presentation we only display output obtained with SPSS. The results are as follows and essentially self-explanatory, given the pertinent discussion in the preceding chapter.

Discriminant

Group Statistics

				Valid N (listwise)	
group		Mean	Std. Deviation	Unweighted	Weighted
.00	coll.asp1	36.8106	13.36800	129	129.000
	coll.asp2	63.3641	14.54808	129	129.000
	coll.asp3	39.5578	15.99565	129	129.000
	coll.asp4	52.8681	10.55085	129	129.000
1.00	coll.asp1	59.2014	12.27452	91	91.000
	coll.asp2	77.5526	11.31437	91	91.000
	coll.asp3	62.7347	13.24867	91	91.000
	coll.asp4	63.1003	9.88728	91	91.000
2.00	coll.asp1	73.8487	9.37505	28	28.000
	coll.asp2	83.4200	9.28923	28	28.000
	coll.asp3	77.8295	10.61901	28	28.000
	coll.asp4	68.7059	7.49891	28	28.000
Total	coll.asp1	49.2083	18.52039	248	248.000
	coll.asp2	70.8348	15.14880	248	248.000
	coll.asp3	52.3832	20.19067	248	248.000
	coll.asp4	58.4108	11.64549	248	248.000

Tests of Equality of Group Means

	Wilks' Lambda	F	df1	df2	Sig.
coll.asp1	.458	144.939	2	245	.000
coll.asp2	.722	47.098	2	245	.000
coll.asp3	.512	116.584	2	245	.000
coll.asp4	.733	44.540	2	245	.000

Pooled Within-Groups Matrices[a]

		coll.asp1	coll.asp2	coll.asp3	coll.asp4
Covariance	coll.asp1	158.395	97.741	148.262	80.703
	coll.asp2	97.741	167.110	124.499	94.820
	coll.asp3	148.262	124.499	210.581	98.518
	coll.asp4	80.703	94.820	98.518	100.268

a. The covariance matrix has 245 degrees of freedom.

Box's Test of Equality of Covariance Matrices

Log Determinants

group	Rank	Log Determinant
.00	4	17.977
1.00	4	17.020
2.00	4	16.025
Pooled within-groups	4	17.516

The ranks and natural logarithms of determinants printed are those of the group covariance matrices.

Test Results

Box's M		25.928
F	Approx.	1.247
	df1	20

Tests null hypothesis of equal population covariance matrices.

Summary of Canonical Discriminant Functions

Eigenvalues

Function	Eigenvalue	% of Variance	Cumulative %	Canonical Correlation
1	1.245[a]	99.4	99.4	.745
2	.007[a]	.6	100.0	.086

a. First 2 canonical discriminant functions were used in the analysis.

Wilks' Lambda

Test of Function(s)	Wilks' Lambda	Chi-square	df	Sig.
1 through 2	.442	198.732	8	.000
2	.993	1.794	3	.616

Standardized Canonical Discriminant Function Coefficients

	Function	
	1	2
coll.asp1	.842	−.177
coll.asp2	−.031	1.420
coll.asp3	.362	−.595
coll.asp4	−.222	−.169

Structure Matrix

	Function	
	1	2
coll.asp1	.975*	.084
coll.asp3	.874*	.089
coll.asp4	.540*	.354
coll.asp2	.552	.794*

Pooled within-groups correlations between discriminating variables and standardized canonical discriminant functions Variables ordered by absolute size of correlation within function.

*. Largest absolute correlation between each variable and any discriminant function

Functions at Group Centroids

	Function	
group	1	2
.00	−1.008	−.026
1.00	.806	.093

Unstandardized canonical discriminant functions evaluated at group means

We note the eigenvalues for the discriminant criterion and Wilks' lambda in the respective preceding output section. Now, for the sake of method comparison, let us define two dummy variables because we have $k=3$ groups to deal with here, as indicated earlier in this section. We can then proceed with a CCA using those dummies as comprising say set A while the four college aspiration measures are contained in the other set B. The relevant results for this comparison are presented next. (See above in this chapter for the necessary software code in SPSS and SAS, as well as for interpretation of the following output.)

```
* * * * * * * * * * * * * * * Analysis of Variance -- design 1 * * * * * * * * * * * * * * *

EFFECT .. WITHIN CELLS Regression

Multivariate Tests of Significance (S=2, M=1/2, N=120)

Test Name        Value   Approx. F  Hypoth. DF   Error DF  Sig. of F
Pillais         .56194    23.73878        8.00     486.00       .000
Hotellings     1.25257    37.73357        8.00     482.00       .000
Wilks           .44213    30.48710        8.00     484.00       .000
Roys            .55460
Note.. F statistic for WILKS' Lambda is exact.
--------------------------------------------------------------------

Eigenvalues and Canonical Correlations

Root No.     Eigenvalue      Pct.    Cum. Pct.   Canon Cor.    Sq. Cor
1                 1.245    99.410       99.410         .745       .555
2                  .007      .590      100.000         .086       .007
--------------------------------------------------------------------
```

We observe that the two eigenvalues in the second column of the last presented output panel are identical to the discriminant criterion eigenvalues that were obtained above with the conventional DA method (see above DA output section that is displayed immediately after Box's M Test result).

```
Dimension Reduction Analysis

Roots      Wilks' L.              F Hypoth. DF     Error DF     Sig. of F

1 TO 2      .44213    30.48710        8.00          484.00        .000
2 TO 2      .99266      .59887        3.00          243.00        .616

--------------------------------------------------------------------
```

In the second column of the last panel, we see identical Wilks' lambda statistics to those found in the respective output section of the discriminant function analysis carried out earlier (see section titled "Wilks' lambda" in the DA output presented previously).

11.5 Generality of Canonical Correlation Analysis

As we alluded to at the beginning of this chapter, CCA can be considered a very general multivariate statistical framework that unifies many of the methods covered so far in this book. To elaborate this relationship, first we note that the canonical correlation concept generalizes the notion of bivariate correlation. Indeed, the latter is obtained from the former in the special case of $p = q = 1$ variables, that is, a single measure in each of the sets A and B (say x in A and y in B). In this case, the canonical correlation for the sets A and B, Corr (Z_1, W_1) in our notation, is identical to the bivariate correlation of their elements, that is, identical to Corr(x, y).

Second, the multiple correlation coefficient of main relevance in RA is also a special case of canonical correlation. Indeed it is obtained when the set A consists of $p = 1$ variable, which is the dependent measure, and the set B consists of q variables that are the predictors in the pertinent regression model. In fact, the multiple correlation coefficient is then identical to the first canonical correlation. We note that there is only $t = 1$ canonical correlation in that case, since $\min(p, q) = 1$ holds.

Third, since various ANOVA designs can be obtained as appropriate special cases of RA, so too can these designs and corresponding ANOVAs be furnished as special cases of CCA. Last but not least, we had seen in the preceding section that discriminant function analysis is a special case of CCA as well. Because DA is "reverse" MANOVA (see discussion in Chapter 10), we can conceptualize the latter also as a method that is covered by the general statistical framework of CCA. Hence, CCA is a very general multivariate analysis method, which subsumes a number of others that are widely used in the behavioral and social sciences.

12

An Introduction to the Analysis of Missing Data

Incomplete data sets, or data matrices that contain missing data values, may be considered the rule rather than exception in the social, behavioral, and educational sciences. Missing data pervade these disciplines as well as many other scientific fields, and most empirical studies conducted in them—no matter how well designed—usually yield incomplete data sets. For these reasons, issues that concern how to deal with missing data have been for decades of special interest in these sciences. In fact, over the past 30 years or so, this interest has led to missing data analysis becoming a major field of research, both in statistics and in the areas of its applications.

This chapter aims at introducing the reader to the problems of missing data, indicating how some earlier methods of "handling" them can be inadequate, and discussing a principled, modern approach to dealing with missing data—one embedded within the framework of maximum likelihood (ML). We emphasize, however, that the analysis of incomplete data is a broad research field, and thus a single chapter like this cannot possibly cover all topics of relevance in it. For more comprehensive discussions, therefore, we refer the readers to a number of excellent and more detailed treatments on missing data, such as Allison (2001), Little and Rubin (2002), and Schafer (1997) (see also Schafer & Graham, 2002), all of which have notably influenced this chapter.

12.1 Goals of Missing Data Analysis

In this chapter, we refer to data sets without missing data as complete data sets. Although initially it may appear that this notion is suitable only for the case of no missing data at all, it turns out that it is highly useful also with data sets containing missing data. Indeed, for a given data set with missing data, we can conceptualize the complete data set as a corresponding one containing no missing values, from which the given data set is obtained as a result of making unavailable the values of some subjects on some variables. The concept of a complete data set will also play an instructive role for the didactic goals of this chapter. In particular, as a first such instance, recall that in most statistical analyses of complete data sets (i.e., those with no missing data), the aim typically is (a) to estimate unknown parameters (with standard errors); (b) to evaluate the plausibility, or goodness of fit, of an underlying model; and possibly (c) to examine parameter restrictions, or competing models, which reflect substantively interesting hypotheses. For example, if we consider an application of regression analysis, the goal would usually be to fit a model under consideration to data, to estimate the partial regression coefficients and intercept (as well as residual variance), and to evaluate how well the model fits the data (e.g., obtain its R^2 index and assess associated residual plots). In addition, one is also likely to be concerned with examining several competing versions of the model, for example, those resulting after dropping some predictors—that is, test the restriction of their coefficients being 0—or adding some other explanatory variables.

The analysis of incomplete data is no different in this respect. That is, when analyzing a data set with missing values, the goal is to estimate parameters and obtain standard errors, to evaluate fit of a model of concern, and possibly examine parameter restrictions (or competing models) (cf. Little & Rubin, 2002). In other words, the fact that the data set contains missing values is considered an undesirable circumstance, but not an issue of primary interest in and of itself. Hence, a meaningful approach to dealing with missing data is not to try to come up with (single) numbers as replacements for the missing values, (i.e., in some sense to "recreate" the missing data), but to use all available data for the particular estimation and modeling aims that a researcher has. With this in mind, in the remainder of the present chapter we will not be concerned with either (a) filling-in the "gaps" in a given data set that represent the missing values, or (b) in some sense optimally estimating the latter. Put another way, we will adopt the standpoint that if a value on a variable is missing for a particular subject, then there is no way of getting to know it or compensate for the fact that it is irreversibly lost. Rather, the following discussion is based on the premise that in the face of missing data a reasonable attitude is to aim at activities (a) through (c) mentioned above, that is, estimate parameters, evaluate models, and possibly test parameter restrictions using all data that have been actually collected.

12.2 Patterns of Missing Data

As in the previous chapters, we presume here that as a result of carrying out a particular study, a rectangular data matrix is available. However, in difference to nearly all prior discussions in the book, we assume that some entries of the data matrix are missing, that is, represent missing values. In such cases, it is helpful to separate (a) the pattern of missing data from (b) the mechanism that has created them. The pattern of missing data only describes the entries that are observed in the data matrix and the entries that are missing. By way of contrast, the missing data mechanism is concerned with the relationships between the fact that certain values are missing, on the one hand, and the observed values as well as the actual missing values, on the other hand.

One can in effect distinguish between several different patterns of missing data. (We note that the following list is not to be considered exhaustive) (cf. Little & Rubin, 2002.) When there are missing values only on a single variable, as may happen in designed experiments, one is dealing with a univariate nonresponse pattern. When the variables in a data matrix, denoted $y_1, y_2, \ldots, y_p$ ($p > 1$), are such that any subject with a missing value on a given measure, say y_i, has also a missing value on each following variable (i.e., on all y_k, with $1 \leq i \leq k \leq p$), then a monotone nonresponse pattern is present. This can occur in some repeated measure studies where subjects who are lost to a follow-up do not provide data after a certain assessment occasion through the end of the investigation. Another kind of missingness pattern of relevance in survey and some educational studies is item and unit nonresponse. In surveys, missingness may be confined to a particular variable (which may or may not be the same across all studied subjects), referred to as item nonresponse; unit nonresponse pattern is present when a whole set of subjects comprising a higher level unit (e.g., class or school) exhibit missing data, typically because their unit has not been contacted or has not provided responses on certain variables. Furthermore, another possibility for a missing pattern is one that is referred to as file-matching. In it, two sets of variables are never jointly observed, that is, there is no subject who provides data on variables from both sets. Last but not least, many variables of main interest in the behavioral, social, and educational sciences are not directly observable, such as intelligence, motivation, aspiration, or parental style, but can only be inferred from manifest behavior indicating them (Raykov & Marcoulides, 2006). These variables, typically referred to as latent variables, can be considered as having missing data for all studied subjects (whether in a sample or population). For this reason, that type of missingness may be called latent variable pattern. This example also suggests that some data analytic and modeling problems or circumstances could be conceived of as dealing in part with missing data even though data are fully recorded (in the example, data may be available on all indicators of the latent variables).

The preceding discussion indicates a fundamental assumption that we will make throughout this chapter. Accordingly, missing values are considered flagging true values for studied subjects and variables, which values, however, are not available (Schafer & Graham, 2002). That is, each missing datum will be considered formally hiding a true and meaningful value for the subject and variable under consideration, which however is "masked" and there is no access to it. As it is readily realized, while in most studies with missing data in the social, behavioral, and educational disciplines this assumption appears plausible, in some settings it may not be so, in particular in surveys with special question series that are not all meaningful to present to all subjects (e.g., questions about pregnancy, or certain questions in some opinion polls). In those cases, the discussion in the remainder of this chapter will not be applicable, and it may be more appropriate to conceptualize strata of a studied population, which are defined by pattern of missingness (Little & Rubin, 2002).

12.3 Mechanisms of Missing Data

Understanding the reasons why certain values may be missing in a given data set can provide a great deal of help in properly handling them. Specifically, the missing data mechanism is crucial for making a decision regarding how to handle an incomplete data set. Until the 1970s, researchers tended to pay little attention to the missingness mechanism, and the widespread attitude toward missing values was largely as an undesirable circumstance that one needed to compensate for. Some 30 years ago, however, primarily through work by Rubin and colleagues, this attitude began to change.

To discuss this shift in attitude, it is helpful to introduce some further notation. For a given empirical study, denote by $Y = [y_{ij}]$ the complete data set—that is, the data matrix with no missing values on all studied variables ($i = 1, \ldots, n$, $j = 1, \ldots, p$; n being sample size, p the number of observed variables). In addition, designate by $M = [m_{ij}]$ an indicator matrix that consists of 0's and 1's, whose entries are defined as follows: $m_{ij} = 1$ if y_{ij} is missing, and $m_{ij} = 0$ if y_{ij} is observed ($i = 1, \ldots, n$, $j = 1, \ldots, p$). Thus, when faced with a data set containing missing values, a researcher has essentially access to the data actually observed plus the indicator matrix M (even though the latter is not separately provided).

The 1970s breakthrough in the way to handle missing data occurred because one began to treat the missing data indicators m_{ij} as random variables that were associated with a certain distribution ($i = 1, \ldots, n$, $j = 1, \ldots, p$). As a consequence, missing data were viewed as a source of uncertainty that needed to be "averaged" over (Little & Rubin, 2002). To a large degree, this shift in attitude was initiated by formal consideration of the mechanism creating the missing values. Specifically, the missing data mechanism was characterized by the conditional probability P(Missing|Y, ϕ), where ϕ denotes, in general, a vector of unknown parameters. In particular, depending on what can be said about this probability of missingness, given knowledge of the complete data (and unknown parameters), three main types of missing data result: (a) missing completely at random (MCAR), (b) missing at random (MAR), and (c) nonignorable missingness (NIM). Before we discuss each of these types in more detail and give formal definitions for them later, we mention here that MCAR is rarely found in empirical research unless it is created by design, while MAR is the kind of missingness that may be considered desirable and is the one that the majority of modern analytic methods have been developed to deal with. By way of contrast, NIM is a highly undesirable type of missing data, which is hard to deal with and typically requires additional assumptions (in need of their own justification) in order to be properly dealt with.

To illustrate this discussion, let us consider a couple of examples. In the first, suppose one is interested in studying the relationships between age and income. To simplify matters for a moment, assume that some data are missing only on the income variable but all subjects provide data on age. Then, if the probability of missing on income is unrelated to both age and income, the data are MCAR. That is, data are MCAR if the fact that a value is missing is unrelated to both the observed and unobserved data. Alternatively, if the probability of missing on income is related to age but not to income, then data are MAR. In other words, an MAR type is present if it is only the observed data that contain information related to the fact that some subjects have missing data. Lastly, if the probability of missing on income is related to income even for persons with the same age, then data are NIM. In that case, missingness on income is related to the actual values of individual income for the subjects refusing to answer the pertinent question.

In the second example, let us assume that one is concerned with examining the relationship between age and alcohol use. Denote these two variables by X and Z, respectively, and let Z denote the number of drinks a person consumes per week. For simplicity, suppose the following small sample data resulted from such an empirical study, whereby a question mark is used to denote any missing value (and ID designates subject identifier) (see Table 12.1).

Since this data set contains missing values, the actually available empirical information consists not only of (a) the five observed values on X and the three observed values on Z; but also of (b) the facts that the last two subjects did not provide data on Z, while all preceding ones did and that all subjects supplied data on X. That is, there are altogether 18 pieces of information that this study provides us with: 8 observed and 10 missing and nonmissing indicators. Hence, in this study, the data set may actually be viewed as follows (see Table 12.2):

TABLE 12.1

Data From a Study of Age and Alcohol Use (Missing Data Denoted by '?')

ID	X	Z
1	25	3
2	33	5
3	37	4
4	42	?
5	46	?

TABLE 12.2

Available Empirical Information From the Example Study on Age and Alcohol Use (cf. Table 12.1)

ID	X	Z	M_X	M_Z
1	25	3	0	0
2	33	5	0	0
3	37	4	0	0
4	42	?	0	1
5	46	?	0	1

We stress that the last two added columns, labeled M_X and M_Z, are additional individual variables that the researcher did not plan to collect originally, but were generated by the fact that the study under consideration contained missing data (i.e., were created due to some subjects not providing data on all variables; these two variables, M_X and M_Z, are the two columns of the earlier mentioned indicator matrix M for the present example). We also note that in this study the previous assumption that the 1's in the columns of M_X and M_Z essentially hide (or mask) true underlying meaningful values is plausible. Obviously, in a study with no missing data, the variables M_X and M_Z do not provide useful information (beyond the fact that there are no missing values in the associated data set).

To give a more formal discussion of the main types of missing data, in the rest of this section we assume for reasons of clarity that in general data are given on a vector of variables $\underline{X}$ and a variable of focal interest Z, with some missing values on Z. (The vector $\underline{X}$ may be empty; we will later generalize to the case where data may be missing on both Z and variables from $\underline{X}$.)

12.3.1 Missing Completely at Random

The above discussion indicated that data are MCAR when subjects with missing data represent a random subsample from the available overall sample. For example, somewhat loosely speaking, if persons with missing data on an alcohol consumption measure (Z in the above notation) are "the same as" or "just like those" with no missing values on it, then data are MCAR. In this case, subjects with missing data on alcohol consumption are not above average (or below average) on age and on alcohol use. However, if persons with missing data on Z are above (or below) average on alcohol consumption, and/or are younger (older) on average than those with data present on Z, then the data set is not MCAR.

As a somewhat more formal definition, denoting by P(.) the probability of an event (that in parentheses), data are MCAR when P(Z missing) is unrelated to both Z and $\underline{X}$, i.e., if P(Z missing) is unrelated to all observed as well as all unobserved data. We note that the fact that data are MCAR does not preclude the case where missingness on Z is related to missingness on another variable(s). For example, even if subjects with missing alcohol consumption measure (Z) were also missing on age (X), MCAR is still possible (Allison, 2001). What is of ultimate relevance for MCAR is that the probability of missingness is not related either to the observed or the unobserved data.

To be more precise, denote by M the statement "data are missing," by $\underline{X}_{obs}$ and Z_{obs} the observed data on $\underline{X}$ and Z, and by $\underline{X}_{mis}$ and Z_{mis} the actual data missing on these variables, respectively. Then a data set is MCAR if and only if

$$P(M|\underline{X}_{obs}, Z_{obs}, \underline{X}_{mis}, Z_{mis}, \phi) = P(M|\phi). \qquad (12.1)$$

Equation 12.1 states that neither the observed data nor the missing data affect the probability of a particular missing value. In other words, data are MCAR when formally assuming knowledge of the missing data and of the observed data (as well as the unknown parameters), we cannot predict any better the probability of missingness than without such knowledge (Schafer, 1997).

Having defined MCAR, the next question refers to how one could determine if he or she is dealing with a case of MCAR in a particular empirical study. Since this question is fundamentally concerned with data that are missing, there is no conclusive statistical test of it, that is, there is no sufficient condition that could be tested and provide an answer to the question. However, there is a necessary condition for MCAR. This condition is referred to as observed at random (OAR). A data set is OAR if there are no systematic distributional differences on fully observed variables between those subjects (Group 1) with data present, on the one hand, and those persons (Group 2) with missing data, on the other hand. Since this is only a necessary condition for MCAR, the latter implies OAR. Yet, data may be OAR but still not MCAR; this is because MCAR requires no relationship between missingness and the actually unavailable values, while OAR in no way refers to the latter.

For this reason, if OAR does not hold, MCAR cannot be the case either. Alternatively, if OAR is not rejected, one may or may not be dealing with an MCAR situation. Hence, if in a given data set, persons with missing data on Z (Group 1) have a different mean or variance, or in general a different distribution, on the other variable(s) with no missing values than those subjects with observed data on Z (Group 2), then that data is definitely not MCAR since they are not OAR. This example indicates that a relatively simple test for an aspect of OAR is that of mean (and/or variance) differences across Groups 1 and 2. If this test is nonsignificant, then data may be OAR. Generally, to test fully for OAR, one needs to examine distributional differences across these groups (e.g., using the Kolmogorov–Smirnov test). If no such differences are found, data are OAR (within the limits of statistical "certainty"), and hence possibly also MCAR, although one cannot be sure of the latter.

An instance where MCAR holds is a situation where data are missing by design. The latter is the case if a variable(s) is measured only on a random subsample of an original sample from a studied population, but other than that there are no missing values. This may be sometimes the case, for example, in disciplines where collecting data may be expensive on certain variables, e.g., in the biomedical and health sciences. Then due to the random choice of subjects not measured on these variables, they represent a random subsample from an original sample, and hence the resulting data set possesses the feature of MCAR (assuming all subjects provided data on all remaining variables).

The final question that arises now is how to analyze MCAR data. First of all, as indicated repeatedly in the literature, if the percentage of missing data is small (up to 5% say) and sample size is large, one may use listwise deletion (complete-case analysis [CCA]; see below). That is, one may then delete from further consideration all cases with at least one missing value. The resulting parameter estimates are with good properties except for some loss of power (efficiency). When this loss is not deemed problematic, this approach available by default in many widely circulated statistical software can be recommended. If the percentage of missing data is larger, however, one can use a method for handling MAR data sets (see below; MAR is a weaker condition than MCAR, as discussed next.)

12.3.2 Missing at Random

The condition of MCAR is based on a very strong assumption. Not only is missingness required in it to be unrelated to the unobserved data but also to be unrelated to the observed data. In many empirical studies in the social and behavioral sciences, however, cases with missing data cannot be considered a random subsample from a given sample, that is, MCAR does not hold. The reason may actually be the fact that whether or not data are missing on a given variable depends on another observed variable(s)—that is, data are MAR; as indicated earlier in this chapter, MAR is the condition when missingness say on Z does not depend on Z (the missing values on Z) but may instead depend on other observed variables, for example, those in $\underline{X}$.

More generally, and using the above notation, a given data set exhibits the MAR property if P(Z is missing) is unrelated to Z_{mis} but may be related to X_{obs} and/or Z_{obs}. Formally, MAR holds if and only if

$$P(M|\underline{X}_{obs}, Z_{obs}, \underline{X}_{mis}, Z_{mis}, \phi) = P(M|\underline{X}_{obs}, Z_{obs}, \phi). \tag{12.2}$$

In other words, MAR is the case if assumed knowledge of the actually missing data (and unknown parameters) does not improve prediction of the probability of missingness over and above knowledge of all observed data.

Comparing Equations 12.1 and 12.2, we note that if data are MCAR they are for sure MAR as well, but not conversely. That is, the MCAR condition is sufficient but not necessary for MAR; conversely, MAR is necessary but not sufficient for MCAR. The reason is that MCAR is a special case of MAR.

If data are not MAR, that is, the probability of missingness depends also on unavailable data (whether on $\underline{X}$ or Z), they are called not-MAR (NMAR). In such a case, data cannot be MCAR either (since MCAR is a restricted form of MAR). That is, data being NMAR is a sufficient condition for not being MCAR. This represents another way in which one could ascertain whether a given data set exhibits the MCAR property. To illustrate the notion of NMAR, consider the case of deterministic censoring (Little & Rubin, 2002). In it, a cutoff is fixed beforehand and any value that is say above it, is set as missing. For example, if starting from a complete data set one declares as missing all positive values on a prespecified variable, one will be implementing deterministic censoring. Since then the probability of missingness depends on the actually missing data, Equation 12.2 is violated and therefore the MAR condition cannot hold (like MCAR that cannot hold then either). The resulting data set is therefore NMAR.

As another empirical example, consider the earlier study of age (X) and alcohol consumption (Z), with data missing on Z but not on X. When would this data set fulfill the NMAR condition? As indicated above, it would satisfy the MAR condition if among persons with the same age there is no relationship between alcohol consumption and the probability of them having a missing value on alcohol use (i.e., if among same-aged persons there is no relationship between how much a person drinks and the probability of him or her refusing/omitting to provide data on this variable Z). Conversely, data are not MAR (i.e., they are NMAR), if among persons with the same age missing values are predominantly found for those who drink more (or, alternatively, less) than average.

Having discussed at length the MAR condition, the natural question that arises at this point is how to test for it. To address this issue, let us mention first that if data are missing by design, they are MCAR, in which case they are also MAR. Second, since the MAR condition involves in a fundamental way reference to missing data (i.e., data that are not available), it is not possible to devise a statistical test of it as data needed for this purpose are missing. We note that in this respect the MAR and MCAR conditions are similar—they are not statistically testable, since they refer to information that is missing and hence one cannot evaluate the probability of missingness and make assertions about it as needed in their definitions. However, as mentioned earlier, we can disprove MCAR by showing that data are not OAR. Also, we can disprove MAR in a particular setting by showing that the probability of missing depends then on the missing data Z_{mis}. Needless to say, showing this will also disprove MCAR, as it is a special case of MAR. We stress that even if means and variances are the same across the above defined Groups 1 and 2 with available and with missing data, it may still be the case that their relevant distributions differ with regard to higher order moments, which discrepancies may be much more difficult to sense statistically. In such cases, showing that the probability of missing depends on the actually unavailable data proves that the data set cannot be MCAR. Hence, this is one more way of examining the MCAR condition (despite the fact that there is no statistical test for MCAR or MAR).

Given the lack of ways to formally test for MAR, this discussion raises the query about what a researcher can do in order to enhance its plausibility. Suppose the variable Z is of focal interest to them and contains missing values. To make the MAR condition more plausible, the following two-step procedure can be followed (Little & Rubin, 2002): (a) collect data on other variables, denoted $W_1, \ldots, W_k$, for both respondents and nonrespondents with respect to Z, whereby the W's are predictive of the missing values on Z; and (b) include these variables $W_1, \ldots, W_k$ as covariates (explanatory, predictor variables) in the subsequent analysis. As a result, step (b) reduces in general the association between missingness and the variable of main interest, Z, thus making the MAR assumption more plausible.

12.3.3 Ignorable Missingness and Nonignorable Missingness Mechanisms

The condition of MCAR and especially the weaker condition of MAR afford in practical terms the researcher with the possibility to disregard the mechanism that has generated the missing data and focus only on his/her specific modeling goals. More precisely, this is possible to do when the missing data mechanism is ignorable. This will be the case when the following two conditions are fulfilled: (a) data are MAR and (b) the parameters governing the missing data process are unrelated to the parameters to be estimated (Little & Rubin, 2002). Hence, a data set follows a nonignorable missing mechanism (i.e., data are NIM), if either they are NMAR, or condition (b) is violated, or both. Most of the time in empirical research, condition (b) is likely to be fulfilled (although in general one does not know for sure if it is so). As an example of when condition (b) would not be fulfilled, the following one could be used: a subject is missing on Z whenever his/her score on X is higher than the mean of X, while one is also interested in the variable means. Most of the time in empirical settings, however, the missingness mechanism will follow its own laws and likely not be related to parameters of interest to the researcher. Therefore, from a practical standpoint, one may focus on ensuring that data are MAR, or at least that this condition is plausible, in order to treat the missingness mechanism as ignorable. That is, in most social and behavioral studies, the ignorability condition can be treated "the same as" the MAR condition.

The notion of ignorable missingness is of particular relevance for purposes of statistical modeling in the context of missing data. The reason is that if the missing data mechanism is ignorable, there is no need to model it (Little & Rubin, 2002). That is, no information is lost then by ignoring and not being concerned with this mechanism, which information would be contained in the data. Hence, in such cases, one would not be sacrificing efficiency of statistical inference by disregarding the mechanism having generated the missing data. Then the researcher can focus on his/her original interest in modeling only the relationships among the variables in $\underline{X}$ and Z.

Alternatively, if the data are NIM (e.g., if they are NMAR), then the missing data mechanism is not ignorable and needs to be modeled as well if one wishes to carry out efficient inferences. To model this mechanism is usually very hard to do in a credible way, since it requires knowledge of how missing data were generated in the first place, which knowledge is rarely available even partially, if ever at all. As a practical way out in such cases, especially when the MAR condition is violated to a limited degree, one may consider still assuming MAR and using methods appropriate with MAR (see end of next section and Section 12.5). In fact, as shown in some initial studies, methods for dealing with missing data that assume the MAR condition possess some robustness against violations of it (Collins et al., 2002; Shafer & Graham, 2002). Further research is, however, needed in order to examine the extent of this robustness, which will also respond to the query as to when their results can be trusted with violations of the MAR condition.

12.4 Traditional Ways of Dealing With Missing Data

The problems resulting from the presence of missing data in empirical studies have long been of concern to methodologists and substantive researchers. Various ad hoc methods have been developed during the past century, which, however, have been shown more recently to be suboptimal in general relative to newer, principled approaches to handling missing data. For completeness of the present chapter, we briefly discuss some of those earlier ad hoc approaches.

12.4.1 Listwise Deletion

This approach is also known under the name "complete case analysis" (CCA). Following it, one disregards all subjects in a given data set, which exhibit at least one missing value. Needless to say, this leads in general to very small samples to carry subsequently the actual statistical analyses on, and therefore sacrifices efficiency of statistical inference. At least as relevant, however, is another potentially major drawback of the method—the general lack of representativeness of the resulting sample for the studied population.

When data are MCAR, as mentioned earlier in the chapter, CCA would likely be a reasonable method of handling missing data if the original sample size is large and a relatively small percentage of values are missing. Even then, however, some lack of efficiency ensues, while the resulting estimates may still have desirable properties (such as unbiasedness, consistency, and large-sample normality). A convenient feature of CCA in empirical research is that it is a nonparametric approach since it proceeds without making any assumptions about the data distribution.

However, most of the time, in practice data are not MCAR. Under these circumstances, CCA leads to biased estimates for the parameters of interest in the originally sampled population, because the complete-case-sample is no longer a random sample from the population. For this reason, the resulting means and covariance matrix cannot be generally trusted as estimates of these parameters in the population of concern, since the mean and covariance structure (matrix) of the studied variables is distorted in the complete-case-sample.

12.4.2 Pairwise Deletion

This method has also been popularized under the name "available case analysis" (ACA). To proceed with it, for any parameter of interest one utilizes the data from all subjects who have complete records on all variables of relevance when estimating that parameter. For example, when one is interested in estimating a correlation coefficient between two variables, this method will utilize the data from all subjects who have available data on these two variables, since this is all data that one needs for estimating their correlation. In this way, it is possible that for some parameters of concern one uses a larger sample size than CCA would. For this reason, ACA is in general more efficient than CCA but like the latter does not ensure representativeness for the studied population of the sample from which a parameter of interest is being estimated. While with MCAR data the ACA method yields estimates that are each possessing desirable properties (other than some loss in efficiency, for the same reasons as with the CCA method), when data are not MCAR its estimates are biased. In addition, for essentially the same reasons as CCA, the ACA approach leads to distorted mean and covariance structures of the set of studied variables. Furthermore, with ACA it is possible that some estimated correlations or covariances lead to nonpositive-definite matrices—that thus cannot be correlation or covariance matrices of real-valued variables—which, may be particularly problematic when one is interested in fitting models to means and covariance matrices as is not infrequently the case in social, behavioral, and educational research (with multinormal data). As yet another downside, the sample base with ACA in general changes from variable to variable (e.g., from one interrelationship index to another), thus giving rise to the serious problem of finding out what the pertinent sample size for subsequent analyses would be.

12.4.3 Dummy Variable Adjustment

When one is interested in regressing Z on the variables in $\underline{X}$, and data are missing on one of the latter, say X_1, some researchers have suggested the creation of a new variable, denoted X^*, defined as follows. For all subjects with available data on X_1, define their value on X^* as identical to that on X_1; for subjects with missing values on X_1, however, define their value on X^* as equal to c, with c being an arbitrary constant. (The same could be considered doing when data are missing on Z, leading to the creation of a dummy variable Z^*.) The dummy variable adjustment (DVA) method then proceeds by regressing Z on the predictors in $\underline{X}$, to which also X^* is added (or Z^*, if possible, in case of missing data on Z). As can be readily realized, the DVA procedure produces biased estimates of partial regression coefficients of the variables in $\underline{X}$, since it changes the model fitted—that actually analyzed has at least one more predictors than the model initially intended to fit.

12.4.4 Simple Imputation Methods

What the DVA method accomplishes in its first step is to deal in a particular way with the "gaps" in one's data set, which are created by the missing data. In this sense, the DVA approach may be considered a precursor to procedures that aim at coming up with a compensating individual score(s) for each missing value. These are called imputation procedures. They aim at filling in for each missing value, that is, drawing from an appropriate distribution of the missing values and filling in the resulting draw (random number) for the missing datum in the original data set. A special characteristic of the so-called simple imputation methods (SIMs) is their feature of filling in only one value per missing datum. (An alternative method using a conceptually similar idea, multiple imputation [MI], is mentioned later in the chapter.)

One of the most popular SIMs in the past has been mean substitution (MS), also referred to as unconditional fill-in. When using it for a given variable, one substitutes all missing values on it with the mean of the remaining cases on that measure. By doing this, however, obviously MS distorts the empirical distribution of this variable. Except for its mean, in general, all other ordinary statistics are not consistent (but distorted) if they are computed from the filled-in sample, as is the ultimate goal of a SIM. In particular, the variance obtained from the so-imputed variable tends to be smaller than the corresponding population variance (since the imputed value does not make any positive contribution to the variance estimate). Alternatively, their correlations will tend to be larger since the degree of linear relationship will tend to be overstated by this variance deflation tendency. That is, while the mean structure is preserved, the covariance and correlation structure (matrix) is distorted relative to the population covariance or correlation matrix, respectively.

A slightly more sophisticated version of this imputation procedure is the so-called conditional fill-in, or regression-based imputation. In it, one substitutes a missing value with a predicted value obtained from the regression of cases with available data on a given measure, on the remaining variables/chosen predictors of that measure. (This approach becomes quite complex with missing data on independent variables as well, but for the present purposes we will not go into these details.) While this is an improvement upon MS, the underestimation feature of the latter mentioned above still remains, although generally not being as pronounced.

Yet another variant of simple imputation is the hot-deck imputation (HDI). Using it, for a given subject with a missing value on a variable of concern, one first finds out "similar" subjects on observed variables, who have the same or very similar values on the latter. As a second step, one substitutes the missing value in question with the value on the critical variable that those similar subjects have (or with their mean if they differ, or with the result of a more complicated HDI procedure then). While this may appear as a fairly meaningful approach at first sight, it can be shown that it also distorts variable relationships. This is due to the built-in tendency of predicted values to be less variable than what may be expected from the missing values.

This discussion highlights a generic problem with the HDI methods. The problem consists in their feature of underestimating variability, which results from the tendency of filled-in values to be less variable than the missing values. As a consequence, the parameter estimate standard errors are underestimated, and in this sense, the sample size effect is overstated. This leads to a spurious tendency to reject tested null hypotheses and proclaim too many significant results in a given study.

The preceding indicates more generally a fundamental flaw of all simple imputation methods. It results from the fact that they lead subsequently to analyzing the so-imputed data set as if it were a complete data set, which it obviously is not (Little & Rubin, 2002). As a result, no adjustment for or any incorporation of the uncertainty brought about by the presence of missing data is carried out. This has as consequences underestimated standard errors, too narrow confidence intervals, and overestimated test statistics leading to spurious significance findings. These problems with SIMs become more serious when the amount of missing values increases and/or with larger number of parameters to be estimated.

12.4.5 Weighting Methods

These methods are typically used in studies where unit nonresponse is present. They accomplish different adjustments that aim at reducing bias resulting from CCA. The essence of the methods is to produce weights by which complete cases are multiplied, in order to "account" for cases with missing data (which cases are deleted prior to the analysis). The conceptual idea behind these methods is that after weighting of the complete cases, the resulting sample more closely resembles the studied population with respect to distribution of important characteristics. These methods furnish weights that are subject-specific and therefore yield the same weight for all variables measured for each complete case. A major problem with weighting methods is that they do not provide principled ways of obtaining appropriate standard errors for parameter estimates of interest. This results from their feature of not ensuring adequate control of sampling variance, which is a consequence of the fact that whereas the weights are treated as fixed once obtained, they are actually worked out from a given sample while the associated sampling error is not properly accounted for.

In conclusion of this section, all methods discussed in it show consequential limitations that make them less trustworthy than has been assumed in the past. Apart from CCA that would yield parameter estimates and standard errors with good properties in case of MCAR and a small portion of missing values, the above procedures for dealing with missing values have potentially serious drawbacks that preclude them from being generally recommended (Allison, 2001). We turn next to an alternative method that could instead be more generally recommended to consider applying, and which is straightforwardly utilized under its assumptions.

12.5 Full Information Maximum Likelihood and Multiple Imputation

When data are MAR and the assumption of multinormality is plausible, one can employ the principle of maximum likelihood (ML), widely utilized in statistical applications. In the context of dealing with incomplete data sets, this utilization of ML has come to be also known as direct ML, raw ML, or full information maximum likelihood (FIML). Its essence is the use of all available data on all subjects and all variables of concern in a given analysis. (To emphasize this relevant feature, in the rest of the chapter we refer to this method as FIML.) In particular, when one is interested in evaluating a specific model as a means of description and explanation of an incomplete data set, FIML accomplishes fitting it to all the data that are available from all subjects on the variables participating in this model.

To highlight this property of FIML as a method of handling missing data, it is instructive to recall that for a given data set, the likelihood function (LF) is the following product of individual datum probability density function (mass function) values:

$$L = f(\underline{x}_1, \underline{\theta})\, f(\underline{x}_2, \underline{\theta}) \ldots f(\underline{x}_n, \underline{\theta}), \tag{12.3}$$

where

n is sample size
$\underline{\theta}$ is the vector of unknown parameters
$\underline{x}_1, \underline{x}_2, \ldots, \underline{x}_n$ are the individual observations

(Roussas, 1997; cf. Raykov, 2005). We emphasize that the LF in Equation 12.3 does not need all individual data records to be of equal length (i.e., the LF does not need all subjects to be observed on all variables of interest in an analysis). In other words, the ith subject contributes to the LF function in Equation 12.3 the value $f(\underline{x}_i, \underline{\theta})$ that is correspondingly obtained from the assumed probability density function (mass function) $f(.,.)$ for any dimensionality of the available data record for him or her, that is, regardless of how many variables have been observed on that person ($i = 1, \ldots, n$). (Note that with different number of observations available, the actual formula of $f(\underline{x}, \underline{\theta})$ differs as well.) Hence, from a conceptual viewpoint, ML would be applicable regardless of whether there are missing data or not (of course, as long as the assumed variable distribution is correct). Theoretical developments (Little & Rubin, 2002), which show that when data are MAR in order to use ML it is sufficient to maximize the likelihood of the observed data only, further corroborate the utility of resulting ML estimates. That is, in the presence of data that are MAR, to accomplish ML estimation one only needs to maximize the likelihood of the observed data as a function of the unknown parameters, and there is no need to model or state the form of the missing data generation process (missing mechanism).

From a practical perspective, it is worthwhile raising the question to what degree the assumption of MAR is essential for an application of FIML. There has not been sufficient research bearing upon this question, which would provide a conclusive answer to this query. To date, according to a relatively limited number of studies (mostly of simulation nature), FIML has been found to be somewhat robust against violations of the MAR condition. This robustness research continues and is highly desirable to be extended further, to cover more conditions of its violation. Until then, when use of FIML is considered, it may be recommended that one becomes first involved in the above mentioned activities to enhance the plausibility of MAR—specifically, fully observing covariates that are predictive of the missing values, and including them in pursued models.

Before we move on to illustrative demonstrations of FIML in the context of multivariate analysis, we mention another modern and principled approach to handling missing data. This is multiple imputation (MI) that may be considered a substantially improved version of the SIMs discussed earlier in the chapter. Unlike SIMs, however, MI commences by producing (under the assumption of MAR) multiple imputed data sets by repeatedly drawing values for each missing datum from an appropriate distribution. In this way, at the end of its first phase, MI affords the researcher with a number of complete data sets that account for the uncertainty brought about by the presence of missing data in the initial sample. Each complete data set is then analyzed by a statistical method of the researcher's choice, typically the one to use if the data set were complete in the first place, in order to address a research question of concern. The results of these analyses are combined into parameter estimates and standard errors (confidence intervals) in the third phase, following specific algorithms. For further discussion of this in general fairly computer-intensive yet highly recommendable and promising method, we refer readers to Allison (2001), Little and Rubin (2002), and Schafer (1997). In the remainder of this chapter, we concern ourselves instead with demonstrating applications of the less computationally intensive method of FIML in some multivariate analysis settings.

12.6 Examining Group Differences and Similarities in the Presence of Missing Data

Multiple-group studies are quite frequent in the social, behavioral, and educational sciences and are commonly concerned with examining group differences and similarities in means, variances, covariances, and possibly correlations of collected measures of interest. Most of the time, however, even in well-designed investigations, researchers cannot avoid the presence of missing data. In such situations, the natural question is how to carry out proper group difference analyses. Modern methods of analyzing incomplete data sets provide principled approaches to addressing this query. In particular, when coupled with specific confirmatory factor analysis procedures (see Chapter 9), FIML can be used to conduct likelihood ratio tests of mean group differences as well as of group differences in covariance matrices or correlation matrices and related follow-up tests. These tests and issues will be the topic of concern next (cf. Raykov & Marcoulides, 2008).

To introduce the empirical context of this section, consider a study of 106 male and 197 female university freshmen on $p = 4$ fluid intelligence tests, with more than 10% of the participants not having provided data on all tests. (The data are contained in the file ch12ex1.dat available from www.psypress.com/applied-multivariate-analysis, with the four consecutive measures denoted "test_1" through "test_4" and the gender variable named "gender" with value 0 for males and 1 for females.) The first two of these measures tapped into one's inductive reasoning ability, while the last pair of tests assessed their figural relations ability. We will be interested in examining whether there are gender differences in fluid intelligence, as evaluated by the four measures, in the presence of this marked amount of missing data. In addition, as a follow-up, we wish to see whether there may be mean differences on each of the tests, and whether the average of the gender mean difference on induction tests is the same as the average of the gender mean difference on the figural relations measures. Furthermore, we would like to examine whether the covariance structure and the correlation structure of these four fluid intelligence tests are the same across groups. That is, our concern will be with testing whether there are gender differences in the four-dimensional mean vectors, in each of the tests, in a contrast of the elements of their group difference vector, and in the 4×4 covariance as well as correlation matrices for both genders.

As is well known, when one is interested in mean differences on multiple observed variables, a traditional approach is to consider using the method of multivariate analysis of variance (MANOVA) (see Chapter 4). When applied on an incomplete data set like this one, however, conventional and widely circulated software implementations of MANOVA will automatically invoke an analysis of the complete-case subsample (CCA), that is, will analyze the subsample resulting after listwise deletion. Yet, as indicated earlier in the chapter, this will generally lead to a substantial loss of efficiency, in particular in the present empirical study that in addition cannot be really considered having a particularly large sample size. We are therefore interested in alternative methods that would use all available information in a sample at hand, such as FIML. The approach outlined next not only permits the study of group mean differences in the presence of missing data but also is applicable for examining group differences in variances, covariances, and correlations.

In order to apply FIML for the aims of this section, one can utilize a special confirmatory factor analysis approach (Raykov, 2001 and references therein) in the context of the multiple-group approach discussed in Section 9.8. To this end, in the general case, we formally conceptualize a model with $p > 2$ observed variables and $q = p$ latent factors in each of G groups under consideration ($G > 1$), whereby each observed variable loads on only one factor and there are no residual terms. Specifically, denoting by $y_{1g}, \ldots, y_{pg}$ the observed variables in group g ($g = 1, \ldots, G$), which in this section are assumed to be normally distributed (see end of section), and by $f_{1g}, \ldots, f_{pg}$ their corresponding factors, the model we adopt is defined as follows (cf. Equations 9.1 and 9.15 in Chapter 9):

$$\underline{y}_g = \underline{a}_g + \Lambda \underline{f}_g + \underline{e}_g, \tag{12.4}$$

where $\underline{y}_g = (y_{1g}, \ldots, y_{pg})'$, $\underline{f}_g = (f_{1g}, \ldots, f_{pg})'$, $\underline{a}_g = (a_{1g}, \ldots, a_{pg})'$ is a vector of intercepts, and for our aims $\underline{e}_g = 0$ is set while Λ is the $p \times p$ identity matrix; in addition, to achieve model identification, we assume that $M(\underline{f}_1) = 0$, that is, all factors have zero mean in group 1, and that $\underline{a}_1 = \underline{a}_2 = \ldots = \underline{a}_G$, that is, each observed variable's mean intercept is the same across all groups (Muthén & Muthén, 2006). In an empirical setting, this model can be readily fitted to data using the popular program M*plus* (see Chapter 9 for a brief introduction; below in this section, we provide the needed software code).

The gain we achieve by using the model defined in Equation 12.4 and immediately after it is that we now have available all benefits of confirmatory factor analysis and thus can easily apply FIML to an incomplete data set from a multiple-group study, under the assumptions of data MAR and multinormality. In particular, a direct implication of Equation 12.4 is that the observed variable means in the first group are parameterized in its mean intercepts, $\underline{a}_1$, while the mean differences between the first and each subsequent group are parameterized in the factor means of the latter group, $M(\underline{f}_g)$ ($g = 2, \ldots, G$). Furthermore, the covariance matrix of the observed variables in the gth group, $\underline{y}_g$, is parameterized in the factor covariance matrix in that group, $\Sigma_{ff,g}$ ($g = 1, \ldots, G$). Moreover, if for model identification purposes we fix the latent variances at 1 while freeing each measure's single factor loading, the correlation matrix of the observed variables in the gth group, $\underline{y}_g$, will be parameterized in the factor covariance matrix in that group, $P_{ff,g}$ (Raykov, 2001, Footnote 1) ($g = 1, \ldots, G$). With this in mind, by testing hypotheses developed in terms of the elements of the mean vector, covariance, or correlation matrix of the factors in the G groups under consideration, we will in fact be testing the same hypotheses about their counterpart observed mean vector, covariance, or correlation matrix elements; in addition, when carrying out the latter tests, we will be able to directly employ FIML. This is the essence of an FIML application in the context of missing data for purposes of group difference testing in means, variances, covariances, and correlations, as well as for carrying out follow-up analyses. (The outlined method is obviously applicable in the same way in case of no missing values.) We exemplify this approach next on the empirical study underlying this section.

12.6.1 Examining Group Mean Differences With Incomplete Data

We start by studying mean differences across the two gender groups in the presence of missing data. To this end, we utilize the model defined in Equation 12.4 and immediately following discussion. Accordingly, for illustration purposes, we are interested in the factor means in the second group that comprises female freshmen. To estimate these means and carry out tests responding to our concerns, we can use the following M*plus* input file. (Explanations for its new lines, relative to the command files utilized in Chapter 9, are provided subsequently. We use an exclamation mark to signal the beginning of a clarifying comment in a line).

```
TITLE:       TESTING GROUP DIFFERENCES WITH MISSING DATA
DATA:        FILE = CH12EX1.DAT;
             !NEED TO ANALYZE THE RAW DATA
VARIABLE:    NAMES = TEST_1 TEST_2 TEST_3 TEST_4 GENDER;
             MISSING = ALL(99);
             GROUPING = GENDER (0 = MALES 1 = FEMALES);
ANALYSIS:    TYPE = MISSING H1;
             !NOTE 'H1', TO REQUEST ALL MODEL FIT INDICES
MODEL:       F1 BY TEST_1@1; TEST_1@0;
             !SYMBOL '@' FIXES PARAMETER TO VALUE NEXT
             F2 BY TEST_2@1; TEST_2@0;
             !VARIABLE NAME REFERS TO ERROR VARIANCE
             F3 BY TEST_3@1; TEST_3@0;
             F4 BY TEST_4@1; TEST_4@0;
```

In this input file, the second line of the VARIABLE command section declares the uniform symbol "99" as indicating missing value throughout the raw data file. (This symbol should be carefully chosen so that it is not used for any subject to denote a legitimate, i.e., nonmissing, value on any variable in any group.) We note that an application of FIML requires access to the raw data, unlike analysis of complete data sets—under the normality assumption—that can proceed also knowing only the observed variable covariance matrix and means. A new command line is subsequently added, ANALYSIS, which requests from the software use of FIML by stating next the subcommand TYPE = MISSING. To ask for all model fit indices, one adds at its end "H1," which symbolically represents the saturated model. The MODEL command section defines each of the latent factors, F1 through F4, as identical to the corresponding observed variable, TEST_1 through TEST_4. This is accomplished by fixing the loading of each observed measure on its factor at 1 and fixing at 0 the variance of the pertinent error term. (This fixing of the only factor loading per measure at 1 is not necessary and carried out by default by the software; we include it here explicitly to emphasize the fact that the factor loading matrix Λ in the defining Equation 12.4 is the identity matrix. Note also the used reference to error term by the name of the pertinent observed variable, at the end of each of the four lines in this command section.)

The default option in M*plus* in the multiple-group case of interest here is: (a) to set equal across groups all observed mean intercepts (i.e., set $\underline{a}_1 = \underline{a}_2$ for the intercept vectors in the above Equation 12.4); and (b) to fix at 0 the factor means in the first listed group (that is, set $M(\underline{f}_1) = 0$), while free the factor means of all remaining groups (here, of female freshmen, i.e., the vector $M(\underline{f}_2)$ contains only free parameters). Therefore, the group mean differences of concern, $M(\underline{y}_2) - M(\underline{y}_1)$, are estimated in the factor mean differences $M(\underline{f}_2) - M(\underline{f}_1)$, and thus in the factor means in group two in this study, that is, in the elements of the vector $M(\underline{f}_2)$. Hence, the group mean difference on the induction tests is estimated in the means of F1 and F2 in the second group (females), whereas the group mean difference on the figural relations tests is estimated in the means of F3 and F4 in that group.

This M*plus* input file produces the following output (only nonredundant parts are provided next; clarifying comments are inserted in the current font, after output statements or sections that are not self-explanatory).

```
INPUT READING TERMINATED NORMALLY
```

There were no syntax errors in the command file submitted to the software.

```
TESTING GROUP DIFFERENCES WITH MISSING DATA

SUMMARY OF ANALYSIS

Number of groups                                    2
Number of observations
   Group MALES                                    106
   Group FEMALES                                  197

Number of dependent variables                       4
Number of independent variables                     0
Number of continuous latent variables               4
```

```
Observed dependent variables

  Continuous
   TEST_1  TEST_2  TEST_3  TEST_4

Continuous latent variables
   F1      F2      F3      F4

Variables with special functions

  Grouping variable      GENDER

Estimator                                              ML
Information matrix                               OBSERVED
Maximum number of iterations                         1000
Convergence criterion                           0.500D-04
Maximum number of steepest descent iterations          20
Maximum number of iterations for H1                  2000
Convergence criterion for H1                    0.100D-03

Input data file(s)
  CH12EX1.DAT

Input data format FREE

SUMMARY OF DATA

    Group MALES
      Number of patterns               7

    Group FEMALES
      Number of patterns               6
```

There were seven patterns of missing values found in the male group and six patterns of missingness in the female group.

```
COVARIANCE COVERAGE OF DATA

Minimum covariance coverage value         0.100

  PROPORTION OF DATA PRESENT FOR MALES

        Covariance Coverage

            TEST_1      TEST_2      TEST_3      TEST_4
            ________    ________    ________    ________
TEST_1       0.962
TEST_2       0.915       0.943
TEST_3       0.915       0.943       0.943
TEST_4       0.821       0.821       0.821       0.849
```

```
PROPORTION OF DATA PRESENT FOR FEMALES

       Covariance Coverage

              TEST_1      TEST_2      TEST_3      TEST_4
              ______      ______      ______      ______
TEST_1         0.970
TEST_2         0.883       0.914
TEST_3         0.883       0.914       0.914
TEST_4         0.817       0.792       0.792       0.843
```

The above two matrices indicate the percentage of complete cases that would be available when looking at each variable variance or interrelationship index. For instance, there were a little over 15% of cases with missing data on the fourth intelligence test, while there were just under 21% of subjects who had a missing value on it or on the second fluid measure, or on both of them.

```
THE MODEL ESTIMATION TERMINATED NORMALLY
```

The numerical optimization routine underlying an application of FIML to the analyzed data has converged to a final solution (presented next), as is needed to be the case in order to proceed with interpretation of the following output:

```
TESTS OF MODEL FIT

Chi-Square Test of Model Fit

          Value                          0.000
          Degrees of Freedom                 0
          P-Value                       0.0000
```

Since the fitted model has not imposed any restrictions beyond those indicated immediately after its defining Equation 12.4, it has as many parameters as there are observed variable means, variances, and covariances, as can be readily worked out (Raykov & Marcoulides, 2006). For this reason, it is a saturated model and therefore fits the data perfectly. (This is also evident in different aspects of fit presented in the following few output lines pertaining to model fit.)

```
Chi-Square Contributions From Each Group

          MALES                              0.000
          FEMALES                            0.000
```

```
Chi-Square Test of Model Fit for the Baseline Model

          Value                                 1245.208
          Degrees of Freedom                          12
          P-Value                                 0.0000

CFI/TLI
          CFI                                      1.000
          TLI                                      1.000

Loglikelihood

          H0 Value                             -3553.269
          H1 Value                             -3553.269

Information Criteria

          Number of Free Parameters                   28
          Akaike (AIC)                          7162.537
          Bayesian (BIC)                        7266.522
          Sample-Size Adjusted BIC              7177.720
            (n* = (n + 2) / 24)

RMSEA (Root Mean Square Error Of Approximation)

          Estimate                                 0.000
          90 Percent C.I.                          0.000   0.000

SRMR (Standardized Root Mean Square Residual)

          Value                                    0.000

MODEL RESULTS

             Estimates        S.E.        Est./S.E.

Group MALES

F1          BY
   TEST_1       1.000         0.000           0.000

F2        BY
   TEST_2       1.000         0.000           0.000

F3          BY
   TEST_3       1.000         0.000           0.000

F4          BY
   TEST_4       1.000         0.000           0.000
```

The last displayed values are the fixed at 1 single factor loadings of each intelligence test on its corresponding factor. (These loadings represent the diagonal elements of the matrix Λ in the model defining

Equation 12.4. The entries in the third column across the last four lines should not be interpreted as they are meaningless in this case of parameters fixed at 1.)

```
F2        WITH
   F1            149.153     21.480          6.944

F3        WITH
   F1             49.334      9.079          5.434
   F2             51.051      9.330          5.471

F4        WITH
   F1            151.521     21.820          6.944
   F2            154.343     22.305          6.920
   F3             49.604      9.330          5.317
```

The last six full lines contain the estimates of the male group covariances between the four administered intelligence tests (first column), followed by their standard errors (second column), and associated *t*-values (last column). The latter values represent the ratio of estimate to standard error, and with large samples may be considered following a standard normal distribution. As seen from these *t*-values, each of the six observed covariances is significant (which could be expected from intelligence tests) since none of their *t*-values falls within the range $(-1.96, +1.96)$ that would indicate lack of significance.

```
Means
   F1              0.000      0.000      0.000
   F2              0.000      0.000      0.000
   F3              0.000      0.000      0.000
   F4              0.000      0.000      0.000

Intercepts
   TEST_1         28.750      1.223     23.501
   TEST_2         30.984      1.253     24.729
   TEST_3         27.653      0.608     45.510
   TEST_4         30.437      1.271     23.944
```

The last four full lines contain the male group means of the four fluid measures, and the next four full lines present their variances.

```
Variances
   F1            157.848     21.887      7.212
   F2            165.045     22.926      7.199
   F3             37.750      5.353      7.053
   F4            167.919     23.600      7.115
```

```
Residual Variances
    TEST_1            0.000        0.000        0.000
    TEST_2            0.000        0.000        0.000
    TEST_3            0.000        0.000        0.000
    TEST_4            0.000        0.000        0.000
```

As indicated after Equation 12.4 defining the fitted model, error variances are set to equal 0. (As indicated above, the entries in the third column across the last four lines should not be interpreted, as they are meaningless here as well.)

```
Group FEMALES
F1         BY
    TEST_1            1.000        0.000        0.000

F2         BY
    TEST_2            1.000        0.000        0.000

F3         BY
    TEST_3            1.000        0.000        0.000

F4         BY
    TEST_4            1.000        0.000        0.000
```

By default, M*plus* assumes that the factor loading matrix is the same in all analyzed groups unless statements to the opposite are made in the command file, which we have not added in the one of the fitted model (see above input file).

```
F2         WITH
    F1              138.615       14.674        9.447

F3         WITH
    F1               55.606        6.626        8.392
    F2               56.630        6.739        8.403

F4         WITH
    F1              136.706       14.679        9.313
    F2              139.445       14.943        9.332
    F3               56.095        6.764        8.293
```

Similarly here, the last six full lines contain the estimates of the female group covariances between the four administered fluid intelligence tests, their standard errors, and associated *t*-values that are also all significant.

```
Means
    F1          5.613      1.503      3.735
    F2          4.950      1.536      3.223
    F3          3.164      0.747      4.234
    F4          3.321      1.560      2.129
```

As indicated earlier, the last four full lines contain the group mean-differences on the four fluid tests. If one were not interested in the multivariate group difference, or the latter was known or previously found to be significant (see next model fitted below), looking at their *t*-values one could conclude that there are significant mean group differences on each of the four measures if one were to use a conventional .05 significance level. Alternatively, one may wish to effect some significant level adjustment to protect the overall Type I error rate from inflation, such as a Bonferroni correction, before conducting these tests. (See Chapter 4 and below in this section where the Bonferroni correction is carried out and the last results correspondingly interpreted, as may be generally recommended for univariate tests following a significant multivariate outcome.)

```
Intercepts
    TEST_1          28.750      1.223     23.501
    TEST_2          30.984      1.253     24.729
    TEST_3          27.653      0.608     45.510
    TEST_4          30.437      1.271     23.944
```

Each one of these intercepts is set equal to the corresponding intercept in the above male group, and like the latter represents the observed mean of the pertinent fluid test in that group.

```
Variances
    F1          149.393     15.096      9.896
    F2          153.196     15.585      9.829
    F3           35.923      3.713      9.676
    F4          156.029     15.950      9.782
```

The last four lines contain the variances of the fluid measures in the current group, that is, for female freshmen. (Note that the restriction of equal mean intercepts only affects the mean structure of the four variables, and thus has no implications for their covariance structure in the currently considered group.)

```
Residual Variances
    TEST_1          0.000       0.000       0.000
    TEST_2          0.000       0.000       0.000
```

```
TEST_3          0.000      0.000      0.000
TEST_4          0.000      0.000      0.000
```

Fitting this saturated model to the data from the present empirical study provides the needed baseline, namely the unrestricted model, for carrying out the likelihood ratio test of no overall mean group difference in the four fluid tests. To accomplish the latter, according to the likelihood ratio theory (Roussas, 1997; see also Bollen, 1989), we now need to fit the restricted model and compare its chi-square with that of the just fitted model. The restricted model results from the fitted model by adding the constraint that reflects the tested null hypothesis, H_0: $\underline{\mu}_1 = \underline{\mu}_2$, where $\underline{\mu}_1$ and $\underline{\mu}_2$ denote the means of the four fluid tests in the male and female groups under consideration, respectively. Obviously, this null hypothesis is equivalent to the following one, $^*H_0$: $\underline{\mu}_2 - \underline{\mu}_1 = 0$, which we immediately note as being phrased in terms of the group mean differences, that is, in terms of the factor means in the female group of the last fitted model (since $\underline{\mu}_1$ is set by default). Hence, the restricted model is obtained from the just fitted model by adding the restriction of vanishing factor means in the female group, i.e., $M(\underline{f}_2) = 0$. This is accomplished by introducing a single line reflecting the restriction in the last used M*plus* input file, leading to the following command file:

```
TITLE:          TESTING GROUP DIFFERENCES WITH MISSING
                DATA
DATA:           FILE = CH12EX1.DAT; ! NEED RAW DATA!
VARIABLE:       NAMES = TEST_1 TEST_2 TEST_3 TEST_4 GENDER;
                MISSING = ALL (99);
                GROUPING = GENDER (0 = MALES 1 = FEMALES);
ANALYSIS:       TYPE = MISSING H1; ! DON'T FORGET 'H1'!
MODEL:          F1 BY TEST_1@1; TEST_1@0;
                F2 BY TEST_2@1; TEST_2@0;
                F3 BY TEST_3@1; TEST_3@0;
                F4 BY TEST_4@1; TEST_4@0;
MODEL FEMALES:  [F1-F4@0];
```

The only difference between this and last fitted model's M*plus* command file, is found in the final line of the current one. This line begins with a reference to the female group, MODEL FEMALE, and then fixes at 0 the factor means in it; as just indicated, this fixing implements the restriction of the tested null hypothesis of no multivariate group mean difference on the four intelligence tests. We note that factor means are referred to in this parameterization by inclusion of the names of the factors in question within brackets.

The last presented command file yields the following output (only nonredundant portions are displayed next):

```
THE MODEL ESTIMATION TERMINATED NORMALLY

TESTS OF MODEL FIT

Chi-Square Test of Model Fit
          Value                              28.965
          Degrees of Freedom                      4
          P-Value                            0.0000
```

Since the previously fitted model was the saturated model that had a perfect fit, this chi-square value is the likelihood ratio test of the restrictions imposed in the current model, that is, of the tested null hypothesis of no group differences on the four fluid measures considered. (Look at the lower section titled "Loglikelihood" and observe that the current chi-square value is twice the difference of the two loglikelihoods presented there) (Roussas, 1997.) Hence, already here, we can conclude that the null hypothesis of no group mean difference when the four intelligence tests are considered together can be rejected because this chi-square value is significant. We provide next for completeness the remaining output sections of the present model, and note in passing that the following parts are not of real interest to us given our current concern with testing this particular null hypothesis.

```
Chi-Square Contributions From Each Group
          MALES                                   19.054
          FEMALES                                  9.912

Chi-Square Test of Model Fit for the Baseline Model
          Value                                 1245.208
          Degrees of Freedom                          12
          P-Value                                 0.0000

CFI/TLI
          CFI                                      0.980
          TLI                                      0.939

Loglikelihood
          H0 Value                             -3567.751
          H1 Value                             -3553.269

Information Criteria
          Number of Free Parameters                   24
          Akaike (AIC)                          7183.503
          Bayesian (BIC)                        7272.632
          Sample-Size Adjusted BIC              7196.517
          (n* = (n + 2) / 24)
```

```
RMSEA (Root Mean Square Error Of Approximation)
          Estimate                              0.203
          90 Percent C.I.                       0.138  0.275

SRMR (Standardized Root Mean Square Residual)
          Value                                 0.095

MODEL RESULTS
             Estimates        S.E.    Est./S.E.

Group MALES
F1        BY
   TEST_1          1.000      0.000        0.000

F2        BY
   TEST_2          1.000      0.000        0.000

F3        BY
   TEST_3          1.000      0.000        0.000

F4        BY
   TEST_4          1.000      0.000        0.000

F2        WITH
   F1            160.721     23.664        6.792

F3        WITH
   F1             56.701     10.539        5.380
   F2             57.559     10.652        5.404

F4        WITH
   F1            160.553     23.552        6.817
   F2            162.050     23.801        6.809
   F3             54.191     10.392        5.215

Means
   F1              0.000      0.000        0.000
   F2              0.000      0.000        0.000
   F3              0.000      0.000        0.000
   F4              0.000      0.000        0.000

Intercepts
   TEST_1         32.359      0.779       41.542
   TEST_2         34.194      0.784       43.612
   TEST_3         29.822      0.389       76.641
   TEST_4         32.688      0.781       41.839

Variances
   F1            171.020     24.334        7.028
   F2            175.157     24.869        7.043
   F3             42.189      6.227        6.775
   F4            174.363     24.801        7.030
```

```
Residual Variances
    TEST_1          0.000       0.000       0.000
    TEST_2          0.000       0.000       0.000
    TEST_3          0.000       0.000       0.000
    TEST_4          0.000       0.000       0.000

Group FEMALES

F1         BY
    TEST_1          1.000       0.000       0.000

F2         BY
    TEST_2          1.000       0.000       0.000

F3         BY
    TEST_3          1.000       0.000       0.000

F4         BY
    TEST_4          1.000       0.000       0.000

F2         WITH
    F1            141.996      15.291       9.286

F3         WITH
    F1             57.491       6.971       8.247
    F2             58.263       7.039       8.277

F4         WITH
    F1            138.744      15.094       9.192
    F2            141.070      15.278       9.234
    F3             57.112       6.966       8.198

Means
    F1              0.000       0.000       0.000
    F2              0.000       0.000       0.000
    F3              0.000       0.000       0.000
    F4              0.000       0.000       0.000

Intercepts
    TEST_1         32.359       0.779      41.542
    TEST_2         34.194       0.784      43.612
    TEST_3         29.822       0.389      76.641
    TEST_4         32.688       0.781      41.839

Variances
    F1            153.349      15.806       9.702
    F2            156.137      16.118       9.687
    F3             36.850       3.884       9.487
    F4            156.896      16.125       9.730
```

```
Residual Variances
   TEST_1          0.000        0.000        0.000
   TEST_2          0.000        0.000        0.000
   TEST_3          0.000        0.000        0.000
   TEST_4          0.000        0.000        0.000
```

After having found that the multivariate group difference test is significant, we may wish to return to the appropriate part of the previously fitted model (without the restriction of all factor means being zero in the female group), and carry out as follow-up analyses tests of group mean differences on each of the four intelligence tests, possibly after some adjustment of significance level. (We note that these univariate tests are not independent of one another, since the fluid measures are interrelated; e.g., Chapter 4.) As noted when discussing the output of that unrestricted model, each fluid measure demonstrates considerable group mean difference. In particular, if one invokes a Type I error protection, for example, uses Bonferroni's procedure, given that one is involved in five tests in total (the multivariate plus the four univariate follow-up ones), one would need to use as a significance level for these univariate tests $\alpha' = .05/5 = .01$. With this in mind, one then finds out that all fluid tests but the last one show significant mean group differences. (Recall that the relevant two-tailed cutoff of the standard normal distribution at that level is 2.58. Note also that the multivariate test would be significant if it were to be performed at the same .01 level, since the chi-square cutoff for 4 degrees of freedom is then 13.277, thus smaller than the chi-square value of the restricted model.)

We move now to our next concern with testing a contrast on the group mean differences. Accordingly, as mentioned earlier in this section, we are interested in testing whether the average group mean difference on the induction tests is the same as the average group mean difference on the figural relations tests. In other words, we are interested in testing the contrast

$$\frac{\Delta_1 + \Delta_2}{2} - \frac{\Delta_3 + \Delta_4}{2} = 0, \tag{12.5}$$

where Δ_1 and Δ_2 are the group mean differences on each of the pair of intelligence tests, and Δ_3 and Δ_4 denote these differences on each measure in the couple of figural relations tests. Since these group mean differences are parameterized in the factor means in the female group, Equation 12.5 actually represents a contrast in these latter means, that is, in the factor means of the previously fitted model (unrestricted model). This contrast can be tested by extending appropriately the previous command file associated with that model. The needed input file is thus as follows (its new features are explained subsequently):

```
TITLE:      TESTING GROUP DIFFERENCES WITH MISSING DATA
            TESTING A CONTRAST ON GROUP MEAN DIFFERENCES
DATA:       FILE=CH12EX1.DAT; !NEED RAW DATA!
VARIABLE:   NAMES=TEST_1 TEST_2 TEST_3 TEST_4 GENDER;
            MISSING=ALL(99);
            GROUPING=GENDER (0=MALES 1=FEMALES);
ANALYSIS:   TYPE=MISSING H1;
MODEL:      F1 BY TEST_1@1; TEST_1@0;
            F2 BY TEST_2@1; TEST_2@0;
            F3 BY TEST_3@1; TEST_3@0;
            F4 BY TEST_4@1; TEST_4@0;
MODEL FEMALES:
            [F1-F4] (P1-P4);
MODEL CONSTRAINT:
            NEW (CONTR_1);
            CONTR_1=P1+P2-P3-P4;
OUTPUT:     CINTERVAL;
```

In this command file, with the added line in the MODEL FEMALES section, we attach special parametric symbols to the factor means in the female group—that is, to its means on the four intelligence measures of concern here—which factor means as we recall from above are free model parameters (by software default). Accordingly, the induction test means in this group are denoted P1 and P2, and the figural relations means are denoted P3 and P4. In the following section MODEL CONSTRAINT, we first introduce a new parameter with the subcommand NEW (CONTR_1), which should represent the contrast being tested (hence its name). In the next line, we define this contrast in a way that is identical to the left-hand side of Equation 12.5 (after multiplying by 2, for simplicity). With the last line of the command file, we ask for confidence intervals (by default provided at 95%- and 99%-condifence level) for all model parameters. We will be particularly interested in the confidence interval associated with the new parameter, CONTR_1. Specifically, we will use the duality between hypothesis testing and confidence interval (e.g., Chapter 4) to test this contrast, by examining if its confidence interval covers 0 (at a pre-specified confidence level). This command file produces the following output:

```
THE MODEL ESTIMATION TERMINATED NORMALLY

TESTS OF MODEL FIT

Chi-Square Test of Model Fit
          Value                          0.000
          Degrees of Freedom                 0
          P-Value                       0.0000
```

Note that the fit of the model is unchanged, relative to the first fitted model in this section. The reason is that here we have actually not changed the unrestricted (saturated) model. This is because we have not introduced any restriction in that model, but only added a new parameter—the contrast we were interested in testing. Since this addition of a new parameter has no impact upon the mean and covariance structure of the model, it remains to be a saturated model and thus its fit is still perfect. For these reasons, no part of the output associated with this model is changed relative to the first fitted model in this section, so we can skip nearly all output parts and move straight to its final section that contains the estimate and confidence intervals of the contrast of interest, along with the confidence intervals of all model parameters (and their estimates provided in the middle of the five numeric columns; the omitted output sections can obviously be obtained readily by submitting to the software the above input file along with the raw data in file ch12ex1.dat available from www.psypress.com/applied-multivariate-analysis).

```
New/Additional Parameters

   CONTR_1          4.077       1.308       3.118
```

The contrast estimate is 4.077, with a standard error of 1.308 and a *t*-value of 3.118 that is significant. Hence, we reject the null hypothesis of the contrast being 0, and conclude that the average group mean difference on induction tests is not the same as the average group mean difference on figural relations tests. By inspecting the associated confidence interval (see final row of next part containing all model parameters' confidence intervals), we can actually make a stronger statement, as we will see shortly.

```
CONFIDENCE INTERVALS OF MODEL RESULTS
                    Lower     Lower     Esti-     Upper     Upper
                     .5%      2.5%      mates     2.5%       .5%

Group MALES
F1       BY
   TEST_1           1.000     1.000     1.000     1.000     1.000

F2       BY
   TEST_2           1.000     1.000     1.000     1.000     1.000

F3       BY
   TEST_3           1.000     1.000     1.000     1.000     1.000

F4       BY
   TEST_4           1.000     1.000     1.000     1.000     1.000
```

```
F2       WITH
   F1            93.825  107.052  149.153  191.255  204.482

F3       WITH
   F1            25.948   31.539   49.334   67.129   72.720
   F2            27.017   32.763   51.051   69.338   75.084

F4       WITH
   F1            95.317  108.754  151.521  194.289  207.726
   F2            96.891  110.626  154.343  198.060  211.795
   F3            25.573   31.318   49.604   67.890   73.635

Means
   F1             0.000    0.000    0.000    0.000    0.000
   F2             0.000    0.000    0.000    0.000    0.000
   F3             0.000    0.000    0.000    0.000    0.000
   F4             0.000    0.000    0.000    0.000    0.000

Intercepts
   TEST_1        25.599   26.353   28.750   31.148   31.901
   TEST_2        27.756   28.528   30.984   33.440   34.211
   TEST_3        26.088   26.462   27.653   28.844   29.218
   TEST_4        27.163   27.946   30.437   32.929   33.711

Variances
   F1           101.470  114.949  157.848  200.747  214.225
   F2           105.993  120.111  165.045  209.980  224.098
   F3            23.962   27.259   37.750   48.241   51.537
   F4           107.131  121.664  167.919  214.175  228.707

Residual Variances
   TEST_1         0.000    0.000    0.000    0.000    0.000
   TEST_2         0.000    0.000    0.000    0.000    0.000
   TEST_3         0.000    0.000    0.000    0.000    0.000
   TEST_4         0.000    0.000    0.000    0.000    0.000

Group FEMALES
F1       BY
   TEST_1         1.000    1.000    1.000    1.000    1.000

F2       BY
   TEST_2         1.000    1.000    1.000    1.000    1.000

F3       BY
   TEST_3         1.000    1.000    1.000    1.000    1.000

F4       BY
   TEST_4         1.000    1.000    1.000    1.000    1.000
```

```
F2       WITH
   F1         100.819  109.855  138.615  167.375  176.411

F3       WITH
   F1          38.538   42.618   55.606   68.593   72.673
   F2          39.270   43.421   56.630   69.839   73.989

F4       WITH
   F1          98.895  107.935  136.706  165.478  174.517
   F2         100.955  110.157  139.445  168.734  177.936
   F3          38.672   42.838   56.095   69.353   73.519

Means
   F1           1.742    2.668    5.613    8.558    9.484
   F2           0.994    1.940    4.950    7.959    8.905
   F3           1.239    1.699    3.164    4.629    5.090
   F4          -0.697    0.264    3.321    6.378    7.338

Intercepts
   TEST_1      25.599   26.353   28.750   31.148   31.901
   TEST_2      27.756   28.528   30.984   33.440   34.211
   TEST_3      26.088   26.462   27.653   28.844   29.218
   TEST_4      27.163   27.946   30.437   32.929   33.711

Variances
   F1         110.508  119.804  149.393  178.981  188.277
   F2          13.051  122.648  153.196  183.743  193.340
   F3          26.360   28.647   35.923   43.200   45.486
   F4         114.945  124.767  156.029  187.290  197.112

Residual Variances
   TEST_1       0.000    0.000    0.000    0.000    0.000
   TEST_2       0.000    0.000    0.000    0.000    0.000
   TEST_3       0.000    0.000    0.000    0.000    0.000
   TEST_4       0.000    0.000    0.000    0.000    0.000

New/Additional Parameters
   CONTR_1      0.709    1.514    4.077    6.641    7.446
```

To obtain the confidence interval at the 95%-confidence level for a parameter of interest, look in the corresponding row to a parameter of interest in this output section, and take as lower and upper endpoint of this interval the entries in the columns titled "Lower 2.5%" and "Upper 2.5%," respectively; similarly, for the 99%-confidence interval take the entries in the columns "Lower .5%" and "Upper .5%." As an alternative to the above significance test, we can now look at the final row of this output part and see that the 95%-confidence interval for the contrast of interest to us here is (1.514, 6.641); since it does not contain the zero point, we conclude that this contrast is significant. (It would have been also declared as significant, had we chosen to perform the test at the $\alpha = .01$ level; see Chapter 4.) Because of this confidence interval being entirely above the 0 point (as its lower endpoint is larger than 0), we can also infer—at the same level of confidence—that the average gender mean difference in induction tests appears to be more pronounced than the average gender difference in figural relations tests.

We note in passing that contrasts of the kind considered in this subsection may be of interest to examine in intervention studies where one may be concerned with assessing whether there may be a transfer of the treatment, training, or instruction effect from certain dimensions to related ones. (For instance, in the present empirical setting, if a training was aimed at enhancing inductive reasoning ability while instead of the gender groups a researcher had a control and experimental group, one might be interested in finding out whether the training might have had any transfer to the figural relations ability—the analyzed data might then be considered consistent with such a transfer if the tested contrast had turned out to be nonsignificant, assuming a nonzero treatment effect.)

We conclude this subsection on examining group mean differences by stating that the analyzed data from the empirical study under consideration suggests gender mean differences when the four fluid intelligence tests are considered simultaneously, with these discrepancies found predominantly in the two induction and first figural relations measures. Thereby, the average gender difference on the induction tests was found to be stronger than the average gender difference on the figural relations tests. We move next to studying gender differences and similarities in the covariance and correlation structure of these fluid measures.

12.6.2 Testing for Group Differences in the Covariance and Correlation Matrices With Missing Data

As outlined earlier in the chapter, the approach followed for purposes of examining group mean differences can be readily used to test also for group differences in the covariance and correlation matrices when analyzing incomplete data (cf. Raykov & Marcoulides, 2008). To accomplish this test for covariance matrices, we can employ again the likelihood ratio test. To this end, in a first step we need to fit the model defined in Equation 12.4 and immediately following discussion, which is the baseline (unrestricted) model. (We already fitted that model at the beginning of the last subsection.) In a second step, we fit the same model with the added restriction of the covariance matrix of the observed variables being the same across the two groups. Since as shown before that covariance matrix is identical to the covariance matrix of the factors, in this step we need to impose the constraint of each element of this matrix being the same in both groups. This is accomplished with the following M*plus* command file: (We discuss its new features subsequently.)

```
TITLE:       TESTING GROUP DIFFERENCES WITH MISSING DATA
             EXAMINING GROUP DIFFERENCES IN THE COVARIANCE
             MATRIX
DATA:        FILE = CH12EX1.DAT; !NEED RAW DATA!
VARI:        NAMES = TEST_1 TEST_2 TEST_3 TEST_4 GENDER;
             MISSING = ALL(99);
             GROUPING = GENDER (0 = MALES 1 = FEMALES);
ANALYSIS:    TYPE = MISSING H1; !DON'T FORGET 'H1'!
MODEL:       F1 BY TEST_1@1; TEST_1@0;
             F2 BY TEST_2@1; TEST_2@0;
             F3 BY TEST_3@1; TEST_3@0;
             F4 BY TEST_4@1; TEST_4@0;
             F1-F4 (P1-P4);
             F1 WITH F2-F4 (P5-P7);
             F2 WITH F3-F4 (P8-P9);
             F3 WITH F4 (P10);
MODEL FEMALES:
             F1-F4 (P11-P14);
             F1 WITH F2-F4 (P15-P17);
             F2 WITH F3-F4 (P18-P19);
             F3 WITH F4 (P20);
MODEL CONSTRAINT:
             P1 = P11; P2 = P12; P3 = P13; P4 = P14; P5 = P15;
             P6 = P16; P7 = P17; P8 = P18; P9 = P19; P10 = P20;
```

With the last four lines of the MODEL command section, we assign parametric symbols to the four factor variances and six factor covariances. Specifically, with the first of these lines—"F1–F4 (P1–P4);"—we assign to the variances of the first through fourth factors the references P1 through P4, respectively. In the next line, we assign the symbols P4 through P7 correspondingly to the covariances of the first with the second, third, and fourth factors. With the following two lines, we then assign the references P8 through P10, respectively, to the remaining three factor covariances. In the MODEL FEMALES section, we do the same in the female group, assigning to their respective variances and covariances the references P11 through P20. In this way, each entry of the covariance matrix of the four fluid tests has been assigned a parametric reference in the male and in the female group. Finally, in the MODEL CONSTRAINT section we set each element of this matrix in the male group to be equal to the same element of this covariance matrix in the female group, which is precisely the null hypothesis of no gender differences in the covariance matrix (structure) that is to be tested.

The last presented command file leads to the following output. To carry out the likelihood ratio test of the hypothesis of gender identity in the covariance structure, since the unrestricted model of relevance is the saturated model associated with perfect fit we only need the chi-square value associated with the presently discussed restricted model; to save space, we dispense with the remaining parts of the output. (These parts can be obtained by submitting to the software the above input file.)

```
THE MODEL ESTIMATION TERMINATED NORMALLY

TESTS OF MODEL FIT

Chi-Square Test of Model Fit
        Value                      13.473
        Degrees of Freedom             10
        P-Value                    0.1983
```

Due to the chi-square value of the restricted model being nonsignificant, we conclude that the null hypothesis of covariance matrix equality may be retained. Thus, there is no evidence in the analyzed data for gender discrepancies in the covariance structure (matrix) of the four fluid intelligence measures.

From this finding, it can be concluded that also the correlation matrix of these four tests is identical across the two groups. The reason is that covariance matrix identity across groups, with its implication of group equality in each variable variance and covariance for any pair of variables, is sufficient for identity of all correlation coefficients, and hence is sufficient for lack of group differences in the correlation matrix. Nonetheless, since obviously the latter identity may hold also when the covariance matrices are not the same across groups, it may be of interest in its own right to be in a position to test for correlation matrix identity across groups also in cases when the test of covariance matrix equality across them had not been carried out beforehand. In this way, one would then be able to answer the question whether there are any gender differences in the standardized indices of linear relationship between any two of the observed variables (the four fluid tests here).

As indicated earlier, for the sake of testing on its own group identity in the correlation matrix, we can still use the same general approach to studying group differences with missing data after employing a different way of identifying the model defined in Equation 12.4. Accordingly, we fix factor variances rather than first loading of observed measures (indicators) upon their factors as done before. Other than that, and observing that we do not need to do anything then about the factor variances, we can use the same idea as the one underlying the preceding test for group identity in covariance matrices. Thus, we need the following M*plus* input file for the restricted model of interest here (the unrestricted model remains the same saturated model that we fitted first in the present section; explanations of new features of the following command file are provided subsequently):

```
TITLE:  TESTING GROUP DIFFERENCES WITH MISSING DATA
        EXAMINING GROUP DIFFERENCES IN THE CORRELATION
        MATRIX
DATA:   FILE=CH12EX1.DAT; !NEED RAW DATA!
VARI:   NAMES=TEST_1 TEST_2 TEST_3 TEST_4 GENDER;
        MISSING=ALL(99);
        GROUPING=GENDER (0=MALES 1=FEMALES);
ANALY:  TYPE=MISSING H1;
MODEL:  F1 BY TEST_1*; TEST_1@0;
        !NOTE FREEING OF ALL FACTOR LOADINGS
        F2 BY TEST_2*; TEST_2@0;
        F3 BY TEST_3*; TEST_3@0;
        F4 BY TEST_4*; TEST_4@0;
        F1-F4@1;
        !NOTE FIXING OF LATENT VARIANCE, FOR IDENTIFIABILITY
        F1 WITH F2-F4 (P1-P3);
        F2 WITH F3-F4 (P4-P5);
        F3 WITH F4 (P6);
MODEL FEMALES:
        F1 BY TEST_1*;
        F2 BY TEST_2*;
        F3 BY TEST_3*;
        F4 BY TEST_4*;
        F1 WITH F2-F4 (P7-P9);
        F2 WITH F3-F4 (P10-P11);
        F3 WITH F4 (P12);
MODEL CONSTRAINT:
        P1=P7; P2=P8; P3=P9; P4=P10; P5=P11; P6=P12;
        !NOTE FEWER CONSTRAINTS ARE BEING IMPOSED, SINCE
        !DIAGONAL OF CORRELATION MATRIX CONSISTS OF
        !THE CONSTANT 1 (SEE MAIN TEXT)
```

With the first four lines in the MODEL command in the first group (i.e., males), we free the loading of each measure on its pertinent factor, but with the fifth line we accomplish model identification by fixing at 1 the factor variances. (In this way, their covariances become identical to their correlations.) We then assign parametric symbols P1 through P6 to the off-diagonal correlation matrix elements in the males group. In the MODEL FEMALES section, we do the same for the respective factor loadings, and assign the reference symbols P7 through P12 to the off-diagonal elements of the correlation matrix in the females group. In the MODEL CONSTRAINT section, we finally set these correlations group invariant. We stress that here we test fewer in number equality constraints since we are imposing the latter only on the off-diagonal elements of the correlation matrix of interest—its diagonal is the constant 1, unlike that of the covariance matrix that is filled with stochastic quantities in any sample.

The last input file yields the following output; again, to save space, only the section pertaining to its chi-square value is presented next since all we are essentially interested in here is the likelihood ratio test of the null hypothesis of no gender differences in the correlation structure of the four fluid measures.

```
THE MODEL ESTIMATION TERMINATED NORMALLY

TESTS OF MODEL FIT

Chi-Square Test of Model Fit
          Value                    12.142
          Degrees of Freedom            6
          P-Value                  0.0588
```

From this output section, in the same way as when testing covariance matrix (structure) identity, we see that the null hypothesis of group equality in the correlation matrices cannot be rejected either. That is, also the correlation matrix shows no gender differences. We stress that the chi-square value now is different from the one of the preceding model fitted to test covariance structure identity. The reason has in essence been already indicated before: here the observed variable standard deviations are not set equal across groups—in fact, they are parameterized in the loadings of each measure upon its factor, which as mentioned above are now free parameters. This is in difference to the case when tested for group identity was the covariance matrix, since then the observed variable variances (and hence their standard deviations as well) were held equal across groups in order to introduce the group equality of the diagonal elements of the covariance matrix. The latter identity obviously need not be imposed when testing group equality in correlation matrices, since the main diagonal elements of the latter are all equal to the constant 1. This fact has also as a consequence that of correspondingly fewer (by 4) degrees of freedom of the last fitted model.

By way of summary of the analyses in this section, we have found no evidence for gender differences in the covariance as well as correlation structures of the four analyzed intelligence tests.

In conclusion of this chapter, we reiterate that the outlined modeling approach based on the ML method not only permits the study of group mean difference and similarities in the presence of missing data but is also applicable when one is interested in examining group differences and similarities in variance, covariances, and correlations. Furthermore, the procedure discussed in the last section is straightforwardly applicable also when of interest is the testing of group differences and similarities on latent variables rather than observed ones (cf. Chapter 9).

13

Multivariate Analysis of Change Processes

13.1 Introduction

Behavioral, social, and educational scientists are oftentimes interested in examining developmental processes in individuals and populations. This allows them to study individual change over time as well as inter-individual differences and similarities in associated patterns of growth or decline. For example, educational scientists may be interested in investigating growth in ability exhibited by a group of subjects in repeatedly administered measures following a particular treatment program, and relating them to various characteristics of the students and their environments. Researchers may also be interested then in comparing the rates of change in these variables across several subpopulations, and/or examining the correlates and predictors of growth in ability (e.g., in an effort to find out which students exhibit the fastest development). Similarly, scientists studying human aging may be interested in examining cognitive functioning decline in late life, comparing its patterns across gender, socioeconomic, or education-related subpopulations of elderly, and ascertaining its antecedents and correlates.

As discussed in Chapter 5, these frequently arising concerns in longitudinal studies can be addressed using traditional analysis of variance (ANOVA) and analysis of covariance (ANCOVA) frameworks. However, as mentioned in Chapter 5, the ANOVA methods have important limitations resulting from assumptions that often can be, and in fact may well be, violated in empirical research: (a) homogeneity of variance–covariance matrices across levels of between-subject factors, and especially (b) sphericity (implying equicorrelation of repeatedly assessed variables). Furthermore, when interested in examining correlates and predictors of change via ANCOVA, one also needs to make the assumptions of (c) perfectly measured covariate(s) and (d) regression homogeneity across groups. Particularly assumption (c) is often hard to consider tenable in social and behavioral research, since it is rare that one has access to measures that do not contain sizeable error. In addition, and no less importantly, the repeated measure ANOVA methods covered in Chapter 5 are in part indifferent to time, in the sense of producing the same statistical test results after "reshuffling assessment occasions" (e.g., letting time "run backward"). Last but not least, with missing data, readily available software implementations of these conventional procedures typically invoke listwise deletion, which leads to potentially serious loss of power and less representative, subsequently analyzed samples from studied populations.

All of the above-mentioned potentially serious limitations of traditional repeated measure ANOVA-based methods for studying change can be dealt with using an alternative methodology when large samples of studied subjects are available. This methodology falls under the general framework of latent variable modeling (LVM) that over the past two decades is becoming increasingly popular across the social, behavioral, and educational sciences. LVM is based on a more comprehensive than the traditional understanding of the concept of latent variable, which plays an instrumental role for its methods. Accordingly, a latent variable can be viewed as a random variable with individual realizations in a given sample (or population, for that matter), which however are not observed (e.g., Raykov & Marcoulides, 2006). Models utilized in applications of LVM are typically developed in terms of latent variables, and can be designed to reflect important aspects of developmental processes in these disciplines.

The present chapter aims at providing an introductory discussion of the utility of the LVM methodology for studying change within a multivariate context. (For a discussion of some LVM applications to the study of growth or decline in settings with single longitudinally followed variables, see Raykov & Marcoulides, 2006.) The models of concern in the rest of this chapter are less restrictive than the above-mentioned repeated measure ANOVA methods (Chapter 5), since the former do not make any of the above assumptions (a) through (d). In addition, these models are also readily applicable in empirical settings with missing data (e.g., via use of full information maximum likelihood or multiple imputation—under their assumptions of data missing at random and multinormality; see also Chapter 12), without sacrificing any available empirical information. Throughout the remainder of the chapter, we illustrate discussed change models by making repeated use of the increasingly popular LVM program M*plus* (Muthén & Muthén, 2006; for a brief introduction to M*plus*, see Chapter 9, or Raykov & Marcoulides, 2006).

13.2 Modeling Change Over Time With Time-Invariant and Time-Varying Covariates

Repeated measure investigations typically focus on developmental processes on at least one longitudinally followed characteristic, and are usually concerned with explaining variability in patterns of growth or decline on them in terms of related variables, often called covariates. These covariates may or may not change during the course of the study and/or may be evaluated repeatedly or only once. When they are constant for each subject, or assessed at a single occasion (for various reasons), they are referred to as time-invariant covariates. Well-known examples are gender, political or religious affiliation, attained educational level, or geographic region, to name a few. If covariates are not constant over the study period and are evaluated more than once along with the main characteristics of interest, they are referred to as time-varying covariates. Any measure that does not remain constant across time can play the role of a time-varying covariate in a longitudinal study as long as repeated assessments of it are taken, and it is of substantive interest for the researcher to include that variable in an explanatory capacity with respect to individual differences in the patterns of change over time on measures of major interest.

The LVM methodology provides a readily available means for modeling temporal development in variables of focal interest in the presence of time-invariant and time-varying covariates. To discuss the potential of the methodology in this regard, denote by Y_{it} the value of the ith subject on a repeatedly assessed variable of main concern in a given study, which is obtained at time point t ($i = 1, \ldots, n$; $t = 1, \ldots T$, $T > 2$; n denoting sample size). Throughout this chapter we presume that the same measures, without any modification or alteration, are administered at all assessment occasions. We also assume until otherwise indicated that all subjects are evaluated at the same points in time; that is, for each given occasion, any variability in the timing of measurement across subjects is assumed not to be of substantive relevance, interest, or consequence. (In such cases, the empirical study is said to yield time-structured data [Willett & Sayer, 1994, 1996].) Denote the repeated measurements, at the same occasions, of a time-varying covariate by C_{it} ($i = 1, \ldots, n; t = 1, \ldots T$). In difference to a considerable part of literature on traditional repeated measure ANOVA as well as mixed models (Pinheiro & Bates, 2000), we will also assume the availability of multiple indicators of a latent variable, η, that cannot be directly observed and is (a) related to the characteristic of main interest, Y; (b) assessed only once; as well as (c) treated as a time-invariant covariate. For ease of discussion, and to set up the context of an example introduced next, we assume that there are two fallible indicators, K_1 and K_2, of η, with linear relations to η. (We note, however, that the following approach is readily extended to the case of more indicators, or more time-invariant/time-varying covariates that can each be fallible as well, or more than one repeatedly assessed variable of focal interest.)

13.2.1 Intercept-and-Slope Model

One of the simplest models for studying change assumes that up to an error term the repeated assessments of a variable of main concern follow a linear trend for each individual, that is,

$$Y_{it} = a_i + b_i\,(t - 1) + e_{it} \tag{13.1}$$

where

a_i and b_i are correspondingly his or her intercept and slope, respectively
e_{it} is an error that is assumed normally distributed with zero mean and unrelated to time of assessment

and the values $(t - 1)$ are taken as such of the predictor in order to enhance interpretability of parameters (in particular, of the intercept that is the expectation of an individual's position at begin of study; $i = 1, \ldots, n$; $t = 1, \ldots T$; cf. Raykov & Marcoulides, 2006). Obviously, Equation 13.1 can be rewritten as

$$Y_{it} = 1\,.\,a_i + (t - 1)\,.\,b_i + e_{it} \tag{13.2}$$

(with "." denoting multiplication, to underscore the idea behind this expression). Across all subjects and measurement occasions, the set of Equation 13.2 is readily seen as that of a restricted confirmatory factor analysis (CFA) model (see discussion in Chapter 9), where a_i and b_i play the role of individual factor scores while the associated constants 1 and $(t-1)$ represent fixed factor loadings ($i = 1, \ldots, n$; $t = 1, \ldots T$). Since a_i and b_i are unknown individual scores, they can be viewed as realizations of latent variables (factors), referred to below as intercept and slope factors, respectively. Hence, the CFA approach available from Chapter 9 can be used to deal with this model of change, which for the above reasons is called the intercept-and-slope model.

As an example of the intercept-and-slope model, in case of five repeated assessment occasions the factor loading matrix associated with it would be as follows:

$$\Lambda = \begin{bmatrix} 1 & 0 \\ 1 & 1 \\ 1 & 2 \\ 1 & 3 \\ 1 & 4 \end{bmatrix}$$

We note that fixing the loadings of the five repeated assessment occasions on the first factor to the value of 1 ensures that it is interpreted as initial true status, that is, as baseline point of the underlying developmental process being investigated. We stress that from a conceptual viewpoint, this approach forms the methodological basis for most of the discussion in the present chapter.

13.2.2 Inclusion of Time-Varying and Time-Invariant Covariates

In the context of interest, in this section dealing with time-invariant and time-varying covariates, the model of concern is readily obtained from the preceding discussion and can be defined with the following equations:

$$\begin{aligned} Y_{it} &= a_i + b_i\,(t-1) + c_t\,C_{it} + e_{it} \\ K_{1i} &= \kappa_1 + \lambda_1\eta_i + \delta_{1i} \\ K_{2i} &= \kappa_2 + \lambda_2\eta_i + \delta_{2i} \\ a_i &= \alpha_0 + \beta_0\eta_i + g_i \\ b_i &= \alpha_1 + \beta_1\eta_i + h_i, \end{aligned} \tag{13.3}$$

where δ_{1i}, δ_{2i}, g_i, and h_i are residual terms assumed normal, with zero mean and unrelated to the predictors in their equations, while α_0, α_1, κ_1, and κ_2 as well as β_0, β_1, λ_1, and λ_2 are unknown parameters (intercepts and slopes, respectively, in the corresponding equations; $i = 1, \ldots, n$; $t = 1, \ldots T$). We note that the last two equations in Equation 13.3 represent regressions of the intercept and slope factors on the time-invariant latent covariate η. Hence, Equation 13.3 define a latent variable model whose measurement part is a restricted factor analysis model (in particular, as far as the repeated measures are concerned), while its structural part contains two regressions among latent variables.

The model in Equation 13.3 can be viewed as a multivariate model for change since it is concerned with the development over time in two repeatedly assessed variables—the one of focal interest, Y, and the time-varying covariate, C (cf. Muthén & Muthén, 2006). We also note that as an extension of this model, one could have in it a fallible time-varying covariate. This is accomplished by including in the model an unobserved covariate that is measured with at least a pair of indicators at each assessment occasion (or by a single indicator, if assuming known error variance or imposing alternatively additional parameter restrictions to attain overall model identification). Furthermore, there is no limit on the number of repeatedly assessed variables of focal interest, time-varying covariates, and time-invariant covariates (which covariates, in case fallible, may have any positive number of indicators, assuming model identification). For a given empirical study, this model can be readily fit using the LVM program M*plus* (Muthén & Muthén, 2006), as is demonstrated next.

13.2.3 An Example Application

To illustrate this discussion, consider a study of $n = 400$ high school students who were administered a college aspiration measure and a school motivation measure repeatedly in 9th through 12th grades. In addition, at the start of the study, a pair of measures tapping into distinct aspects of parental style—parental dominance and encouragement for academic progress—were administered to all participants. A researcher is interested in (a) examining the patterns of change over time in college aspiration, in relation to school motivation that is simultaneously assessed; and (b) explaining individual differences in these patterns in terms of parental style. That is, he/she is concerned with evaluating whether parental style, as assessed with this pair of indicators, can explain why students may differ in their starting position and rate (gradient) of change in college aspiration, after controlling for parallel development in school motivation. We emphasize that here we are dealing with (a) a repeatedly measured focal variable, namely college aspiration, which is assessed simultaneously with (b) a time-varying covariate, school motivation, and (c) a fallible time-invariant covariate, parental style, which is evaluated by two indicators. That is, we are interested in fitting the latent variable model defined in Equation 13.3 for $T=4$ (i.e., assessment occasions $t=0, 1, 2, 3$). Preliminary analysis of the observed variable distribution suggested that the assumption of normality (conditional on covariates) could be considered plausible.

Before we go into detail regarding how to conduct the modeling of concern, we note that on the assumption of multinormality, an application of the maximum likelihood method of parameter estimation and model testing on the raw data is equivalent to fitting the latent variable model to the covariance matrix and variable means using the earlier discussed fit function in Equation 9.13 (see Chapter 9), which is routinely available in widely circulated software for modeling with latent variables, such as LISREL, EQS, M*plus*, AMOS, and SAS PROC CALIS. The reason for this equivalence is that with multinormal data, the empirical means and covariance matrix represent sufficient statistics (Roussas, 1997). For this reason, it can be shown that maximizing the likelihood of the raw data is tantamount to fitting the model to these sample statistics using that fit function (i.e., choosing values of its parameters so that the distance between the model and the means as well as covariance matrix is minimized [Bollen, 1989]). We emphasize that analysis of change cannot meaningfully proceed by fitting the model only to the covariance structure of a set of observed variables, since this would disregard the empirical means and their dynamic. As a result, different patterns of change over time (e.g., increase over time, decline, or growth followed by decline or vice versa) could be equally consistent with a given covariance matrix resulting in a repeated measure context, which would make the purpose of modeling void.

Returning to the above empirical example, its analytic goals are accomplished with the following M*plus* command file (explanatory comments on new lines are provided subsequently; the analyzed data are provided in the file ch13ex1_mcm.dat available from www.psypress.com/applied-multivariate-analysis, where the 10 variables follow in free format the order given in the VARIABLE command next).

```
TITLE:    MODELING CHANGE IN THE CONTEXT OF A TIME-
          VARYING COVARIATE AND A TIME-INVARIANT LATENT
          COVARIATE
DATA:     FILE = CH13EX1_MCM.DAT;
          TYPE = MEANS COVARIANCE;
          NOBS = 400;
```

```
VARIABLE:  NAMES = COLLASP1 COLLASP2 COLLASP3 COLLASP4
           PARSTYL1 PARSTYL2 MOTIVN1 MOTIVN2 MOTIVN3
           MOTIVN4;
MODEL:     INTERCEP SLOPE | COLLASP1@0 COLLASP2@1 COLLASP3@2
           COLLASP4@3;
           PARSTYLE BY PARSTYL1 PARSTYL2;
           INTERCEP SLOPE ON PARSTYLE;
           COLLASP1 ON MOTIVN1;
           COLLASP2 ON MOTIVN2;
           COLLASP3 ON MOTIVN3;
           COLLASP4 ON MOTIVN4;
```

This command file makes a reference to data to be analyzed as presented by a sequence of means and covariance matrix for the observed variables. To this end, after stating the name of the file containing these statistics, the second line of the DATA command indicates that the data to be analyzed come in this form, followed by a statement of the sample size. The first line of the MODEL command is a software built-in statement that invokes fitting the intercept-and-slope model to the variables stated after the vertical bar (assumed measured at time points 0, 1, 2, and 3, as here). The next six lines define the latent variable PARSTYLE as indicated by PARSTYL1 and PARSTYL2, include the regression of the intercept and slope factors on it, and the regressions of each college aspiration measure on the simultaneous assessment of motivation. This command file yields the following output (with added comments following main sections):

```
THE MODEL ESTIMATION TERMINATED NORMALLY

TESTS OF MODEL FIT

Chi-Square Test of Model Fit
          Value                   39.512
          Degrees of Freedom          26
          P-Value                 0.0435
```

The fit of the model can be viewed as acceptable (see also the associated root mean square error of approximation [RMSEA] index presented below, and in particular its left endpoint that is well below the threshold of .05).

```
Chi-Square Test of Model Fit for the Baseline Model
          Value                              2054.279
          Degrees of Freedom                       39
          P-Value                              0.0000
CFI/TLI
          CFI                                   0.993
          TLI                                   0.990

Loglikelihood
          H0 Value                          -8304.061
          H1 Value                          -8284.305

Information Criteria
          Number of Free Parameters                25
          Akaike (AIC)                      16658.121
          Bayesian (BIC)                    16757.908
          Sample-Size Adjusted BIC          16678.581
            (n* = (n+2) / 24)

RMSEA (Root Mean Square Error Of Approximation)
          Estimate                              0.036
          90 Percent C.I.                       0.006  0.058
          Probability RMSEA <= .05              0.843

SRMR (Standardized Root Mean Square Residual)
          Value                                 0.029

MODEL RESULTS
                 Estimates     S.E.   Est./S.E.

INTERCEP |
   COLLASP1      1.000       0.000       0.000
   COLLASP2      1.000       0.000       0.000
   COLLASP3      1.000       0.000       0.000
   COLLASP4      1.000       0.000       0.000

SLOPE    |
   COLLASP1      0.000       0.000       0.000
   COLLASP2      1.000       0.000       0.000
   COLLASP3      2.000       0.000       0.000
   COLLASP4      3.000       0.000       0.000

PARSTYLE BY
   PARSTYL1      1.000       0.000       0.000
   PARSTYL2      0.453       0.065       6.980
```

The measurement model for parental style is satisfactory with respect to the factor loadings that are significant (see last column of output section).

```
INTERCEP ON
   PARSTYLE         0.033     0.005      6.079
SLOPE ON
   PARSTYLE         0.015     0.003      5.690
```

Both individual starting position and rate of change (slope of growth) are significantly related to parental style. Hence, there is evidence that high values on the latent variable parental style are associated with high values on both starting position and gradient of change in college aspiration, even after accounting for individual differences that could be explained by such in contemporary motivation—as we see in the next four lines, at each assessment there is a significant and positive relationship between college aspiration and motivation.

```
COLLASP1 ON
   MOTIVN1          0.214     0.042      5.155
COLLASP2 ON
   MOTIVN2          0.261     0.027      9.784
COLLASP3 ON
   MOTIVN3          0.306     0.030     10.177
COLLASP4 ON
   MOTIVN4          0.352     0.048      7.301
SLOPE WITH
   INTERCEP         0.357     0.062      5.715
MOTIVN1 WITH
   PARSTYLE        -1.012     0.973     -1.040
MOTIVN2 WITH
   PARSTYLE         0.586     1.057      0.554
MOTIVN3 WITH
   PARSTYLE        -0.297     0.964     -0.308
MOTIVN4 WITH
   PARSTYLE        -0.427     1.002     -0.426
```

Motivation shows only weak relationships with parental style at each assessment occasion (see last four lines of the section).

```
Intercepts
   COLLASP1          0.000      0.000      0.000
   COLLASP2          0.000      0.000      0.000
   COLLASP3          0.000      0.000      0.000
   COLLASP4          0.000      0.000      0.000
   PARSTYL1         98.795      1.003     98.514
   PARSTYL2         99.240      0.782    126.878
   INTERCEP          9.878      2.082      4.746
   SLOPE            -1.229      1.233     -0.996
```

The intercept of the regression for starting position (intercept factor) is estimated at 9.88 and is significant, while that for the rate of change (slope factor) is estimated at −1.23 and is nonsignificant. As usual, in regression analysis applications, we are not particularly interested in these intercepts (with data from independent subjects, as in this case). The final part of the output, presented next, contains the factor and error variance estimates, and we see that all of them are positive.

```
Variances

   PARSTYLE        394.975     56.903      6.941

Residual Variances

   COLLASP1          0.551      0.082      6.695
   COLLASP2          0.622      0.058     10.810
   COLLASP3          0.586      0.069      8.486
   COLLASP4          0.699      0.127      5.492
   PARSTYL1          9.103     49.218      0.185
   PARSTYL2        164.014     15.381     10.664
   INTERCEP          1.565      0.155     10.068
   SLOPE             0.438      0.044     10.050
```

13.2.4 Testing Parameter Restrictions

As found in the last output section, the error variances associated with the repeated assessment of the variable of main interest, college aspiration, are all significant and of similar magnitude. This suggests that they may be identical in the studied population. Even before looking at the data, such a hypothesis appears not unlikely to be plausible in many repeated assessments settings, since use of the same measuring instrument is often associated with similar impact of error at all measurement occasions.

In order to test this hypothesis, we can use the likelihood ratio approach. To this end, we consider the present model as the "full" model, and fit next the restricted one—the model resulting from that just fitted, by adding the restriction equivalent to the tested null hypothesis. This is accomplished with the following M*plus* input file:

```
TITLE:        MODELING CHANGE IN THE CONTEXT OF A TIME-
              VARYING COVARIATE AND A TIME-INVARIANT LATENT
              COVARIATE, ADDING THE RESTRICTION OF TIME-
              INVARIANT ERROR VARIANCES (SEE LAST LINE BELOW)
DATA:         FILE=CH13EX1_MCM.DAT;
              TYPE=MEANS COVARIANCE;
              NOBS=400;
VARIABLE:     NAMES=COLLASP1 COLLASP2 COLLASP3 COLLASP4
              PARSTYL1 PARSTYL2 MOTIVN1 MOTIVN2 MOTIVN3
              MOTIVN4;
MODEL:        INTERCEP SLOPE | COLLASP1@0 COLLASP2@1
              COLLASP3@2 COLLASP4@3;
              PARSTYLE BY PARSTYL1 PARSTYL2;
              INTERCEP SLOPE ON PARSTYLE;
              COLLASP1 ON MOTIVN1;
              COLLASP2 ON MOTIVN2;
              COLLASP3 ON MOTIVN3;
              COLLASP4 ON MOTIVN4;
              COLLASP1-COLLASP4 (1);
```

The only new feature of this command file, relative to that of the last fitted model, is its final line. Using the dash as a shorthand (indicating inclusion of all measures between the first and last, which were defined in the VARIABLE command), we refer to the error variances by the names of the observed variables that they pertain to. By assigning the same parameter number, (1), to all of them at the end of the line, we request that these error variances be held constant. If this restriction is found to be plausible, we will retain it in the model that will then represent a more parsimonious—and hence more desirable—means of data description and explanation. To carry out the pertinent likelihood ratio test, we need to examine the fit indices of this restricted model, which are presented next. (At this moment, we are not interested in the remainder of the output for this model, except, of course, ensuring that a meaningful admissible solution has been obtained, which is readily found to be the case by submitting the last command file to the data of interest, and for this reason is dispensed with next.)

```
THE MODEL ESTIMATION TERMINATED NORMALLY

TESTS OF MODEL FIT

Chi-Square Test of Model Fit

        Value                          40.783
        Degrees of Freedom                 29
        P-Value                        0.0719
```

Hence, the likelihood ratio test of stability in error variance has the statistic $40.783 - 39.512 = 1.261$, which is nonsignificant when evaluated against the chi-square distribution with pertinent degrees of freedom, $29 - 26 = 3$. (The relevant cutoff at the conventional significance level of .05 is 7.84, and thus much higher than this test statistic.) We conclude that the hypothesis of time-invariant error impacts upon the measurement of college aspiration can be retained.

A main goal of statistical modeling in a substantive area of application is finding a most parsimonious and tenable model, which is also of substantive interest; such a model permits most precise estimation of and interpretative statements about its parameters. To this end, we examine next whether the impact of the time-varying covariate is the same over time. In order to achieve this goal, we impose in the last fitted model the restriction of identical slopes of the four consecutive regressions of college aspiration upon motivation. This is accomplished with the following command file:

```
TITLE:        MODELING CHANGE IN CONTEXT OF A TIME-
              VARYING COVARIATE AND A TIME-INVARIANT LATENT
              COVARIATE, WITH THE RESTRICTIONS OF TIME-
              INVARIANT ERROR VARIANCES AND SLOPE OF
              RELATIONSHIP OF REPEATED FOCAL MEASURE UPON
              TIME-VARYING COVARIATE
DATA:         FILE = CH13EX1_MCM.DAT;
              TYPE = MEANS COVARIANCE;
              NOBS = 400;
VARIABLE:     NAMES = COLLASP1 COLLASP2 COLLASP3
              COLLASP4 PARSTYL1 PARSTYL2 MOTIVN1
              MOTIVN2 MOTIVN3 MOTIVN4;
MODEL:        INTERCEP SLOPE | COLLASP1@0 COLLASP2@1
              COLLASP3@2 COLLASP4@3;
              PARSTYLE BY PARSTYL1 PARSTYL2;
              INTERCEP SLOPE ON PARSTYLE;
              COLLASP1 ON MOTIVN1 (2);
              COLLASP2 ON MOTIVN2 (2);
              COLLASP3 ON MOTIVN3 (2);
              COLLASP4 ON MOTIVN4 (2);
              COLLASP1-COLLASP4 (1);
```

The new feature in this command file is the addition of the same reference number, (2), to the slopes of the regression of repeated focal measure—college aspiration—upon the time-invariant covariate, motivation (see end of the four lines declaring this regression). In this way, that regression slope is held constant across time; since this constant obviously need not be equal to the common error variance, a different numerical symbol (viz. 2) is assigned to it at the end of the respective four command lines. This input file produces the following output that we include in its entirety, since it represents the most parsimonious tenable model of interest in this section.

```
THE MODEL ESTIMATION TERMINATED NORMALLY

TESTS OF MODEL FIT

Chi-Square Test of Model Fit
          Value                      44.641
          Degrees of Freedom             32
          P-Value                    0.0680
```

Given these fit indices, the likelihood ratio test of the hypothesis of a time-invariant regression slope of repeated measure upon the time-varying covariate has the statistic $44.641 - 40.783 = 3.858$, which is also nonsignificant when evaluated against the chi-square distribution with pertinent degrees of freedom, $32 - 29 = 3$ (the relevant cutoff is 7.84, as mentioned above, and is thus higher than that test statistic). We interpret this finding as suggesting support for the claim that the slopes of the regression of college aspiration on motivation are stable over time. We add, however, that this finding is not sufficient to reconsider motivation as a time-invariant covariate. The latter is only then the case, when a covariate is constant within subject over time (or evaluated only once in a study, for whatever technical or design-related reasons). This constancy cannot obviously be inferred from the test outcome just presented, and we move on to the remainder of the output.

```
Chi-Square Test of Model Fit for the Baseline Model
   Value                              2054.279
   Degrees of Freedom                       39
   P-Value                              0.0000

CFI/TLI
   CFI                                   0.994
   TLI                                   0.992
```

```
Loglikelihood
    H0 Value                                 -8306.625
    H1 Value                                 -8284.305

Information Criteria
    Number of Free Parameters                       19
    Akaike (AIC)                             16651.251
    Bayesian (BIC)                           16727.089
    Sample-Size Adjusted BIC                 16666.800
      (n* = (n+2) / 24)

RMSEA (Root Mean Square Error Of Approximation)
    Estimate                                     0.031
    90 Percent C.I.                              0.000       0.052
    Probability RMSEA <= .05                     0.931
```

In addition to the above-presented tenable chi-square value, the RMSEA value and confidence interval (see in particular its left endpoint) furnish further evidence that the model can be considered a plausible means of data description and explanation.

```
SRMR (Standardized Root Mean Square Residual)
    Value                0.029

MODEL RESULTS

                     Estimates      S.E.     Est./S.E.

INTERCEP |
    COLLASP1           1.000        0.000        0.000
    COLLASP2           1.000        0.000        0.000
    COLLASP3           1.000        0.000        0.000
    COLLASP4           1.000        0.000        0.000

SLOPE |
    COLLASP1           0.000        0.000        0.000
    COLLASP2           1.000        0.000        0.000
    COLLASP3           2.000        0.000        0.000
    COLLASP4           3.000        0.000        0.000

PARSTYLE BY
    PARSTYL1           1.000        0.000        0.000
    PARSTYL2           0.454        0.065        7.005

INTERCEP ON
    PARSTYLE           0.033        0.005        6.105

SLOPE ON
    PARSTYLE           0.015        0.003        5.692
```

```
COLLASP1 ON
   MOTIVN1            0.276      0.025     10.895

COLLASP2 ON
   MOTIVN2            0.276      0.025     10.895

COLLASP3 ON
   MOTIVN3            0.276      0.025     10.895

COLLASP4 ON
   MOTIVN4            0.276      0.025     10.895

SLOPE WITH
 INTERCEP             0.360      0.059      6.064

MOTIVN1 WITH
 PARSTYLE            -1.011      0.973     -1.039

MOTIVN2 WITH
 PARSTYLE             0.584      1.057      0.553

MOTIVN3 WITH
 PARSTYLE            -0.294      0.964     -0.305

MOTIVN4 WITH
 PARSTYLE            -0.425      1.002     -0.424

Intercepts
  COLLASP1            0.000      0.000      0.000
  COLLASP2            0.000      0.000      0.000
  COLLASP3            0.000      0.000      0.000
  COLLASP4            0.000      0.000      0.000
  PARSTYL1           98.795      1.003     98.513
  PARSTYL2           99.240      0.782    126.878
  INTERCEP            6.814      1.269      5.369
  SLOPE               1.054      0.040     26.197

Variances
  PARSTYLE          393.974     56.625      6.958

Residual Variances
  COLLASP1            0.614      0.031     20.000
  COLLASP2            0.614      0.031     20.000
  COLLASP3            0.614      0.031     20.000
  COLLASP4            0.614      0.031     20.000
  PARSTYL1           10.107     48.898      0.207
  PARSTYL2          163.808     15.360     10.665
  INTERCEP            1.549      0.152     10.217
  SLOPE               0.434      0.041     10.465
```

We note that similar substantive interpretations are suggested here as with the first fitted version of this model (which had no time-invariance constraints on any of its parameters). At the same time, however, we also observe that the standard errors are notably smaller on average in the presently considered model version and thus allow more precise statistical inferences. This is due to the fact that the number of free parameters is smaller here than that for the first model version considered in this section. As indicated earlier, this is a main reason why more parsimonious models are generally preferable to less parsimonious models when tenable for an analyzed data set.

13.3 Modeling General Forms of Change Over Time

The models considered in Section 13.2 made an important assumption about the pattern of change over time. This was the assumption of linear growth or decline in the repeatedly followed variables of main concern. Specifically, the linear relationship between assessment occasion and variable value at it was embodied in the basic Equation 13.1 that underlies the model given by Equation 13.3 and all its extensions and versions indicated in the last section. As is well known, however, oftentimes studied developmental processes in the social, behavioral, and educational sciences evolve in a nonlinear fashion, especially across longer periods of time. Although in some cases the linear function in Equation 13.3 may represent a reasonable approximation, in particular after taking into account possibly unequal lengths of time elapsing between successive assessments, in many settings it may not be satisfactory. Then weaker assumptions in the models dealt with above would be desirable. Incorporation of such assumptions is also possible within the framework of LVM and the methodological approach followed in this chapter, as demonstrated in this section.

13.3.1 Level-and-Shape Model

In order to discuss a model for change, which is based on an unrestricted pattern of development over time, we relax the linear relationship assumption in Equation 13.1. To this end, we represent any pattern of temporal change across the T repeated assessments under consideration as follows (McArdle & Anderson, 1990; Rao, 1958; Tucker, 1958):

$$Y_{it} = 1 \,.\, l_i + \gamma_t \,.\, s_i + e_{it}, \tag{13.4}$$

where $\gamma_1 = 0$ and $\gamma_T = 1$ while $\gamma_2, \ldots, \gamma_{T-1}$ are unknown parameters, and the same assumptions about the error term are made as in Equation 13.1. Similar to Equation 13.1, Equation 13.4 can be viewed as representing a restricted CFA model (see Chapter 9) having two factors, with subject scores on them being l_i and s_i $(i = 1, \ldots, n)$. Taking expectation across persons from both sides of Equation 13.4 and some straightforward algebra, one sees that $\gamma_2, \ldots, \gamma_{T-1}$ are correspondingly the ratios of mean change occurring from first assessment (starting position) until the 2nd, ..., $(T-1)$st occasion, to mean change across all measurement points. Similarly, the mean of starting position on the repeatedly followed variable, $M(Y_1)$, is seen to equal the mean of the factor l, $M(l)$, which for this reason is called level factor. In the same way, the mean of the factor with scores s_i $(i = 1, \ldots, n)$ is shown to equal overall mean change across the study period. For this reason, this factor is referred to as shape factor, and the model defined by Equation 13.4 is called level-and-shape model (McArdle & Anderson, 1990; see also Raykov & Marcoulides, 2006).

In a multivariate level-and-shape model, at least two repeatedly followed variables of focal interest are simultaneously analyzed by postulating Equation 13.4 for each one of them. In addition, time-invariant and possibly time-varying covariates may be included, whereby none of them needs to be perfectly measured. For the case of a pair of longitudinally measured characteristics of main concern, denoted Y and Z, and a fallible time-invariant covariate assessed by a pair of indicators, this model's defining equations are as follows (cf. Equation 13.3; for simplicity, and to emphasize similarity to the intercept-and-slope model, we use the same notation where possible with respect to the covariate indicators and associated equation terms):

$$
\begin{aligned}
Y_{it} &= 1 \,.\, l_i + \gamma_t \,.\, s_i + e_{it} \\
Z_{it} &= 1 \,.\, u_i + \pi_t \,.\, v_i + d_{it} \\
K_{1i} &= \kappa_2 + \lambda_2 \eta_i + \delta_{1i} \\
K_{2i} &= \kappa_2 + \lambda_2 \eta_i + \delta_{2i} \\
l_i &= \alpha_0 + \beta_0 \eta_i + g_i \\
s_i &= \alpha_1 + \beta_1 \eta_i + h_i \\
u_i &= \alpha_2 + \beta_2 \eta_i + m_i \\
v_i &= \alpha_3 + \beta_3 \eta_i + w_i
\end{aligned}
\qquad (13.5)
$$

In Equations 13.5, in addition to the assumptions and notation used in Equation 13.3, u and v symbolize the level and shape factors for the longitudinally followed variable Z, with pertinent parameters $\pi_2, \ldots, \pi_{T-1}$ and error term d (assumed also normal and unrelated to these factors; $\pi_1 = 0$ and $\pi_T = 1$), while α_0 through α_3 and β_0 through β_3 denote the intercepts and slopes, respectively, of the regressions of the level and shape factors of both repeatedly assessed variables upon the latent covariate, with associated error terms g through w (assumed normal, with zero mean, and uncorrelated with predictors in pertinent equations). As was the case with the generic model in Equation 13.3), in the current one there is no limit on the number of repeatedly followed variables of main interest, time-varying and time-invariant covariates, as well as on their indicators in case any covariate of either type cannot be measured perfectly (assuming overall model identification). (Like Equation 13.3, time-varying covariates are included in this model by adding them on the right-hand side of any or both of the first pair of Equations 13.5.)

13.3.2 Empirical Illustration

To illustrate the discussed general modeling approach, consider a study of $n=160$ high school seniors with each student measured on several intelligence tests. Two of the tests, dealing with their ability for inductive reasoning with regard to stimuli consisting of symbols and of letters, were presented at four different assessment occasions across the academic year. In addition, at the first measurement point, a figural relations and a culture-fair test were administered, which can be considered indicators of the latent construct figural relations ability. A researcher is interested in studying the patterns of change over time in the inductive reasoning measures, symbolized by Y_{it} and Z_{it} in the notation used above ($n=1, \ldots, 160$; $t=1, \ldots, 4$). Thereby, he/she wishes to account for individual differences in the figural relations ability at the start of the investigation, which is denoted η in Equations 13.5, with K_1 and K_2 designating its markers—the figural relations and culture-fair tests, respectively. (The sample means and covariance matrix are provided in the file ch13ex2_mcm.dat available from www.psypress.com/applied-multivariate-analysis, with the order of variables being the same as that indicated in the VARIABLE command given next. Preliminary examination showed that the assumption of normality could be considered plausible.) In order to respond to these concerns, to fit the model given by Equations 13.5 to the data from this study we use the following M*plus* command file. (IR11 through IR14 denote below the four repeated measures with the first induction test, and IR21 through IR24 those with the second induction test; FR and CR are the two indicators of figural relations ability.)

```
TITLE:        BIVARIATE LEVEL-AND-SHAPE MODEL WITH
              FALLIBLE TIME-INVARIANT COVARIATE
DATA:         FILE = CH13EX2_MCM.DAT;
              TYPE = MEANS COVARIANCE;
              NOBS = 160;
VARIABLE:     NAMES = IR11 FR1 IR21 CF1 IR12 IR22 IR13 IR23
              IR14 IR24;
```

```
MODEL:         LEVEL1 SHAPE1 | IR11@0 IR12*1 IR13*1 IR14@1;
               LEVEL2 SHAPE2 | IR21@0 IR22*1 IR23*1 IR24@1;
               FR BY FR1 CF1;
               LEVEL1 ON FR; LEVEL2 ON FR;
               SHAPE1 ON FR; SHAPE2 ON FR;
```

In this input file, LEVEL1 and LEVEL2 denote the level factors for the two repeatedly presented induction tests, while SHAPE1 and SHAPE2 are their shape factors (see Equation 13.4 and following discussion). Of particular interest are the regressions of each of these four factors upon the latent, time-invariant covariate, figural relations ability (assessed at first occasion). Since we are fitting the two-dimensional level-and-shape model, according to its underlying Equations 13.5, the loadings of the shape factors on the second and third measurement are parameters to be estimated, which fact is symbolized in the first two lines of the MODEL command by asterisks following the name of the measure they pertain to—these are the parameters γ_2, γ_3, π_2, and π_3 in the notation of Equations 13.5. This input file yields the following output, with interpretative comments included after pertinent sections.

```
THE MODEL ESTIMATION TERMINATED NORMALLY

TESTS OF MODEL FIT

Chi-Square Test of Model Fit
     Value                                  35.534
     Degrees of Freedom                         29
     P-Value                                0.1875
```

The model fits acceptably well (see also its RMSEA value further below and associated confidence interval).

```
Chi-Square Test of Model Fit for the Baseline Model
     Value                                2595.709
     Degrees of Freedom                         45
     P-Value                                0.0000

CFI/TLI
     CFI                                     0.997
     TLI                                     0.996

Loglikelihood
     H0 Value                            -4900.723
     H1 Value                            -4882.956
```

```
Information Criteria
   Number of Free Parameters                   36
   Akaike (AIC)                          9873.445
   Bayesian (BIC)                        9984.152
   Sample-Size Adjusted BIC              9870.189
     (n* = (n + 2) / 24)

RMSEA (Root Mean Square Error Of Approximation)
   Estimate                                 0.038
   90 Percent C.I.                          0.000    0.075
   Probability RMSEA <= .05                 0.668

SRMR (Standardized Root Mean Square Residual)
   Value                                    0.013

MODEL RESULTS
              Estimates     S.E.    Est./S.E.

LEVEL1     |
   IR11         1.000      0.000      0.000
   IR12         1.000      0.000      0.000
   IR13         1.000      0.000      0.000
   IR14         1.000      0.000      0.000

SHAPE1     |
   IR11         0.000      0.000      0.000
   IR12         1.031      0.033     31.188
   IR13         1.172      0.036     32.725
   IR14         1.000      0.000      0.000
```

As can be seen by examining the above factor loadings, there is significant improvement in subject performance on the first induction test at second and third assessment relative to starting position (see parameters γ_2 and γ_3 that are significant). There is some drop in performance on this test at the final assessment, however, relative to third occasion, as could be suggested by examining the 95%-confidence (large-sample) interval of γ_3 that is entirely above the point 1. As can be readily obtained, this confidence interval is $(1.172 - 1.96 \times .036, 1.172 + 1.96 \times .036)$, that is, (1.101, 1.243). (This drop in performance could be perhaps explained by lower motivation to take the test at the end of the academic year.)

```
LEVEL2     |
   IR21         1.000     0.000      0.000
   IR22         1.000     0.000      0.000
   IR23         1.000     0.000      0.000
   IR24         1.000     0.000      0.000
```

```
SHAPE2    |
    IR21            0.000     0.000      0.000
    IR22            1.044     0.033     31.798
    IR23            1.142     0.038     29.791
    IR24            1.000     0.000      0.000
```

The same interpretation of the factor loadings apply to the second induction test and subject performance development on it over time.

```
FR          BY
    FR1             1.000     0.000      0.000
    CF1             1.396     0.099     14.044
```

Both the factor loadings and error variances (see pertinent rows in section "Residual Variances" below) are significant, as would be expected from a well-defined measurement process of latent variables.

```
LEVEL1    ON
    FR              1.367     0.124     11.063

LEVEL2    ON
    FR              1.804     0.137     13.140
```

The time-invariant, fallible covariate—figural relations ability—is significantly related to starting position on both induction tests. Specifically, subjects high on figural relations ability perform well at first assessment on both tests, which is not unexpected considering the fact that figural relations and inductive reasoning are interrelated fluid intelligence subabilities (Horn, 1982).

```
SHAPE1    ON
    FR              0.414     0.089      4.633

SHAPE2    ON
    FR              0.156     0.089      1.764
```

While overall mean change on the induction test dealing with symbols is significantly related to figural relations ability, mean change across the four assessments with the induction test using letter-based stimuli is not.

```
SHAPE1    WITH
    LEVEL1        -12.253    11.607     -1.056
```

The correlation between the residuals of starting position on the symbol-based induction test with that of overall change on it is not significant, and the same result is found below with respect to the letter-based induction test (see five lines below). That is, on either induction test, the unexplained by figural relations parts of its initial level and change over the course of the study are only weakly related.

```
LEVEL2    WITH
   LEVEL1        25.249      5.516      4.578
   SHAPE1        -0.182      3.584     -0.051

SHAPE2    WITH
   LEVEL1        -4.521      3.485     -1.297
   SHAPE1        16.172      3.146      5.140
   LEVEL2         1.282     13.314      0.096

Intercepts
   IR11           0.000      0.000      0.000
   FR1           20.783      0.534     38.884
   IR21           0.000      0.000      0.000
   CF1           34.839      0.716     48.637
   IR12           0.000      0.000      0.000
   IR22           0.000      0.000      0.000
   IR13           0.000      0.000      0.000
   IR23           0.000      0.000      0.000
   IR14           0.000      0.000      0.000
   IR24           0.000      0.000      0.000
   LEVEL1        21.569      0.816     26.445
   SHAPE1        12.424      0.580     21.405
   LEVEL2        26.227      0.967     27.118
   SHAPE2        10.613      0.546     19.443

Variances
   FR            33.408      5.115      6.531

Residual Variances
   IR11           6.418     12.111      0.530
   FR1           12.300      1.957      6.284
   IR21          17.580     13.928      1.262
   CF1           16.983      3.313      5.126
   IR12          11.740      1.705      6.886
   IR22           7.393      1.316      5.618
   IR13           9.881      1.705      5.795
   IR23          12.075      1.852      6.521
   IR14          18.357      2.383      7.702
   IR24          12.947      1.805      7.175
   LEVEL1        37.592     13.400      2.805
```

```
SHAPE1         25.259    11.061      2.284
LEVEL2         23.338    15.131      1.542
SHAPE2         17.532    12.764      1.374
```

We conclude this subsection by stressing that perhaps most of the time when a researcher would use the discussed model or closely related models, he/she will be interested primarily in the loading estimates of the shape factor and the regression coefficients of the intercept and shape factors on the employed covariate(s), with which parameters our preceding discussion was mostly concerned.

13.3.3 Testing Special Patterns of Growth or Decline

As elaborated earlier, the generic model in Equation 13.5 allows a general form of growth or decline over time. Consequently, the model permits development in repeatedly assessed variables to occur in any fashion. For instance, as in the example just considered, there may be initial growth followed by decline or vice versa on any longitudinally followed dimension. The fact that the pattern of change is permitted to be completely unrestricted gives this model a higher chance to fit better empirical data from a repeated measure study than those in the preceding section that were based on the linear growth or decline assumption. At the same time, however, it needs to be stressed that with the present model one cannot obtain as a by-product a specific quantitative description of the development that occurs over time in the variables of focal interest, while the latter description is a by-product of the intercept-and-slope model when found tenable for an analyzed data set.

In some settings, therefore, it may be of substantive interest to examine within the model in Equations 13.5 whether there is evidence in the data hinting to a particular pattern of change over time. When this pattern is well defined beforehand, it may be possible to derive specific values of the intermediate parameters $\gamma_2, \ldots, \gamma_{T-1}$ (and the correspondingly ones for other repeatedly assessed dimensions). Then, in a subsequent modeling session, these parameters may be fixed at those values, and the likelihood ratio approach used to test the hypothesis that growth/decline proceeds in that particular manner. For instance, in the last example, one may be interested (before looking at the data) in testing the hypothesis that development over time is linear for both induction tests. To this end, all one needs to do is fix the parameters γ_2 and π_2 at .333, while γ_3 and π_3 at .667, and compare the chi-square values of this restricted model to that of the last fitted one, in which the former is nested. This can be accomplished with the following *Mplus* command file:

```
TITLE:     BIVARIATE LEVEL-AND-SHAPE MODEL WITH FALLIBLE
           TIME-INVARIANT COVARIATE
DATA:      FILE=CH13EX2_MCM.DAT;
           TYPE=MEANS COVARIANCE;
           NOBS=160;
VARIABLE:  NAMES=IR11 FR1 IR21 CF1 IR12 IR22 IR13 IR23
           IR14 IR24;
MODEL:     LEVEL1 SHAPE1 | IR11@0 IR12@.333 IR13@.667
           IR14@1;
           LEVEL2 SHAPE2 | IR21@0 IR22@.333 IR23@.667
           IR24@1;
           FR BY FR1 CF1;
           LEVEL1 ON FR; LEVEL2 ON FR;
           SHAPE1 ON FR; SHAPE2 ON FR;
```

The only difference in this input to the one of the last fitted model lies in the first two lines of the MODEL command. In them, as mentioned above, we fix the intermediate loadings of the shape factor at the values they would have if change over time were to be linear. (This model parameterization is identical to that underlying a two-dimensional intercept-and-slope model; see preceding section.) This command file is associated with the following output:

```
THE MODEL ESTIMATION TERMINATED NORMALLY

   WARNING: THE LATENT VARIABLE COVARIANCE MATRIX (PSI)
   IS NOT POSITIVE DEFINITE. THIS COULD INDICATE A
   NEGATIVE VARIANCE/RESIDUAL VARIANCE FOR A LATENT
   VARIABLE, A CORRELATION GREATER OR EQUAL TO ONE BETWEEN
   TWO LATENT VARIABLES, OR A LINEAR DEPENDENCY AMONG MORE
   THAN TWO LATENT VARIABLES. CHECK THE TECH4 OUTPUT FOR
   MORE INFORMATION. PROBLEM INVOLVING VARIABLE LEVEL1.
```

While the numerical optimization routine underlying the model fitting process converged (which is the message of the first output line cited), it did so to an inadmissible solution. Indeed, the residual variances associated with the disturbance terms pertaining to the level and shape factors on both induction tests are estimated at negative values, as will be seen shortly. (The fact that one of these estimates is nonsignificant is irrelevant.) Since no variance of a real-valued variable—in which all variables studied in the behavioral, social, and educational sciences currently are—can be negative, this solution is improper. Hence, in general, no part of the output can be trusted or interpreted in substantive terms. In particular, one should not interpret the goodness of fit indices or any parameter estimates (except noting, of course, the negative variance estimates of the disturbance terms pertaining to the level and shape factors, in this example). For completeness, we include the rest of the output next, and continue after it the discussion of this important phenomenon of inadmissible solutions.

```
TESTS OF MODEL FIT

Chi-Square Test of Model Fit
    Value                                502.894
    Degrees of Freedom                        31
    P-Value                               0.0000

Chi-Square Test of Model Fit for the Baseline Model
    Value                               2595.709
    Degrees of Freedom                        45
    P-Value                               0.0000

CFI/TLI
    CFI                                    0.815
    TLI                                    0.731

Loglikelihood
    H0 Value                           -5134.403
    H1 Value                           -4882.956

Information Criteria
    Number of Free Parameters                 34
    Akaike (AIC)                       10336.805
    Bayesian (BIC)                     10441.361
    Sample-Size Adjusted BIC           10333.730
      (n* = (n + 2) / 24)

RMSEA (Root Mean Square Error Of Approximation)
    Estimate                               0.308
    90 Percent C.I.                        0.285      0.332
    Probability RMSEA <= .05               0.000

SRMR (Standardized Root Mean Square Residual)
    Value                                  0.149

MODEL RESULTS
                       Estimates     S.E.    Est./S.E.

LEVEL1      |
    IR11                  1.000     0.000      0.000
    IR12                  1.000     0.000      0.000
    IR13                  1.000     0.000      0.000
    IR14                  1.000     0.000      0.000
```

```
SHAPE1   |
   IR11              0.000    0.000    0.000
   IR12              0.333    0.000    0.000
   IR13              1.075    0.041   26.175
   IR14              1.000    0.000    0.000

LEVEL2   |
   IR21              1.000    0.000    0.000
   IR22              1.000    0.000    0.000
   IR23              1.000    0.000    0.000
   IR24              1.000    0.000    0.000

SHAPE2   |
   IR21              0.000    0.000    0.000
   IR22              0.333    0.000    0.000
   IR23              1.102    0.046   23.963
   IR24              1.000    0.000    0.000
FR       BY
   FR1               1.000    0.000    0.000
   CF1               1.383    0.098   14.066

LEVEL1   ON
   FR                1.598    0.124   12.929

LEVEL2   ON
   FR                1.877    0.138   13.604

SHAPE1   ON
   FR                0.202    0.075    2.690

SHAPE2   ON
   FR                0.088    0.068    1.290

SHAPE1   WITH
   LEVEL1           44.907    6.548    6.858

LEVEL2   WITH
   LEVEL1           25.405    5.587    4.547
   SHAPE1            6.924    3.131    2.211

SHAPE2   WITH
   LEVEL1           -1.294    2.553   -0.507
   SHAPE1            8.435    2.011    4.194
   LEVEL2           36.514    5.601    6.520

Intercepts
   IR11              0.000    0.000    0.000
   FR1              20.783    0.534   38.884
   IR21              0.000    0.000    0.000
   CF1              34.839    0.716   48.637
```

```
     IR12                 0.000      0.000      0.000
     IR22                 0.000      0.000      0.000
     IR13                 0.000      0.000      0.000
     IR23                 0.000      0.000      0.000
     IR14                 0.000      0.000      0.000
     IR24                 0.000      0.000      0.000
     LEVEL1              27.153      0.874     31.060
     SHAPE1               7.995      0.486     16.433
     LEVEL2              31.494      0.995     31.646
     SHAPE2               5.946      0.410     14.493

 Variances
     FR                  33.732      5.128      6.578

 Residual Variances
     IR11                97.696     11.657      8.381
     FR1                 11.976      1.941      6.171
     IR21                85.005     10.053      8.456
     CF1                 17.609      3.330      5.288
     IR12                50.952      6.147      8.289
     IR22                39.882      4.839      8.241
     IR13                13.710      2.314      5.926
     IR23                15.719      2.419      6.497
     IR14                18.948      2.673      7.087
     IR24                11.315      1.979      5.716
     LEVEL1             -16.505      8.112     -2.035
     SHAPE1             -32.205      6.595     -4.883
     LEVEL2              -3.488      8.300     -0.420
     SHAPE2             -26.126      5.385     -4.852
```

13.3.4 Possible Causes of Inadmissible Solutions

Inadmissible solutions are a major problem in applications of LVM. When they occur, the entire associated output is rendered useless, except of course for the purpose of attempting to locate the source of the problem. In more general terms, inadmissible solutions are typically those where (a) a variance estimate is negative, (b) an estimated correlation is larger than 1 or smaller than −1, or (c) a resulting (implied) covariance/correlation matrix is not positive definite. In essentially all cases in practice, LVM software will alert its user of such solutions. In particular, M*plus* issues then a warning of the kind provided at the beginning of the last presented output, which describes the nature and possible formal reasons of the problem.

A major cause for the occurrence of an inadmissible solution is a misspecified model. This is a model that postulates relationships that are in a sufficiently strong conflict with the analyzed data. For example, a model may omit an important parameter, or may postulate a relationship between some of its variables (whether latent or observed) that runs counter to their population relationship. The latter may occur when a seriously nonlinear population relationship is actually postulated in a fitted model as linear. We note that in a sense any realistic model, in order to be useful, is likely to be misspecified. This will not always lead to an inadmissible solution, however. If a model is not seriously misspecified, it is not necessarily likely that such a solution will result. These kinds of models, if they in addition fit an analyzed data set sufficiently well, are the ones of interest to researchers and the object of further scrutiny as possible means of data description and explanation that may respond to substantive questions asked.

A seriously misspecified model, on the other hand, makes an assumption that has implications substantially inconsistent with the analyzed data. Not all seriously misspecified models will be associated with an inadmissible solution, however, but some will. The last considered example represents a case in point. As we can readily see from the observed variable means, and from a perusal of the output associated with the unrestricted model fitted above in this section, development over time proceeds in other than a linear fashion. (This is seen, e.g., also from looking at the shape factor loading estimates and values, which are not all increasing.) Yet this model assumed that pattern to be linear, which represents a model misspecification. As it turned out, this misspecification was sufficiently serious to lead to an inadmissible solution.

Not all models associated with inadmissible solutions when fitted to data will be seriously misspecified, however. It is possible that lacking or limited variability in a studied population can lead to such a solution, especially with a sample that is not large, also with models that are not seriously misspecified. A small sample may in its own right be a reason for an inadmissible solution then, since as is well-known sampling error (i.e., lack of population representativeness) is inversely proportional to sample size. Hence, a small sample typically embodies a large sampling error that may lead to an inadmissible solution. A data set that reflects low reliability and validity of measurement, and hence a high degree of measurement-related "noise" may also lead to an inadmissible solution, even with models that may be reasonable approximations to studied phenomena. Last but not least, inappropriate starting values may also cause an inadmissible solution, although this may be considered currently less likely given the software advances in determining relatively "good" start values based on model and analyzed data.

All these possible reasons need to be assessed in a particular empirical setting (where perhaps more than one of them may be in effect), when examining the possible reason for inadmissible solution in a given case. Further development of the considered model, in particular after critical examination of its assumptions, and/or increasing sample size when possible, could contribute to successfully resolving the problem leading to an inadmissible solution. For the last fitted model in the preceding subsection, given the empirical means, one would suggest that the inadmissible solution resulted from the assumption of linear growth in the repeatedly administered induction tests. From this perspective, the model fitted before that one—which postulated an unrestricted pattern of change over time—is therefore to be preferred as a means of data description and explanation.

13.4 Modeling Change Over Time With Incomplete Data

Missing data are the rule rather than exception in social, behavioral, and educational research. Repeated measure studies conducted in these sciences very often yield incomplete data sets. This is due to the fact that usually subject attrition occurs at different points during the course of these longitudinal investigations. The LVM framework employed so far in this chapter is very well suited for an application in such empirical settings. In particular, full information maximum likelihood (FIML) or multiple imputation can be readily utilized within this approach, under their assumptions (Chapter 12). In order to use FIML in such settings, one develops a substantively meaningful model to be fitted, as in the case when there were no missing values, introduces a uniform symbol(s) to designate all missing values in the raw data file, and submits the model to analysis utilizing all collected data. Any of the models discussed in the present chapter, with their possible extensions, can be employed in this way for purposes of studying change processes along multiple repeatedly measured dimensions. Thereby, parameter estimates and standard errors are interpreted in the same way as in the case of no missing data, and parameter restrictions are also tested similarly.

To demonstrate analysis of change with incomplete data, we use FIML to fit a multivariate level-and-shape model to data from a study of the relationship between intelligence, motivation, and aspiration with $n = 160$ senior high school students. At four assessment occasions (one per quarter), an intelligence test and a motivation measure were administered to them, whereby a notable proportion of students were absent on at least one measurement point. In addition, at the first occasion, two college aspiration scales were also administered. (The data on all 10 observed variables are presented in the file ch13ex3.dat available from www.psypress.com/applied-multivariate-analysis, where "99" denotes missing datum. In the input file below, following the position of the variables in the data file, the name V# is used to denote variables of no interest for the current analytic purposes, IQ# to designate the consecutive intelligence test measures, and MOT# those for motivation, while CA1 and CA2 symbolize the pair of college aspiration scales.) A researcher is interested in examining the interrelationships among patterns of change over time in intelligence and motivation, after controlling for individual differences in college aspiration. This goal can be achieved using a two-dimensional version of the level-and-shape model with a fallible time-invariant covariate assessed by a pair of indicators, which is defined in Equations 13.5. This model is fitted using the following M*plus* command file (to save space, we abbreviate to the first 4 letters, at least, as accepted by the software):

```
TITLE:    TWO-DIMENSIONAL LEVEL AND SHAPE MODEL WITH
          FALLIBLE TIME-INVARIANT COVARIATE AND MISSING
          DATA
DATA:     FILE=CH13EX3.DAT;
VARI:     NAMES=IQ1 CA1 MOT1 CA2 V5-V8 IQ2 V10 MOT2 V12-V16
          IQ3 V18 MOT3 V20-V24 IQ4 V26 MOT4 V28-V35;
          USEVARIABLE=IQ1-CA2 IQ2 MOT2 IQ3 MOT3 IQ4 MOT4;
          MISSING=ALL(99);
ANALY:    TYPE=MISSING H1;
MODEL:    LEVEL1 SHAPE1 | IQ1@0 IQ2*1 IQ3*1 IQ4@1;
          LEVEL2 SHAPE2 | MOT1@0 MOT2*1 MOT3*1 MOT4@1;
          COLLASP BY CA1 CA2;
          LEVEL1-SHAPE2 ON COLLASP;
```

After selecting the 10 variables to be analyzed from the original data file with altogether 35 variables (see subcommand USEVARIABLE in the VARIABLE command, which is employed for this purpose), the uniform missing value symbol is defined and FIML requested as analytic method. The level-and-shape model is then defined for both repeatedly followed variables, intelligence and motivation, with the college aspiration factor representing a time-invariant latent covariate that is evaluated by the two subscales administered at initial assessment. In the final line, the level and shape factors for intelligence and for motivation are regressed upon that covariate. This command file yields the following output:

```
THE MODEL ESTIMATION TERMINATED NORMALLY

TESTS OF MODEL FIT

Chi-Square Test of Model Fit
  Value                          34.435
  Degrees of Freedom                 29
  P-Value                        0.2237
```

As seen from this section, model fit is acceptable (see also the RMSEA value below). We therefore can move on to an interpretation of parameter estimates and standard errors (confidence intervals).

```
Chi-Square Test of Model Fit for the Baseline Model
    Value                                   2532.831
    Degrees of Freedom                            45
    P-Value                                   0.0000

CFI/TLI
    CFI                                        0.998
    TLI                                        0.997

Loglikelihood
    H0 Value                               -4845.434
    H1 Value                               -4828.217

Information Criteria
    Number of Free Parameters                     36
    Akaike (AIC)                            9762.869
    Bayesian (BIC)                          9873.575
    Sample-Size Adjusted BIC                9759.612
      (n* = (n + 2) / 24)

RMSEA (Root Mean Square Error Of Approximation)
    Estimate                                   0.034
    90 Percent C.I.                            0.000  0.072
    Probability RMSEA <= .05                   0.710

SRMR (Standardized Root Mean Square Residual)
    Value                                      0.013

MODEL RESULTS
                          Estimates     S.E.   Est./S.E.

LEVEL1 |
    IQ1                      1.000     0.000      0.000
    IQ2                      1.000     0.000      0.000
    IQ3                      1.000     0.000      0.000
    IQ4                      1.000     0.000      0.000

SHAPE1 |
    IQ1                      0.000     0.000      0.000
    IQ2                      1.037     0.033     31.654
    IQ3                      1.169     0.036     32.802
    IQ4                      1.000     0.000      0.000
```

Using the same confidence interval-based approach as in the preceding section, we can see that there is evidence of a considerable increase in the intelligence test results at second and third assessments relative to first occasion, and some drop at final measurement where subject performance remains, however, still higher than starting position (see similar pattern in motivation below, which may well provide an explanation for this finding).

```
LEVEL2 |
    MOT1                  1.000    0.000     0.000
    MOT2                  1.000    0.000     0.000
    MOT3                  1.000    0.000     0.000
    MOT4                  1.000    0.000     0.000

SHAPE2 |
    MOT1                  0.000    0.000     0.000
    MOT2                  1.042    0.033    31.280
    MOT3                  1.139    0.039    29.283
    MOT4                  1.000    0.000     0.000
```

The same pattern of development across the academic year is observed on the motivation measure—initial increase until third assessment and then some drop at final occasion (but still remaining above starting position).

```
COLLASP BY
    CA1                   1.000    0.000     0.000
    CA2                   1.391    0.100    13.855

LEVEL1 ON
    COLLASP               1.354    0.125    10.787

SHAPE1 ON
    COLLASP               0.428    0.091     4.681

LEVEL2 ON
    COLLASP               1.806    0.140    12.932

SHAPE2 ON
    COLLASP               0.168    0.090     1.866
```

The last four lines of this output section show that college aspiration explains significant proportion of variance in initial position on intelligence and on motivation. In addition, while college aspiration is a significant predictor of overall change in intelligence, there is no marked relationship between individual differences on college aspiration and overall increase in motivation.

```
SHAPE1 WITH
    LEVEL1              -15.479   12.959    -1.194
```

```
LEVEL2 WITH
    LEVEL1              25.635     5.564      4.607
    SHAPE1              -0.597     3.633     -0.164

SHAPE2 WITH
    LEVEL1              -4.501     3.529     -1.275
    SHAPE1              16.152     3.180      5.080
    LEVEL2               1.032    13.945      0.074
```

The preceding values are the covariances between the unexplained by college aspiration parts of starting position in intelligence and motivation, and as such are not of particular relevance for us here.

```
Intercepts
    IQ1               0.000      0.000      0.000
    CA1              20.719      0.532     38.921
    MOT1              0.000      0.000      0.000
    CA2              34.731      0.710     48.886
    IQ2               0.000      0.000      0.000
    MOT2              0.000      0.000      0.000
    IQ3               0.000      0.000      0.000
    MOT3              0.000      0.000      0.000
    IQ4               0.000      0.000      0.000
    MOT4              0.000      0.000      0.000
    LEVEL1           21.455      0.809     26.513
    SHAPE1           12.484      0.578     21.600
    LEVEL2           26.112      0.964     27.083
    SHAPE2           10.647      0.549     19.394
```

The final output portion presented next shows the latent covariate and residual variance estimates, and we note that none of them is negative, as would be the case in an admissible solution.

```
Variances
    COLLASP          32.952      5.075      6.493

Residual Variances
    IQ1               2.988     13.376      0.223
    CA1              12.387      1.973      6.279
    MOT1             17.919     14.512      1.235
    CA2              17.017      3.318      5.129
    IQ2              11.217      1.693      6.626
    MOT2              7.469      1.345      5.555
    IQ3              10.108      1.760      5.743
    MOT3             12.379      1.907      6.491
    IQ4              18.535      2.411      7.689
```

```
MOT4          12.861     1.827      7.041
LEVEL1        41.414    15.159      2.732
SHAPE1        28.096    11.934      2.354
LEVEL2        23.374    15.783      1.481
SHAPE2        17.470    13.433      1.301
```

In conclusion of this chapter, we stress that the LVM approach used in it for purposes of studying change represents a widely and readily applicable method for simultaneous analysis of change processes. In addition to focusing on at least two longitudinally followed dimensions, the approach permits inclusion of time-varying as well as time-invariant covariates. Thereby, none of these covariates needs to be perfectly evaluated, and in case of being error-prone needs to be measured by proxies (preferably at least two, unless substantively defensible restrictions on error variance for single indicators or other parameters are introduced in order to achieve model identification). This LVM approach is less restrictive than traditional repeated measure ANOVA or ANCOVA (regardless of whether uni- or multivariate approaches are used). In addition, under the assumptions of multinormality and data missing at random, the method is directly applicable with incomplete data sets, whereby no available sample information is sacrificed. A main feature of the approach is that it is best used with large samples, which may represent a limitation in settings with limited number of studied subjects. Another characteristic of the discussion in this chapter is that it assumed time-structured data. Extensions of its models to the case of individual-specific measurement points is possible, however, as are applications of them in many cases of nonnormal data via corrected test statistics and standard errors (Muthén & Muthén, 2006).

Appendix: Variable Naming and Order for Data Files

This section provides the variable names and their order, where appropriate, within the text (ASCII) files with data available from www.psypress.com/applied-multivariate-analysis. Variables in the command files containing only covariance matrices and means, which are used in Chapters 9, 12, and 13, have the order that is given in the VARIABLE command (see subcommand NAMES=) in the pertinent M*plus* input file found in the main text of the respective chapter.

Each text file with raw data presents as its first row the names of the variables contained in it. This needs to be taken into account and corresponding program-specific actions undertaken when reading the data into the corresponding software (SPSS or SAS).

The following table lists in its first column the name of each downloadable raw data file, as the file is referred to in the respective chapter where it is used, and in the next column provides the variable names (in the order in which they are contained in that file).

TABLE A.1

File Names, and Names of Variables (in Their Order), for Raw Data Used in the Book

Raw Data File Name	Variable Names and Order
ch1ex1.dat	gpa init_ab iq hours_tv
ch3ex1.dat	id Exam_Score Aptitude_Measure Age_in_Years Intelligence_Score Attention_Span
ch3ex2.dat	ir1 group gender sqrt_ir1 ln_ir1
ch4ex1.dat	test.1 test.2 test.3 group
ch4ex2.dat	motiv.1 motiv.2 motiv.3 ses
ch4ex3.dat	aspire.1 aspire.2 aspire.3 group ses
ch5ex1.dat	test.1 test.2 test.3
ch5ex2.dat	test.1 test.2 test.3 ses
ch5ex3.dat	ir.1 ir.2 ir.3 ir.4 gender
ch6ex1.dat	ir.1 ir.2 group
ch6ex2.dat	ir.1 ir.2 fr.2 cf.2 group ir1.group
ch6ex3.dat	ir fr cf group

(continued)

TABLE A.1 (continued)

File Names, and Names of Variables (in Their Order), for Raw Data Used in the Book

Raw Data File Name	Variable Names and Order
ch7ex1.dat	ir.lett fig.rel ir.symb cult.fr raven
ch7ex2.dat	ability.1 ability.2 ability.3 ed.motiv.1 ed.motiv.2 ed.motiv.3
ch8ex1.dat	extravsn neurotic agreeabl algebra geomet trigono
ch8ex2.dat	V1 V2 V3 V4 V5 V6 V7 V8 V9
ch10ex1.dat	test.1 test.2 test.3 test.4 group
ch10ex2.dat	pt.1 pt.2 pt.3 pt.4 group
ch11ex1.dat	ir.symb fig.rel ir.lett cult.fr raven pers.sp1 pers.sp2 vocab
ch11ex2.dat	coll.asp1 coll.asp2 coll.asp3 coll.asp4 group dummy.1 dummy.2
ch12ex1.dat	test_1 test_2 test_3 test_4 gender

Note: First three (or four, where appropriate) symbols of the file name indicate the number of the chapter where that raw data file is used.

References

Agresti, A. (2002). *Categorical data analysis*. New York: Wiley.

Allison, P.D. (2001). *Missing data*. Thousand Oaks: Sage.

Anderson, T.W. (1984). *Introduction to multivariate statistics*. New York: Wiley.

Baltes, P.B., Dittmann-Kohli, F., & Kliegl, R. (1986). Reserve capacity of the elderly in aging-sensitive tasks of fluid intelligence: Replication and extension. *Psychology and Aging, 1*, 172–177.

Bekker, P.A., Merckens, A., & Wansbeek, T.J. (1994). *Identification, equivalent models, and computer algebra*. New York, NY: Academic Press.

Belsley, D.A., Kuh, E., & Welsch, R.E. (1980). *Regression diagnostics: Identifying influential data and sources of collinearity*. New York, NY: John Wiley & Sons. Inc.

Bentler, P.M. (2004). *EQS structural equations program manual*. Encino, CA: Multivariate Software.

Bollen, K.A. (1989). *Structural equations with latent variables*. New York: Wiley.

Box, G.E.P. & Cox, D.R. (1964). An analysis of transformations. *Journal of the Royal Statistical Society, Series B, 26*, 211–246.

Browne, M.W. & Cudeck, R. (1993). Alternative ways of assessing model fit. In K.A. Bollen & J.S. Long (Eds.), *Testing structural equation models* (pp. 136–162). Newbury Park, CA: Sage Publications.

Byrne, B.M. (1994). *Structural equation modeling with EQS and EQS/Windows: Basic concepts, applications, and programming*. Thousand Oaks, CA: Sage Publications.

Byrne, B.M. (1998). *Structural equation modeling with LISREL, PRELIS, and SIMPLIS: Basic concepts, applications, and programming*. Mahwah, NJ: Lawrence Erlbaum Associates.

Byrne, B.M. (2001). *Structural equation modeling with AMOS: Basic concepts, applications, and programming*. Mahwah, NJ: Lawrence Erlbaum Associates.

Campbell, D.T. & Stanley, J.C. (1963). *Experimental and quasi-experimental designs for research*. Chicago, IL: Rand McNally.

Cattell, R.B. (1966). The scree test for the number of factors. *Multivariate Behavioral Research, 1*, 245–276.

Cattell, R.B., Eber, H.W., & Tatsuoka, M.M. (1970). Handbook for the Sixteen Personality Factor Questionnaire (16PF) in Clinical, Educational, Industrial and Research Psychology. Champaign, IL: Institute for Personality and Ability Testing.

Collins, L., Schafer, J.L., & Kam, C.-H. (2001). A comparison of inclusive and restrictive strategies in modern missing data procedures. *Psychological Methods, 6*, 330–351.

Cramer, E. & Nicewander, W.A. (1979). Some symmetric, invariant measures of multivariate association. *Psychometrika, 44*, 43–54.

De Carlo, L.T. (1997). On the meaning and use of kurtosis. *Psychological Methods, 2,* 292–307.
Dunn, G., Everitt, B., & Pickles, A. (1993). *Modeling covariances and latent variables using EQS.* London, UK: Chapman & Hall.
Fisher, R.A. (1936). The use of multiple measurements in taxonomic problems. *Annals of Eugenics, 7,* 179–188.
Fung, W.-K. (1993). Unmasking outliers and leverage points: A confirmation. *Journal of the American Statistical Association, 88,* 515–519.
Harman, H.H. (1976). *Modern factor analysis.* Chicago, IL: University of Chicago.
Hastie, T., Tibshiriani, R., & Friedman, J. (2001). *The elements of statistical learning.* New York: Springer.
Hayashi, K. & Marcoulides, G.A. (2006). Examining identification issues in factor analysis. *Structural Equation Modeling, 13,* 631–645.
Hays, W.L. (1994). *Statistics.* Fort Worth, TX: Harcourt Brace Jovanovich.
Heck, R.H. & Thomas, S.L. (2000). *An introduction to multilevel modeling techniques.* Mahwah, NJ: Lawrence Erlbaum Associates.
Horn, J.L. (1982). The nature of human mental abilities. In B.B. Wolman (Ed.), *Handbook of developmental psychology.*
Hosmer, D.W. & Lemeshow, S. (2000). *Applied logistic regression.* New York: Wiley.
Howell, D.C. (2002). *Statistical methods for psychology.* Pacific Grove, CA: Wadsworth.
Hox, J. (2002). *Multilevel analysis. Techniques and applications.* Mahwah, NJ: Lawrence Erlbaum Associates.
Huber, P.T. (1981). *Robust statistics.* New York, NY: Wiley.
Huitema, B.J. (1980). *Analysis of covariance.* New York: Wiley.
Johnson, R.A. & Wichern, D.W. (2002). *Applied multivariate statistical analysis.* Upper Saddle River, NJ: Prentice Hall.
Jolliffe, I.T. (1972). Discarding variables in a principal component analysis, I: Artificial data. *Applied Statistics, 21,* 160–173.
Jolliffe, I.T. (2002). *Principal component analysis* (second edition). New York, NY: Springer-Verlag, Inc.
Jöreskog, K.G. & Sörbom, D. (1996). *LISREL8 user's reference guide.* Lincolnwood, IL: Scientific Software International.
Kaiser, H.F. (1960). The application of electronic computer to factor analysis. *Educational and Psychological Measurement, 20,* 141–151.
Kaplan, D. (2000). *Structural equation modeling. Fundamentals and extensions.* Thousand Oaks, CA: Sage.
Kirk, R.E. (1994). *Experimental design: Procedures for the behavioral sciences.* Pacific Grove, CA: Brooks/Cole.
Little, R.J.A. & Rubin, D.B. (2002). *Statistical analysis with missing data.* New York: Wiley.
Lord, F.M. (1980). *Applications of item response theory to practical testing problems.* Hillside, NJ: Lawrence Erlbaum Associates.
Marcoulides, G.A. & Drezner, Z. (2001). Specification searches in structural equation modeling with a genetic algorithm. In G.A. Marcoulides & R.E. Schumacker (Eds.), *Advanced structural equation modeling: New developments and techniques* (pp. 247–268). Mahwah, NJ: Lawrence Erlbaum Associates.
Marcoulides, G.A. & Hershberger, S.L. (1997). *Multivariate statistical methods. A first course.* Mahwah, NJ: Lawrence Erlbaum Associates.

Marcoulides, G.A. & Saunders, C. (2006). PLS: A silver bullet? *MIS Quarterly, 30*(2), *iii–ix*.

Marcoulides, G.A. & Schumacker, R.E. (1996). (Eds.). *Advanced structural equation modeling: Issues and techniques*. Hillsdale, NJ: Lawrence Erlbaum Associates.

Marcoulides, G.A., Drezner, Z., & Schumacker, R.E. (1998). Model specification searches in structural equation modeling using Tabu search. *Structural Equation Modeling, 5*, 365–376.

Mardia, K.V. (1970). Measures of multivariate skewness and kurtosis with applications. *Biometrika, 50*, 519–530.

Maxwell, S.E. & Delaney, H.D. (2004). *Designing experiments and analyzing data*. (second edition). Mahwah, NJ: Lawrence Erlbaum Associates.

McArdle, J.J. & Anderson, E. (1990). Latent variable growth models for research on aging. In J.E. Birren & K.W. Schaie (Eds.), *Handbook of the psychology of aging* (third edition, pp. 21–44). New York: Academic Press.

Mills, J.D., Olejnik, S.F., & Marcoulides, G.A. (2005). The Tabu search procedure: An alternative to the variable selection methods. *Multivariate Behavioral Research, 40*, 351–371.

Morrison, D.F. (1976). *Multivariate statistical methods*. San Francisco, CA: McGraw Hill.

Muthén, B.O. (2002). Beyond SEM: General latent variable modeling. *Behaviormetrika, 29*, 81–117.

Muthén, L.K. & Muthén, B.O. (2002). How to use a Monte Carlo study to decide on sample size and determine power. *Structural Equation Modeling, 9*, 599–620.

Muthén, L.K. & Muthén, B.O. (2004). Observed and latent categorical variable modeling using M*plus*. M*plus* short course handout. Alexandria, VA.

Muthén, L.K. & Muthén, B.O. (2006). *Mplus user's guide*. Los Angeles, CA: Muthén & Muthén.

Nesselroade, J.R. & Baltes, P.B. (1979). (Eds.). *Longitudinal research in the study of behavior and development*. New York: Academic Press.

Olson, C.L. (1976). On choosing a test statistic in multivariate analysis of variance. *Psychological Bulletin, 83*, 579–586.

Pedhazur, E.J. (1997). *Multiple regression in behavioral research* (third edition). New York: Wadsworth.

Pinheiro, J. & Bates, D. (2000). *Mixed models with Splus*. New York: Springer.

Rao, C.R. (1952). *Advanced statistical methods in biometric research*. New York: Wiley.

Rao, C.R. (1958). Some statistical methods for comparison of growth curves. *Biometrics, 14*, 1–17.

Raykov, T. (2001). Testing multivariable covariance structure and means hypotheses via structural equation modeling. *Structural Equation Modeling, 2*, 224–257.

Raykov, T. (2005). Analysis of longitudinal data with missing values via covariance structure modeling using full-information maximum likelihood. *Structural Equation Modeling, 12*, 331–341.

Raykov, T. & Marcoulides, G.A. (1999). On desirability of parsimony in structural equation modeling selection. *Structural Equation Modeling, 6*, 292–300.

Raykov, T. & Marcoulides, G.A. (2006). *A first course in structural equation modeling*. (second edition). Mahwah, NJ: Lawrence Erlbaum Associates.

Raykov, T. & Marcoulides, G.A. (2008). Studying group differences and similarities in observed variable means and interrelationship indices in the presence of missing data using latent variable modeling. Manuscript submitted for publication.

Raudenbush, S.W. & Bryk, A.S. (2002). *Hierarchical linear models*. Thousand Oaks, CA: Sage.

Rencher, A.C. (1995). *An introduction to multivariate statistics*. New York: Wiley.

Rencher, A.C. (1998). *Multivariate statistical inference and applications*. New York: Wiley.

Roussas, G.G. (1997). *A course in mathematical statistics*. New York: Academic Press.

Schafer, J.L. (1997). *Analysis of incomplete multivariate data*. London: Chapman & Hall.

Shafer, J.L. & Graham, J.W. (2002). Missing data: Our view of the state of the art. *Psychological Methods, 7*, 147–177.

Schatzoff, M. (1964). *Exact distributions of Wilk's likelihood ratio criterion and comparisons with competitive tests*. Unpublished doctoral dissertation, Harvard University.

Schatzoff, M. (1966). Exact distributions of Wilk's likelihood ratio criterion. *Biometrika, 53*, 347–358.

Shadish, W.R., Cook, T.D., & Campbell, D.T. (2002). *Experimental and quasi-experimental designs for generalized causal inference*. Boston, MA: Houghton Mifflin.

Skrondal, A. & Rabe-Hesketh, S. (2004). *Generalized linear latent variable modeling*. London: Chapman & Hall.

Spearman, C. (1904). General intelligence objectively determined and measured. *American Journal of Psychology, 5*, 201–293.

Steiger, J.H. & Lind, B.C. (1980, June). Statistically based tests for the number of common factors. Paper presented at the Psychometric Society Annual Meeting, Iowa City, IA.

Stewart, D. & Love, W. (1968). A general canonical correlation index. *Psychological Bulletin, 70*, 160–163.

Tabachnick, B.G. & Fidell, L.S. (2007). *Using multivariate statistics*. Boston, MA: Allyn & Bacon.

Tatsuoka, M.M. (1971). *Multivariate analysis*. New York: Macmillan.

Tatsuoka, M.M. (1988). *Multivariate analysis*. New York: Macmillan.

Thurstone, L.L. (1947). *Multiple-factor analysis*. Chicago, IL: University of Chicago.

Timm, N.H. (2002). *Applied multivariate analysis*. New York: Springer.

Tucker, L. (1958). Determination of parameters of a functional relation by factor analysis. *Psychometrika, 23*, 19–23.

Wilcox, R.R. (2003). *Applying contemporary statistical techniques*. New York: Academic Press.

Wilks, S.S. (1932). Certain generalizations in analysis of variance. *Biometrika, 24*, 471–494.

Willett, J.B. & Sayer, A.G. (1994). Using covariance structure analysis to detect correlates and predictors of change. *Psychological Bulletin, 116*, 363–381.

Willett, J.B. & Sayer, A.G. (1996). Cross-domain analyses of change over time: Combining growth modeling and covariance structure analysis. In G.A. Marcoulides & R.E. Schumacker (Eds.), *Advanced structural equation modeling: Issues and techniques* (pp. 125–157). Mahwah, NJ: Lawrence Erlbaum Associates.

Author Index

W

Subject Index

A

B

C

D

E

F

G

H

I

K

L

M

N

O

P

T